GIS The geographic language of our age

GIS

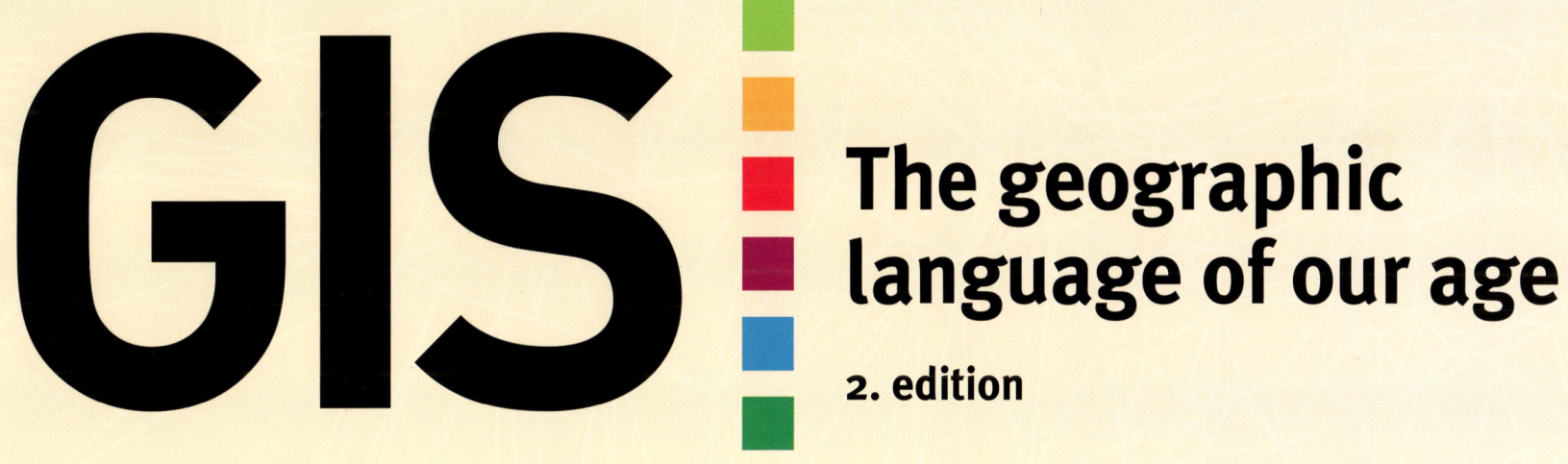

The geographic language of our age

2. edition

Knut Grinderud   Anders C. Haavik-Nilsen   Halvard Bjerke   Øystein Sanderud   Per Gunnar Ulveseth
Øyvind Mauseth   Steinar Nilsen   Magnus Fjetland   Alexander Steffensen   Ingvill Richardsen

FAGBOKFORLAGET

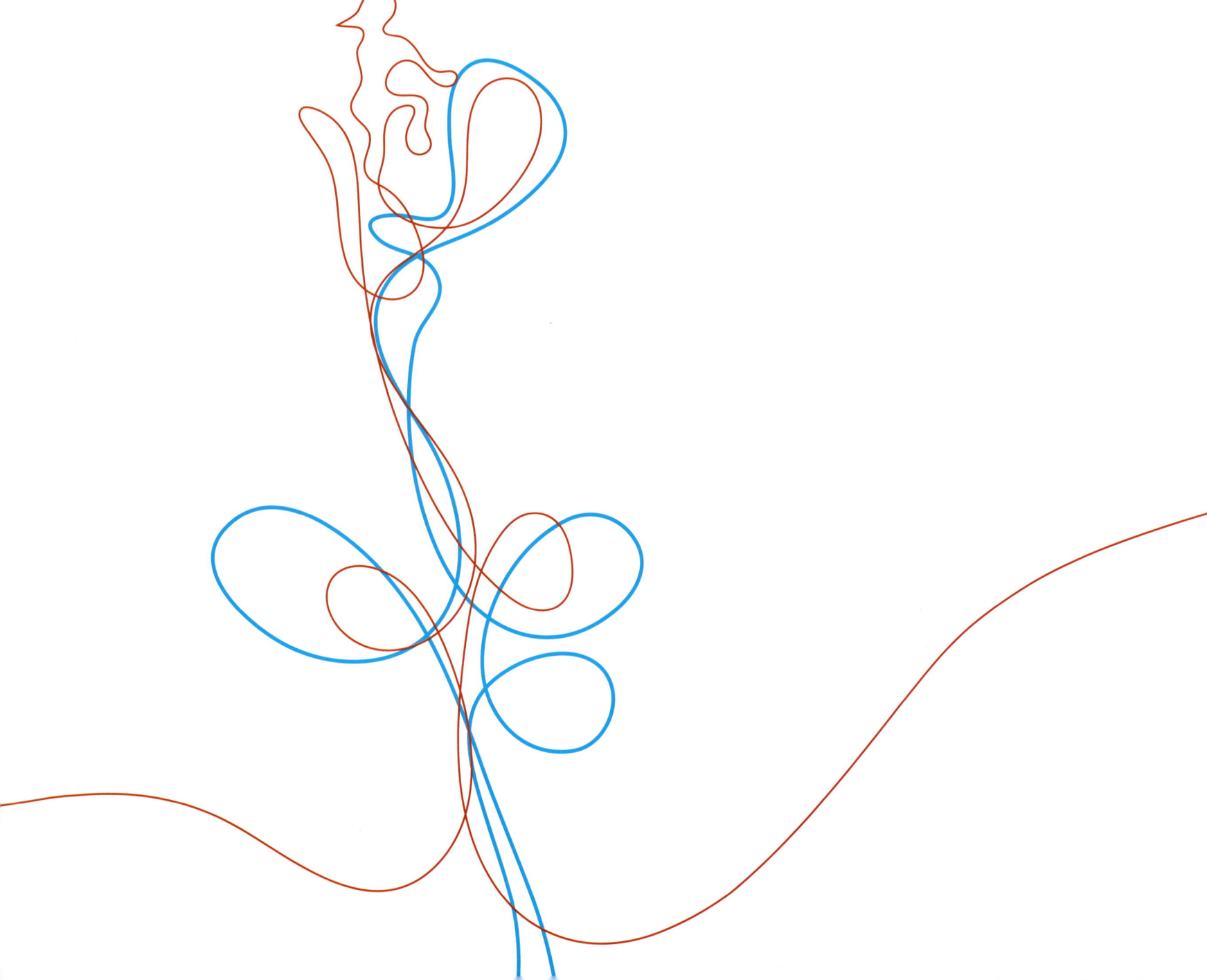

Learning from GIS

The intriguing idea behind this book is that geographical information systems (GIS) can be used as a method of planning and investigating our surroundings. It focuses in an innovative manner on how and why GIS is applied, rather than on how it works. In this way, the tremendous value of GIS is exposed for all to see.

In this book project, the authors also emphasise the fact that applied GIS links together a wide range of professional issues, and demonstrate that governmental, academic and private institutions can use it to increase knowledge by communicating through the language of geographical and social awareness.

Today, there are few books written in Norwegian that provide an overall understanding of geographical information systems. The publications used in academic institutions are highly technical and do not focus on the role of GIS in social development and planning.

The group behind this new book has strong opinions with respect to developing the potential represented by the thinking and methods of GIS. Through this book, they provide a basic understanding that can open new doors in the field, rather than technical details of the systems themselves.

GIS is still thought of as a technical discipline, but it is far more relevant to discuss it as a tool for organising and managing our social and environmental surroundings.

This educational book introduces people to the thinking, methods and challenges that lie behind GIS. Its strong focus on the communicative aspect of geographical information, both through cartography and student works, will help to make GIS both pertinent and relevant to a wider professional field.

Jack Dangermond
ESRI President

Copyright © 2009 by
Vigmostad & Bjørke AS
All Rights Reserved

ISBN: 978-82-450-2011-3

2. edition 2017

After an idea and initiative of Knut Grinderud

Graphic production: John Grieg, Bergen

Graphic design: EinArt – Einar Nilsson and Per Bækken in partnership with Haakon Rasmussen and Knut Grinderud

Typeset by Fagbokforlaget

Editors: Anders C. Haavik-Nilsen and Knut Grinderud

Halvard Bjerke
Anders C. Haavik-Nilsen
Knut Grinderud
Øystein Sanderud
Per Gunnar Ulveseth
Øyvind Mauseth
Ingvill Richardsen
Steinar Nilsen
Magnus Fjetland
Alexander Steffensen

Language consultant: Catriona Turner

Artwork created by authors / graphic designers / publishers, unless otherwise noted.

Inquiries about this text can be directed to:
Fagbokforlaget
Kanalveien 51
5068 Bergen
Tlf.: 55 38 88 00 Faks: 55 38 88 01
e-post: fagbokforlaget@fagbokforlaget.no
www.fagbokforlaget.no

All rights reserved. No part of this publication may be reproduced, stored in a retrieval system, or transmitted, in any form or by any means, electronic, mechanical, photocopying, recording, or other-wise, without the prior written permission of the publisher.

Contents

Foreword *8*

1 Application *10*
Modelling reality *12*, The models' applications *13*, What is GIS? *16*, Users of GIS *19*, GIS in spatial planning *22*, GIS in emergency preparedness *23*, Planning methodology *25*, From national guidelines to local solutions *32*, The northern regions *35*

2 Systems *38*
Makes the earth appear flat *40*, A computer model of reality *46*, Data collection of geographic data *50*, Data quality *60*, Sources of errors *63*, SOSI *64*, Geographic data in databases *64*, Web-based services *67*, Technological trends *69*

3 Geographic data *72*
What is geographic information? *74*, Geographic infrastructure *75*, Basic geodata *77*, Thematic data *88*, Land use planning *123*, Civil protection and emergency preparedness *127*, Statistical data *133*

4 Analysis *138*
Analysis methodology *140*, Spatial statistical analysis *148*, Network analyses *151*, Digital elevation models *155*, Practical analysis example *165*, Sharing the results of analyses *172*

5 Presentation *174*
Cartography *176*, Perception level *178*, Information variables *179*, Graphic elements *180*, Visual variables *180*, The perceptual properties of the visual variables *184*, Cartographic method *185*, Point symbolization *186*, Line symbolization *187*, Area symbolization – choropleth maps *189*, Other maps with area symbolization *192*, Cartographic work process *196*, Map design *199*, Web cartography *201*

6 Military geography *204*
Military working processes and methodical approach *206*, Contribution categories *209*, Military methodology example – planning and decision-making process *210*, Military use of GIS *215*

Summary *224*
Glossary *226*
Norwegian institutions and organizations – formal English names *229*
References *231*
Index *233*

Foreword

All of us have perhaps dreamt of realizing an ambitious idea. This is exactly how the team that produced this book feel. A project like this would never have been possible without a large network of people with different personal qualities and professional backgrounds. The deep commitment shown by everybody who has contributed to the process has enabled us as a team to produce a different kind of textbook – one with many dimensions. As a reader of this book you will notice that we have tried to give the book a two-dimensional common thread.

Concept. We have tried to shift the focus away from a technological perspective. In our opinion, the technology used to produce the answers is of secondary importance. Instead, we have chosen to focus on how geographic information systems (GIS) can constitute a language and aid in the investigations of complex everyday life. We look at how this tool can help planners to ask the right questions that will later lead them to deduce the correct answers. This brings us to the book's other dimension.

Methodology. In this book we aim to structure the challenges arising from various geographical assigments, projects, and tasks into four stages: the collection, processing, analysis, and presentation of geographic information. In all of these stages, human

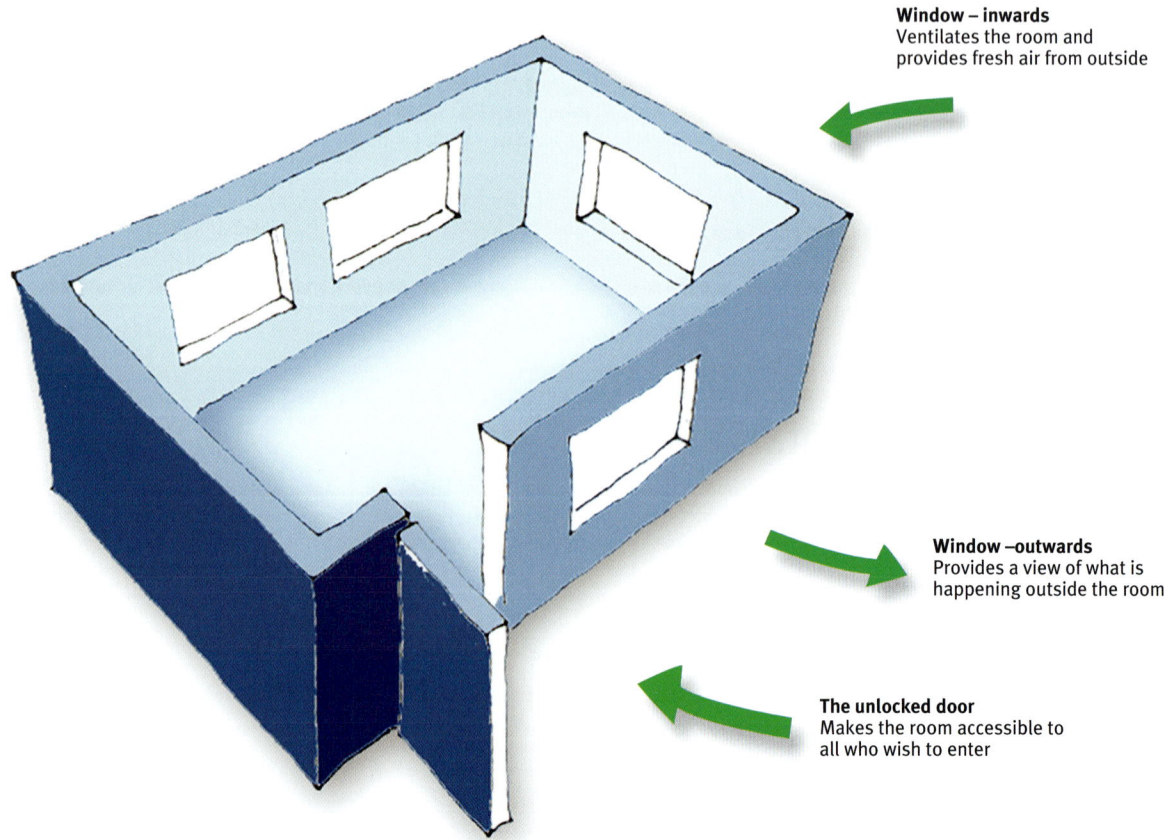

Window – inwards
Ventilates the room and provides fresh air from outside

Window – outwards
Provides a view of what is happening outside the room

The unlocked door
Makes the room accessible to all who wish to enter

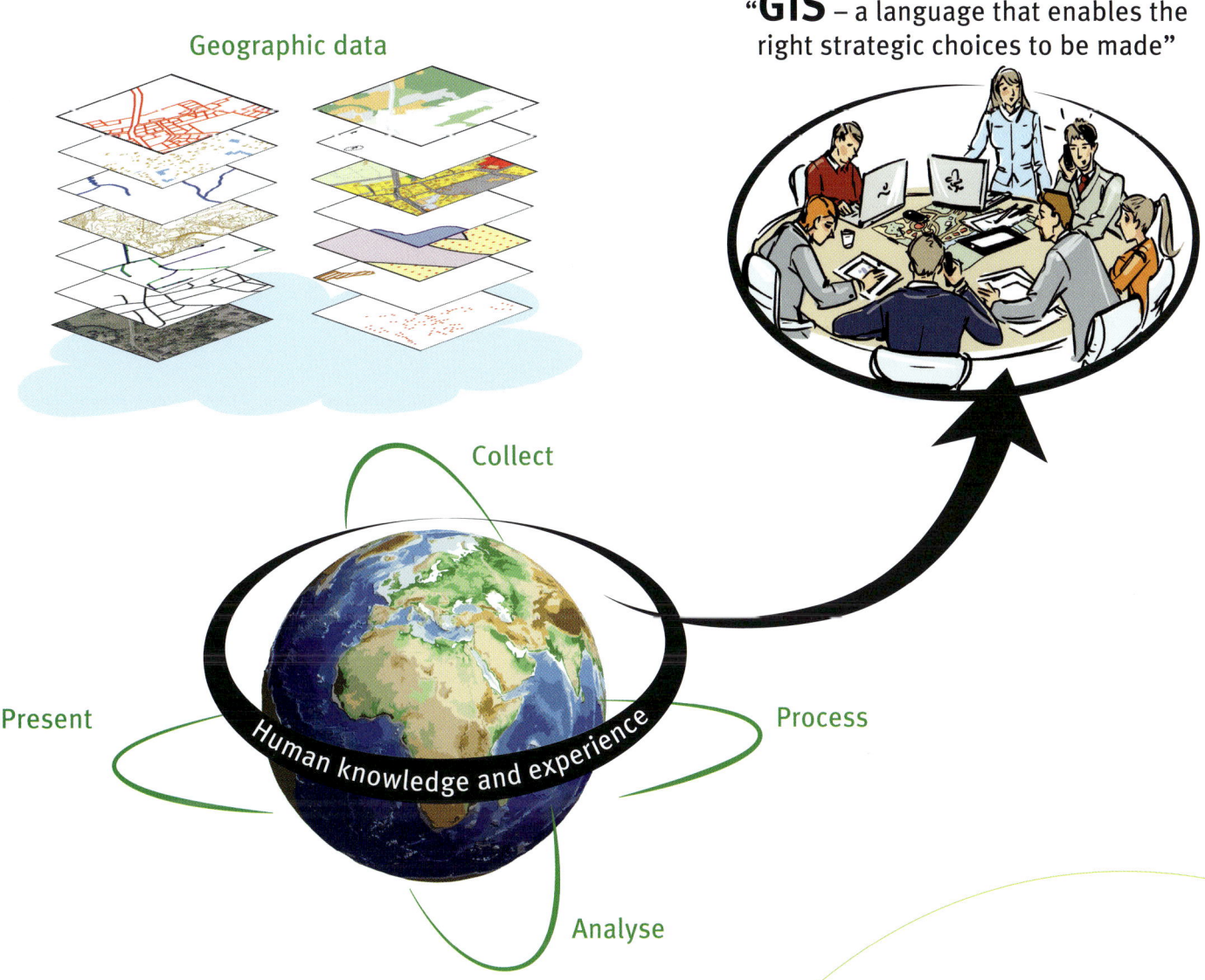

knowledge and experience are relevant factors. We therefore aim to identify different situations in which GIS are important tools for describing the geography around us and for making appropriate strategic decisions when necessary.

We especially thank Captain Ragnar Øien, a master of science in GIS-engieering, who has been of great assistance in checking the academic content of the Norwegian version of this book. We also thank professor Brendan MacBride at the Norwegian Military Academy for his help with translation of Chapters 1 and 6.

This book is written for Norwegian readers and later translated into English. This means that it is used a lot references to Norwegian institutions and agencies, who are not as familiar to international readers. See also Appendix Norwegian institutions and organizations.

We hope that this book will generate new ideas on how GIS can be used in everyday life.

Halvard Bjerke holds a master's degree in geomatics from the Norwegian University of Life Sciences. He is an Assistant Professor of engineering at the Norwegian Military Academy.

Anders C. Haavik-Nilsen holds military qualifications from the Norwegian Military Academy, and military qualifications from the Norwegian Defence Command and Staff College. He also holds a master's degree in civil engineering from the Norwegian University of Science and Technology and a master's degree in management from the BI Norwegian Business School. He has a background in both national and international service in the army, and is the head of the Military Academy's engineering programme.

Knut Grinderud holds a master's degree in engineering from Norway's Institute of Technology (now the Norwegian University of Science and Technology), and a master's degree in business administration in strategic management from the NHH Norwegian school of economics. He works as chief engineer and head of strategy and development in the Norwegian Defence Estates Agency (NDEA). He has previously been responsible for spatial planning, GIS, property registers, and firing range and training area exercises in the NDEA.

The chapter is a revised version of the chapter written by Knut Grinderud and Haakon Rasmus Rasmussen in the 1st edition.

Application

Modelling reality	12
The models' applications	13
What is GIS?	16
Users of GIS	19
GIS in spatial planning	22
GIS in emergency preparedness	23
Planning methodology	25
From national guidelines to local solutions	32
The northern regions	35

Modelling reality

Throughout the ages, humankind has created representations of the world as tools to aid exploration and understanding. People have always sought a greater perspective for their own experiences. Earlier models of the world were not necessarily always accurate, nor did they need to be, since their purpose was merely outward representations of what was imaginable based on limited knowedge of that time. Even to this day there are disagreements about which models best represent alternative needs and perspectives. A model's structural properties can be critical with respect to its usability and relevance, such that an astrophysicist may benefit from using an inaccurate model of the universe that makes it easier to see relationships or a course of events. This manipulation of reality is valid and important, and was both present in earlier times and is still present in society today.

So, our representations and perceptions of our surroundings are expressed also in the maps we produce, and our perception of the world is formed by our maps. Figure 1.1 shows the 'T-O' map dating from the 13th or 14th centuries. The map is oriented with east at the top and consists of a 'T' within an 'O'. The 'T' represents the Mediterranean, the river Don in Russia, and the Nile, which were regarded as the borders between the only continents that had been discovered at the time, namely Europe, Asia, and Africa. The 'O' represent the vast world ocean around the continents. Christianity's absolute status as the valid worldview is reflected in the fact that Jerusalem is located at the centre of the world and that theological places such as 'Paradise' and the mountain where 'Noah's Ark' was grounded are depicted as specific physical locations in the world.

Additionally, the maps in Figure 1.1 reflect the perception that parallel to our physical world there also exists a spiritual world that comes into contact with our world in certain places and at certain times. On early maps, demons and mythical creatures were often drawn at the edge of the world, as shown in Figure 1.2. Such depictions reflected the notion that voyages beyond the world's oceans were a form of suicidal madness. Thus, Christopher Columbus's real achievement should not be seen as the actual journey he undertook but rather his decision to take it defiance of the prevailing worldview.

Exploration and the Age of Enlightenment caused the Middle Age's worldview to collapse. It was replaced by the perception of the world we have generally had to this day. However, today's fast means of transport, the Internet, and global environmental burdens have changed our perception of the earth as a much smaller and more vulnerable place than the huge planet, with its overpowering natural forces, into which the explorers of the 1500s ventured forth.

Albert Einstein's new models for perceiving mass, time, and space, and the astrophysicists who have followed in his wake have shattered our general perceptions of the universe. Their representations of curved space and wormholes that enable time travel are barely understandable today, and probably just as astonishing as a transatlantic flight would have been for people in the 1300s. Yet, is instead the concept of Gaia (Lovelock 1979) – the earth as a living organism – the only valid perception of the world if humankind is to ensure its future existence?

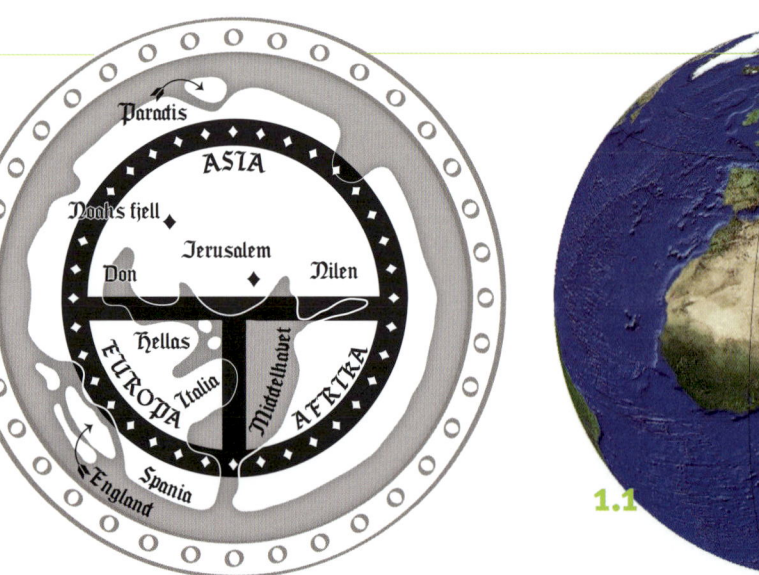

1.1 Images of the world from the Middle Ages and the present: a so-called T-O map (left) and a digital 3D model of the globe draped with a satellite photo (right).

The models' applications

One particularly recognizable example of modelling the world around us is found in British physician Dr John Snow's investigations into how cholera spread in London in the mid-1800s. London was the first city in modern Europe with 1 million inhabitants and as a direct consequence it was the first to experience the challenges associated with large-scale city planning. At the time the occurance of cholera was not understood to be connected with the waterborne bacteria thriving beneath the rapidly expanding city's poorly separated fresh water and sewage systems.

In the 1800s, physicians had two different, incompatible theories about why infectious diseases occurred. The miasma theory, which dominated until towards the end of the 1800s, held that diseases were caused by noxious vapours in the air called 'miasmata'. In line with the miasma theory, Ullevål Hospital in Oslo, built in 1887, was constructed with well-ventilated buildings separated by wide lawns to prevent the spread of poisonous air. The second hypothesis was the germ theory, which held that diseases were caused by small, invisible germs. The germ theory first achieved a breakthrough when the existence of these invisible organisms was proven with the discovery of bacteria and viruses.

Dr John Snow was determined to find the cause of the cholera epidemic that was raging through the streets of London. He was a proponent of the germ theory and had his suspicions about what the cause might be, but no clear evidence. Dr Snow decided to collect information about where those who died from cholera lived and thus construct an abstract representation of London that only contained certain types of information.

1.2

1.2 The Carta marina wall map was completed in 1539 by Olaus Magnus in Venice. The map is regarded as the first map of the Nordic countries. It depicts demons and mythical creatures that sailors might have expected to encounter off the coast of Norway. Source: Magnus (1539).

14 The models' applications

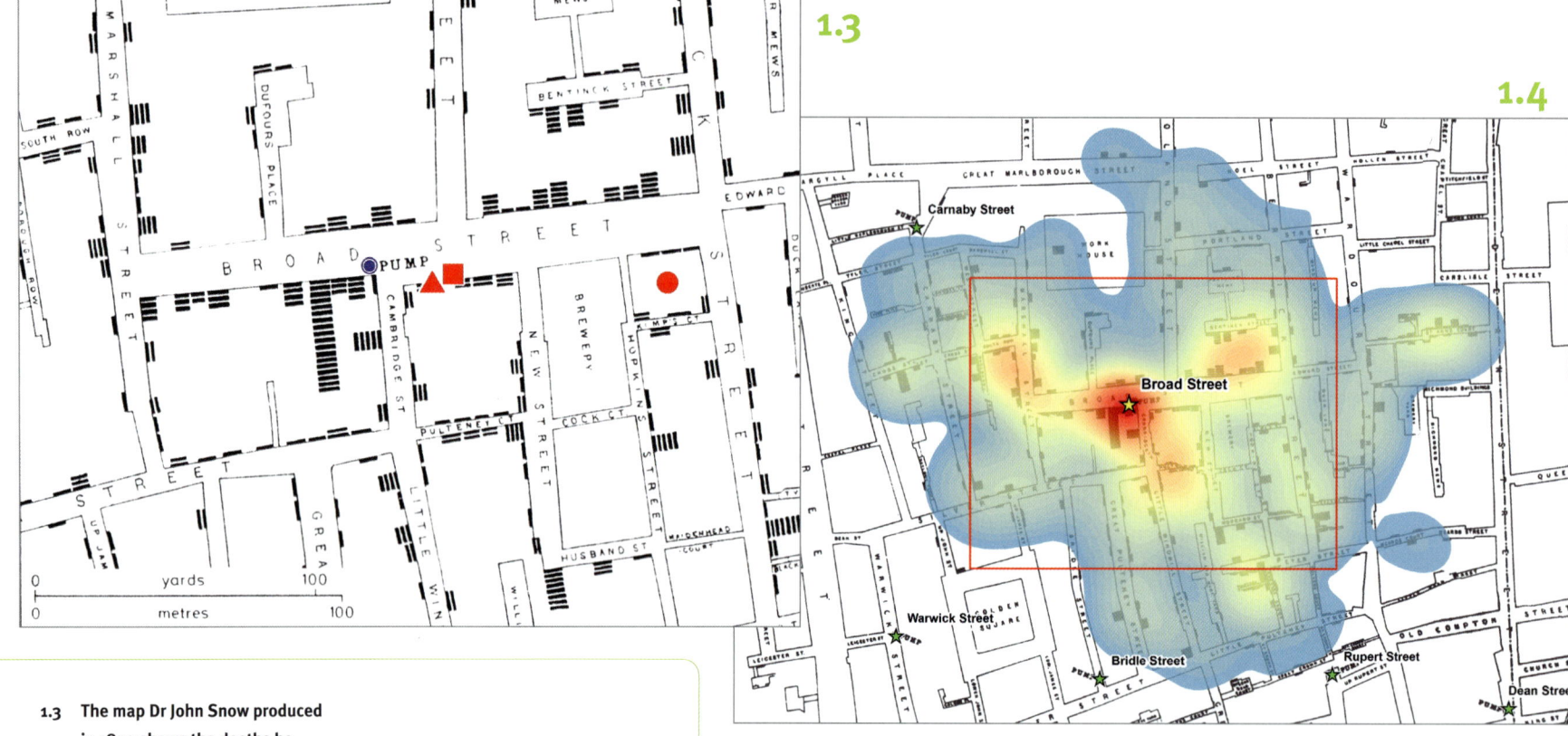

1.3 The map Dr John Snow produced in 1854 shows the deaths he recorded, depicted by black bar symbols. The triangle depicts the mean value, the square the median value and the circle the mode value (mode: most frequently occurring value in a dataset). The deaths were clustered around a particular water pump in Broad Street. Dr John Snow had already hypothesised that the 'organism' that caused cholera was waterborne. He used the map to illustrate his hypothesis. Source: John Snow (1854).

1.4 Gravity analysis of the data from the deaths recorded by Dr Snow. Source background map: John Snow (1854).

The procedure used by Dr Snow was to register where the deaths due to cholera occurred, and then add the results from the records to a map and perform an analysis by calculating the density of the plotted points. Through such an analysis, he was able to record all illnesses in map form and show the area in which the source of infection might be located. Most of those who were infected lived near one particular water pump in Broad Street, as shown in Figure 1.3.

Dr Snow met fierce resistance from physicians and others who explained the disease on the basis of the miasma theory. Nonetheless, in the autumn of 1854, he persuaded the city's authorities to remove the handle of the water pump in Broad Street. The cholera epidemic ended immediately. This outcome strengthened the germ theory as an explanation model and resulted in people finally recognizing the link between cholera and the importance of separating water and sewage. This in turn eventually resulted in a totally new water and sewage system being constructed in London.

Snow's work on the use of geographic information was revolutionary. Few others had incorporated such information into a system in order to be able to draw conclusions from or base decisions on the results of an analysis. For this reason, the London doctor is often referred to as among the first persons to use a *geographic information system*, also known as a *GIS*.

Today, GIS analyses are largely conducted using the same method as used by Dr Snow, though with slightly more advanced tools. Nowadays, these models can be programmed into computers. Storage capacity used to be a major problem because of the high costs involved. However, computers have made it possible to handle large amounts of data and complex issues, and computers' efficiency is continuing to improve.

A case that involved a parallel approach to Dr Snow's data models and map presentations was the outbreak of legionnaires' disease in Fredrikstad in Eastern Norway in the spring of 2005. This was one of the biggest crisis situations managed by Norwegian local authorities in recent years and serves as a good example for comparison with how modern approaches and tools are used.

During the legionnaires' disease epidemic more than 50 people were infected, several of whom died. It was obvious that it was extremely important to find the source of the outbreak quickly because no one could be sure whether the source of the outbreak was still active. In previous outbreaks, such as in Stavanger in 2002, ordinary paper maps had been used to analyse how many people were infected and where they contacted their disease. Fredrikstad Municipality started with an ordinary paper map but quickly realized that managing and then analysing large volumes of data in this way would be a cumbersome task. Together with the Norwegian Institute for Air Research (NILU) and a supplier of GIS software, the municipality's staff therefore chose to solve the task with the aid of digital data and the GIS tool.

To provide a basis for the analysis, data were collected on the infected people's addresses, their movements throughout the entire infection period, the cooling towers and locations of other possible sources of the outbreak, and the spread of the disease based on wind and weather conditions – Snow's data collection approach. There was a huge quantity of complex data, but the municipality had a tool with which to process it. The data provided a basis for simulating the assumed dispersal patterns

1.5 Dr John Snow (1813–1858). Source: Ralph R. Frerichs, UCLA Department of Epidemiology, School of Public Health.

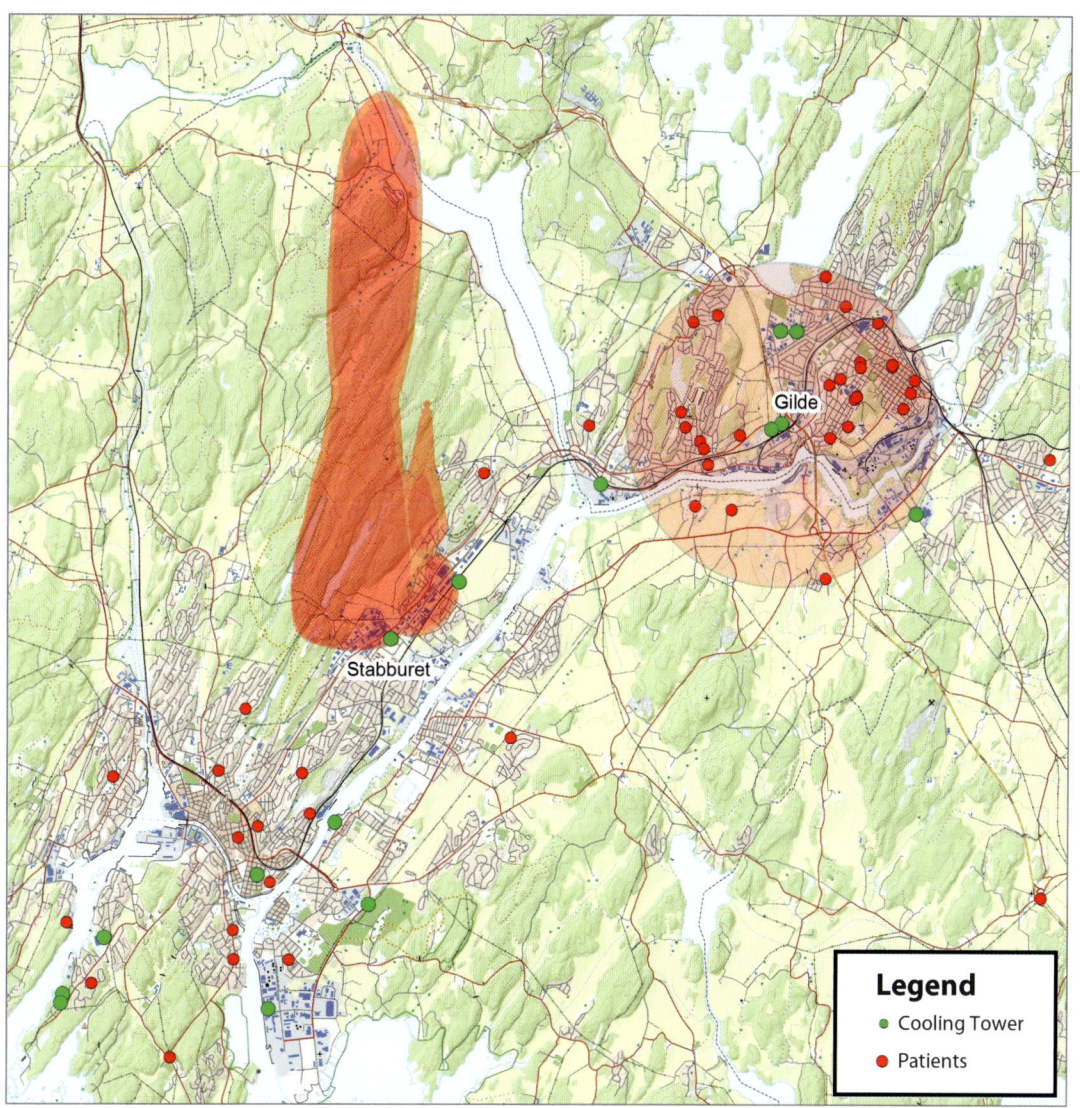

1.6 Buffer analysis and distribution analysis of the outbreak of Legionnaires' disease in Fredrikstad, 2005. The distribution analysis shows a simulated spread of the infection via contaminated water droplets from the factory cooling tower, based on atmospheric dispersion and wind parameters. Source: Fredrikstad Municipality.

from specific cooling towers and finding out which patients were within or outside a dispersal field during a given period of time. A complex data model in a suitable GIS tool enabled the users to take account of air humidity, wind speed, and temperature, making the job of finding the most probable source of the outbreak easier. Part of the analysis is presented in Figure 1.6.

The map analyses alone did not provide the final answer. However, they were crucial for identifying the probable location of the source of the outbreak, based on the movement patterns of the infected people. There are a considerable number of cooling towers and other possible sources of legionnaires' disease in the area, but the map analyses made it easier and faster to find the most likely cooling towers. This meant that DNA profiles taken from the sources could be analysed quickly and compared with those of the patients. The source of the outbreak was thus found more quickly using a GIS. The experience gained from using GIS tools in this case contributed to not only GIS tools but also the use of GIS expertise becoming an important part of emergency planning in Eastern Norway and elsewhere in the country.

What is GIS?

Even though people have used geographic models in the form of maps and globes for a long time it was not until the 1960s that IT-based geographic information systems were developed. The *Canadian Geographic Information System* (CGIS) from 1962 is considered to be the first GIS. It was developed because people working in traditional thematic cartography wanted to make and produce thematic maps more efficiently as well as to be able to store information in connection with major building projects.

It was not until the 1970s that computing software, including GIS software, became more widely available. The first systems called GIS originated from *computer-aided design* (CAD) systems. Over time, as the need for better quality computer graphics and higher plotting resolutions increased, specialized solutions were required. Presenting map data is very demanding as far as computer graphics are concerned and GIS software has since developed into a separate category of software.

There have been many explanations of the concept of geographic information systems. As a result there are many terms – including geographic information science, geographic information systems (GIS), geographic information technology, geographic information processing, and digital geographic information – and even more definitions. The following definitions of GIS have been sourced from various disciplines.

Blankholm provides the following definition of GIS: 'A geographical information system (GIS) is a tool for the collection, storage, analysis, and presentation of data in the spatial plane' (Blankholm, 2004). Bernhardsen (2000) provides an alternative definition: 'A geographic information system is an electronic data processing system that handles geographic data, information on characteristics, and relationships to objects that are uniquely geographically localized.' By contrast, Pickles (1995) states: '[a] GIS is a set of tools, technologies, approaches and ideas that are vitally embedded in broader transformation of science, society and culture.'

Other researchers emphasize that without relevant qualifications, it would be difficult for users to take full advantage of the potential of GIS. Roger Tomlinson states: 'No GIS can be a success without the right people involved. A real-world GIS is actually a complex system of interrelated parts and at the centre of this system is a smart person who understands the whole' (Tomlinson 2003). Duek (1989) has given following definition: 'Geographic Information System – a system of hardware, software, data, people, organizations and institutional arrangements for collecting, storing, analyzing and disseminating information about areas of the earth.'

Clearly, there are different perceptions of what constitutes a GIS. Some researchers define GIS as a tool or science, whereas others make the important point that a true GIS cannot exist in the absence of skilled or competent persons who can make the best use of the tools and who are knowledgeable about how to treat basic data derived from GIS. User interfaces in systems that are available today are becoming more accessible and more user-friendly. This means there are more opportunities to focus on the processes themselves, rather than on having to ensure that the systems function properly, as was the case in the past.

The above-mentioned definitions of GIS have three main elements in common:

- geographic data
- hardware and software
- human knowledge and experience

The interaction of these components enables us to *collect*, *process*, *analyse*, and *present* the geographic information around us with the aid of digital technology. In other words, a GIS can be defined as follows:

> *A geographic information system (GIS) is the combination of geographic data, map systems, methods, and human knowledge and experience that makes it possible to collect, process, analyse, and present the geography around us.*

Figure 1.7 can be used as an educational aid to describe the different elements in the definition of GIS. The elements are emphasized wherever they are mentioned in the chapters in this book. The first chapter (Application) presents background information on what constitutes a GIS. Chapter 2 (Systems) explains how GIS theory is fundamental for recreating reality as accurately as possible. In addition, the chapter provides an overview of some methods for collecting geographic data and how such data are processed before they can be analysed. Chapter 3 (Geographic data) describes the types of geographic data currently available. Chapter 4 (Analysis) presents an approach to working with GIS, followed by examples of analytical methods within GIS, and Chapter 5 (Presentation) focuses on the visualization and presentation of geographic information. The book concludes with Chapter 6 (Military geography), which describes the application of GIS in the Norwegian Armed Forces.

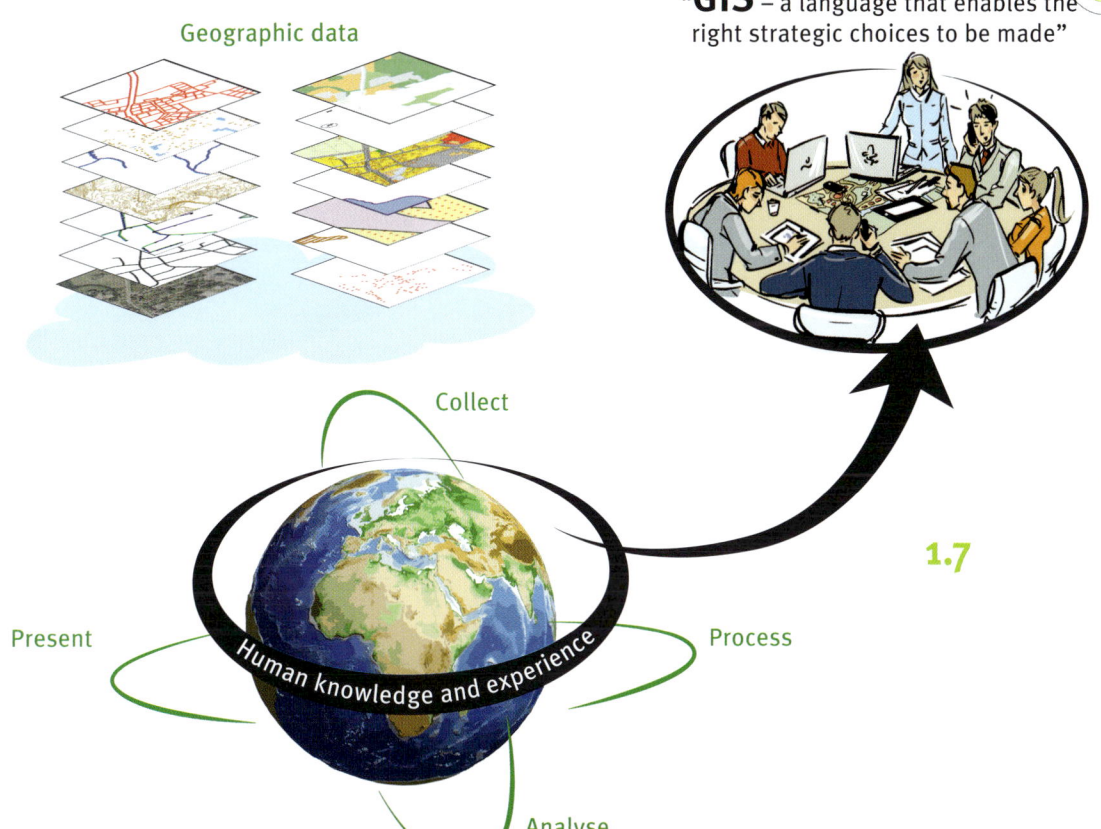

1.7 The four components that form the foundation of a modern GIS. Few people possess the latest expertise in all of these fields speciality, but by combining these individual competences together, GIS can allow clearer visualization of the geographic reality around us.

18   What is GIS?

1.8 Georeferenced, oblique aerial photos. The photos are taken from five different angles enabling the user to see a building from all sides and from above, which means that she can move around a building or zoom out and view the neighbourhood. Oblique photos can also be used to drape a 3D model making it true to life. Source: Blom Geomatics.

1.9 The difference between digital models and physical models Figures a, b and c show a digital terrain model with and without an ortophoto draped over it and a physical model of Akershus Fortress in Oslo, respectively. The digital terrain model's strength is that it is visually precise, can be rotated in all directions, and all unnecessary information can be removed. With an ortophoto draped over it the model provides a very close depiction of the real world. The model's weakness is that the buildings become very dominant, and it takes a lot of work to achieve a correct image of the façades. The physical model's strength is that it provides a better impression of elevations, and the consequences of changes are more visible and robust.

1 – APPLICATION

The book introduces readers to the thinking, methods, and challenges behind a GIS. It also focuses on the communicative aspects of a GIS and how acquired knowledge can be used to increase our understanding of differing environments. In this respect, a GIS can contribute to sustainable development in societies. In summary, **GIS is 'the geographic language of our age'**.

Users of GIS

Geography is a universal language and GIS is the tool that allows people to speak to each other. One can find geographic information technology everywhere, often without even realizing it. Everything from search engines on the Internet when one wants to find the shortest route from A to B or an address, to a taxi control room in which the operators know where all their cars are at any given time, to more

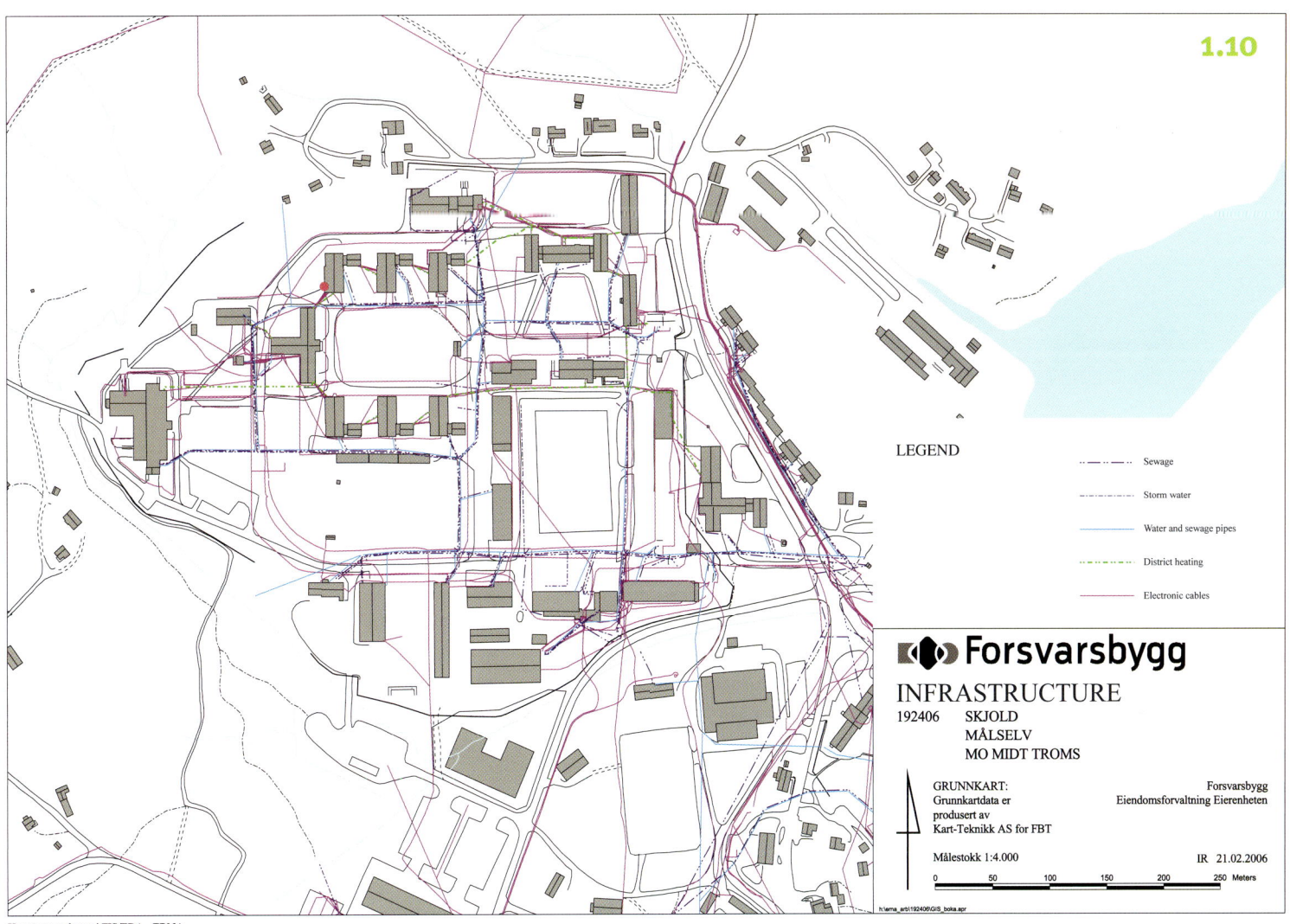

1.10 Map showing the infrastructure of Skjold Military Base. Information about thousands of meters of wires, pipelines, transformers etc. needs to be provided and updated regularly.

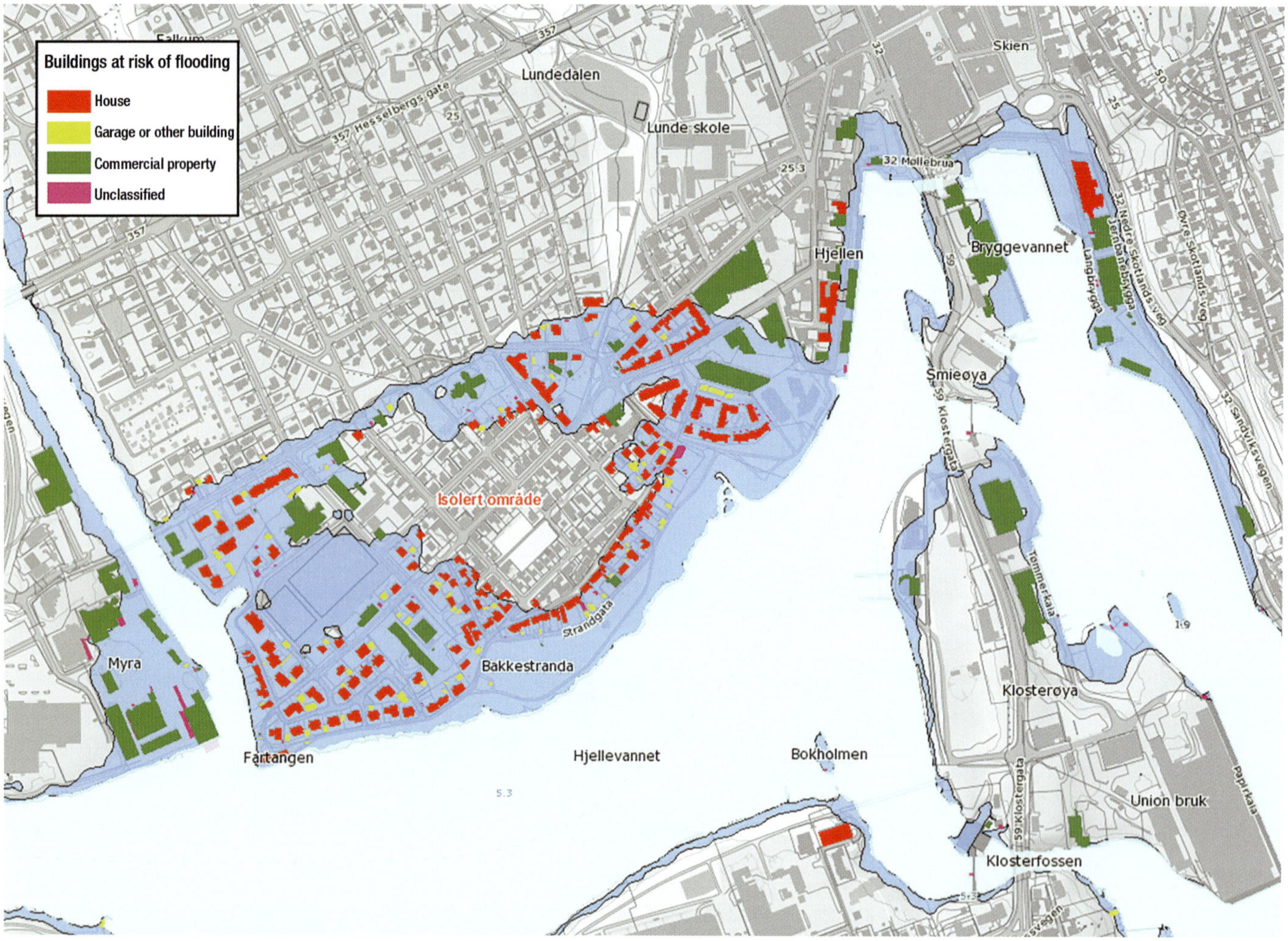

1.11 Flood maps from the Norwegian Water Resources and Energy Directorate (NVE) provide municipalities with a better basis for spatial planning and preparedness. The map shows the estimated height of the water during a 500-year flood in Skien and which buildings would be affected. Flood data: NVE; background map: © Norwegian Mapping Authority.

weighty professional systems designed for special issues. These are only a few examples of the use of geographic information systems.

The number of disciplines that use GIS in some form or other has increased dramatically in recent years. Their needs rely increasingly on geographic data as a reliable basis for decision and planning procedures. Modern GIS can be said to have derived from traditional mapping and surveying disciplines. The first of these to adopt GIS were geodesy, land and sea surveying, remote sensing, photogrammetry, and cartography. These disciplines together with GIS can be gathered under the term *geomatics*, which includes all activities related to collecting, processing, analyzing and presenting spatially localized information.

Examples of other disciplines that are potential users of GIS include geology, geophysics, oceanography, agriculture, biology, environmental sciences, geography, sociology, political science, and anthropology. Other examples within business include the transport industry and the taxi trade.

More and more businesses, organizations, and government agencies are implementing GIS in their internal systems. Statskog (the Norwegian state-owned land and forest enterprise) uses GIS for the management of resources on state properties, whereas the Norwegian Armed Forces use GIS for planning operations and other purposes. In addition to internal systems, geographic information can be made publicly available via the web-based GIS services. Some municipalities have published web-based GIS servic-

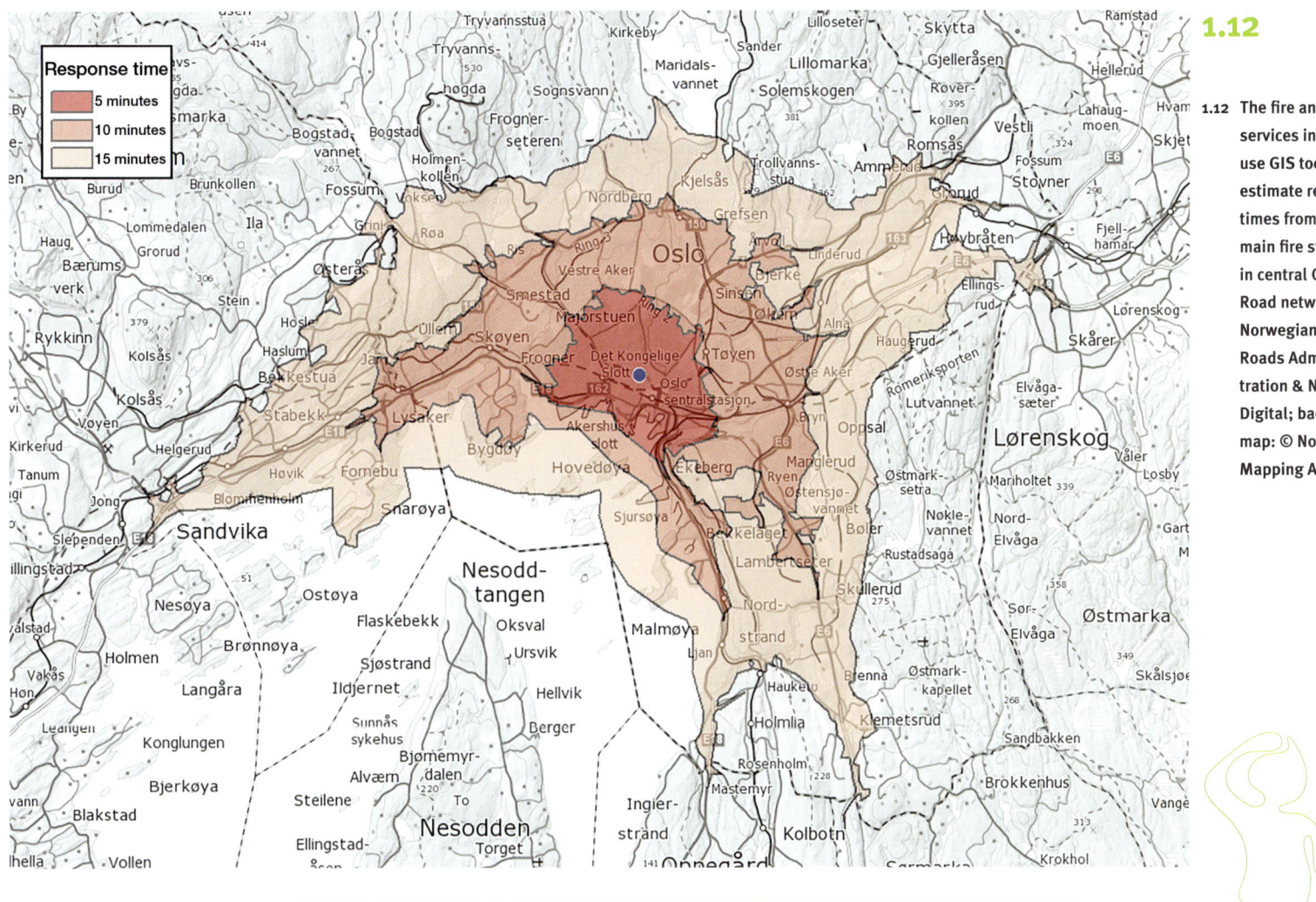

1.12 The fire and rescue services in Oslo can use GIS tools to estimate response times from the main fire station in central Oslo. Road network data: Norwegian Public Roads Administration & Norway Digital; background map: © Norwegian Mapping Authority.

es that citizens can use. These types of online maps also provide limited GIS functions such as searches and simple analyses, but their functionality is expected to develop in line with technological developments.

While recognizing that GIS is a powerful and impressive technology with multiple application in different contexts, there needs to be caution regarding some potential deficiencies. However, one has to be aware of the deficiencies in such technology. Here, it is important to refer back to our definition of a GIS. To take full advantage of the technology, the user requires relevant knowledge and the ability to ask analytical questions related to what it can achieve. The challenges inherent in this include knowing the quality of the basic data being used and ensuring that the focus is on a process delivering results, not on the actual technology itself. Traditionally, GIS, like many other computer-aided processes, has been perceived as a subject for technicians. Better user interfaces for GIS software, together with increased understanding of GIS in general, means that instead the focus can be on what processes can be accomplished, rather than on how they work technically. The input of poor quality or innaccurate data will result in equally poor result in the final onscreen map.

Presenting information is also about interpretation and this awareness needs to be understood. Other challenges include combining data that are not logically related and, not least, how one should present the information. It is quite possible to 'lie' with maps. Sources of errors are described thoroughly in Chapter 2 (Systems).

GIS in spatial planning

GIS is an important tool for both spatial planning and effective resource management. When various types of thematic data and basemap data are combined by a GIS the results can contribute new knowledge and lead to greater understanding.

Thematic data are presented as data layers over a basemap, as shown in Figure 1.13. Thus, various geographic data can be used to provide different images and details of the reality on the ground. Through the use of spatial statistics and analyses, the transformation of various public records onto maps can provide new images of a given reality as well as new knowledge and understanding of geographic patterns

1.14 GIS is an important tool in spatial planning.

and relationships. This in turn can provide the basis for those involved in a planning process to formulate specific questions. Analyses and studies of thematic data sets can be used to demonstrate the consequences of different actions.

The thematic breadth of GIS applications is considerable and includes biodiversity, population age structure, arterial roads with high accident rates, and aquaculture farms along the coast, to name a few examples. In any work related to land use, environments, and resources, it is important to be aware of the diversity in the thematic data available as well as where such information exists. Criteria for success include the exploitation of the information potential of whatever database is used, awareness of data quality, knowledge of what treatment methods a GIS permits, and knowledge of how cartographic communication can be used to disseminate knowledge.

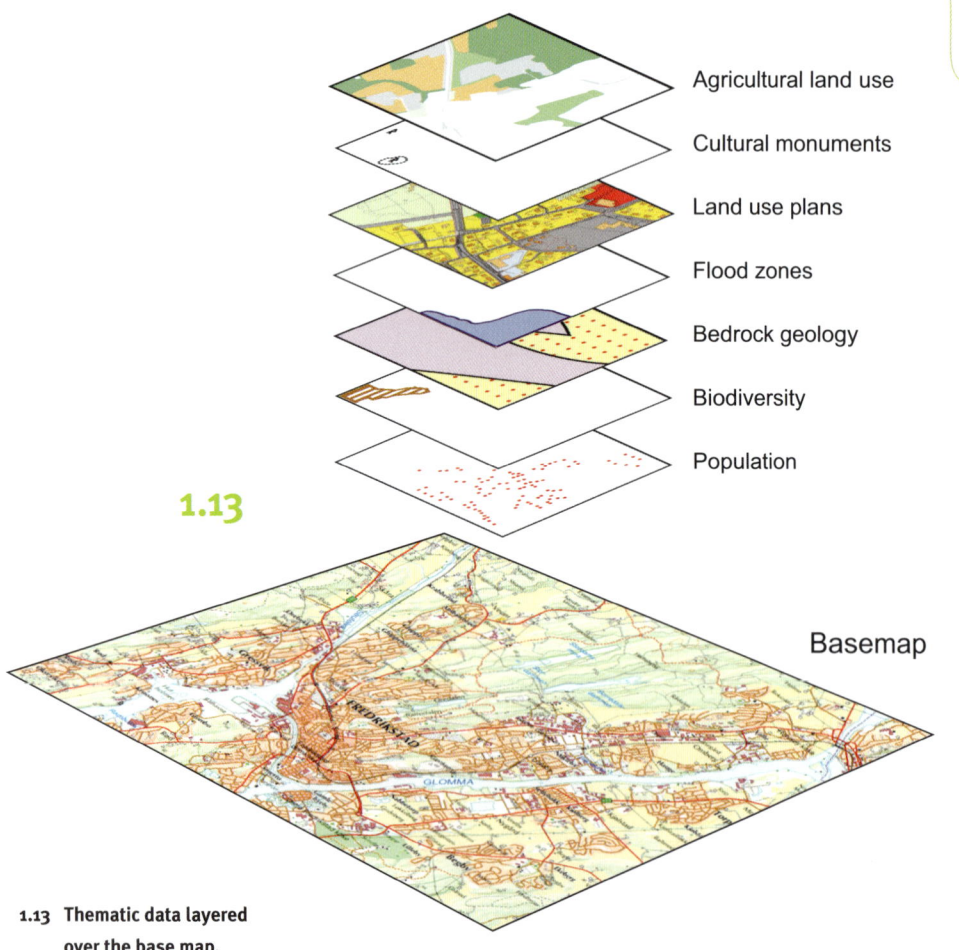

1.13 Thematic data layered over the base map.

GIS in emergency preparedness

In recent years there has been increased focus on civil protection and emergency planning. Climate changes are leading to more frequent occurrences of extreme weather with subsequent flooding and landslides as well as their consequences, such as loss of lives and extensive material damage. Hence, prevention is becoming increasingly important in order to avoid the adverse effects of extreme weather events. Under the Civil Defence Act of 2010, § 14, municipal authorities in Norway should identify what adverse events might occur in their municipality, the likelihood of them happening, and their likely impacts.

In general, municipalities should assess the risks of all types of land use. Their assessments should be compiled in a comprehensive risk and vulnerability (RAV) analysis that should identify unsatisfactory conditions and provide a basis for evaluating countermeasures. The RAV analysis can then be used in further work on a contingency plan in which, in accordance with § 15 of the Civil Defence Act of 2010, the municipality has to include a plan for crisis management, notification lists, an overview of resources, evacuation plans, and plans to disseminate information to the public and media.

In many cases, a GIS can be a tool for public safety and emergency preparedness. For example, during the preparation of a RAV it may be relevant to consider the consequences of a landslide. What mass volume could be moved during a landslide? Is accommodation available for rehousing victims? Which roads will be affected? What detour routes are possible? What are the emergency services' response times? As part of RAV analyses, map products and GIS analyses may help to answer these questions. RAV analyses are also important in

1.15 Skjeggestadbrua in Holmestrand. Photo: Tore Meek / NTB scanpix.

24 GIS in emergency preparedness

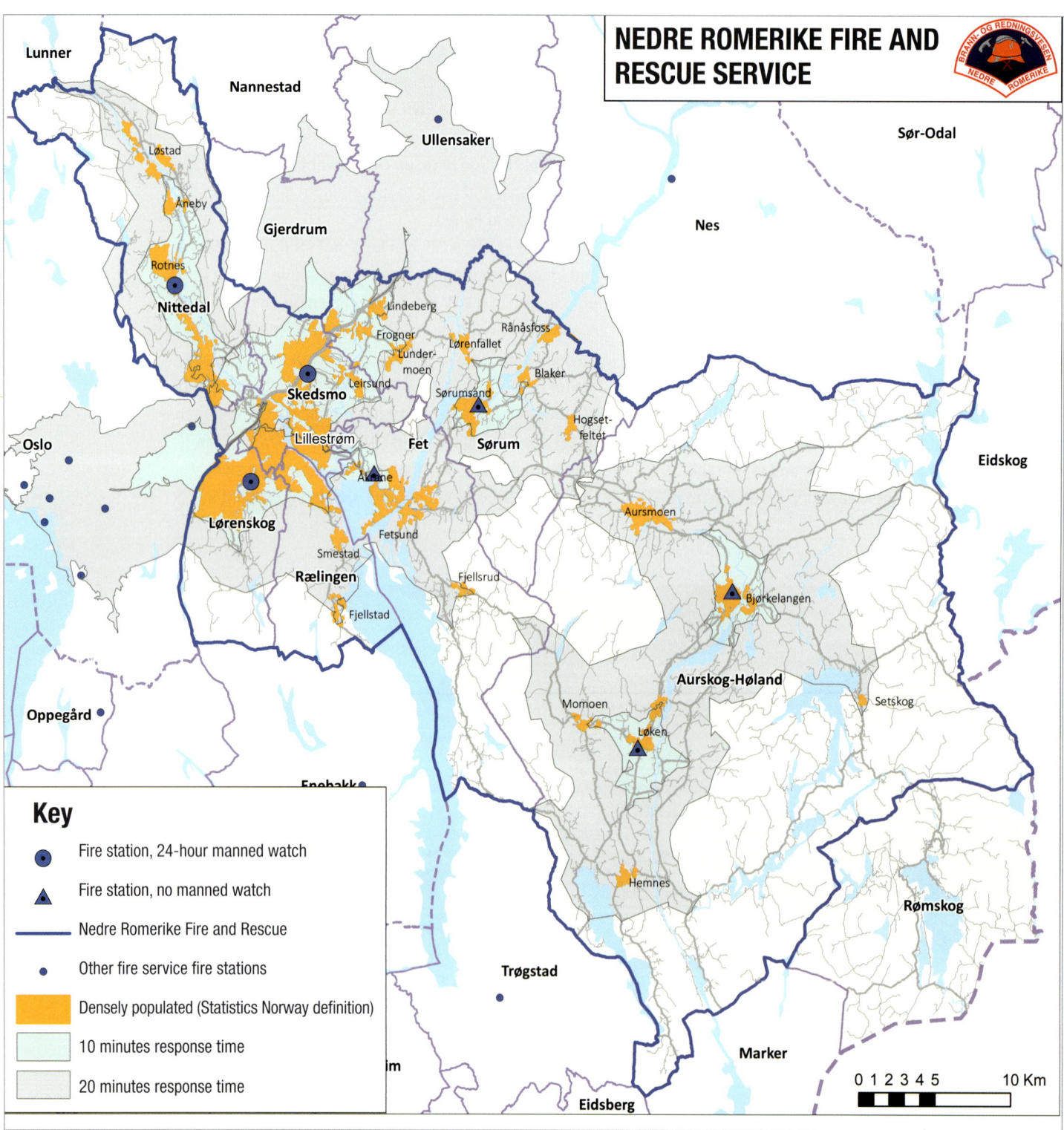

1.16 Commute time analysis by the Norwegian Directorate for Civil Protection (DSB). The analysis shows the response time for various Fire Stations in the area. Source: DSB.

land use planning, for example by ensuring that housing is not constructed in either landslide-prone or flood-prone areas.

In February 2015, excavations in Holmestrand caused an avalanche that resulted in the partial collapse of Skjeggestadbrua (Skjeggestad bridge) on the E18. Due to the extensive damage, the southbound highway over the two parallel bridges had to be dismantled. A more thorough geological analysis of the underlying quick clay deposits, would have led to building deeper and stronger peir foundations into the surrounding bedrock. If municipal plans highlight the potential risks for excavation work, there should be a greater chance of project owners and the general public avoiding such accidents. In this context, GIS can be used in different types of planning to map vulnerable areas and potential impacts.

In order to safeguard society and be well prepared for the unforeseen, joint exercises are conducted by the Norwegian armed forces, police force, fire service, ambulance service, and other civil society actors. The exercises are carried out in various scenarios that include natural disasters, accidents, explosion hazards, and acts of terrorism, where GIS can be used to create better understanding of the problem and more effective coordination efforts to handle such emergencies. By sharing geographic information or products from pre-defined analyses, all involved parties can plan their missions from the same starting point. The use of GIS facilitates interaction across disciplines, whereby different services and departments can give input and provide specialist information to solve a common problem.

Planning methodology

Why do we plan? According to General Dwight D. Eisenhower, who was responsible for planning the allied forces' invasion of France and Germany in 1944–1945, 'Plans are nothing, planning is everything.' The quote highlights that a process is often more important than the end product. Hence, it is important to focus on the methods used to arrive at the end product.

Planning is predictable in the sense that it always reflects contemporary and changing challenges. However, it may seem more fruitful to study the applicability of various methods than the methodology and procedural factors. In such cases, it becomes necessary to find out how to ask the right questions rather than how to find answers. In this perspective, the responsible individual and their ways of acting forms a good basis for more sustainable discussions.

From the numerous different tools and methods available, we have chosen as the starting point for this book the experiences of two different actors: a teaching and research establishment (the Norwegian Military Academy) and a property manager (the Norwegian

1.17 Joint exercise between the Norwegian Armed Forces, police, fire service, ambulance service, and other civil society actors during the exercise Flotex Silver Rein 2012. Photo: Torbjørn Kjosvold, Norwegian Armed Forces.

1.18 Norwegian Armed Forces Special Operations Command, Marinejegerkommandoen (the marine and naval component of the Norwegian Special Operations Forces), and the Emergency Response Unit (a divisions of the police force) training in dealing with counter-terrorism, at Rena military base, in preparation for exercise Gemini 2014. Photo: Didrik Linnerud, Norwegian Armed Forces.

Defence Estates Agency, NDEA). Both actors use GIS to solve various work-related problems. All actors involved in planning have to establish their own types of toolboxes according to their requirements and opportunities. The remainder of this chapter will provide readers with a description of the three different types of toolboxes that can be used as aids to cope with a given situation.

PLANNING IN THE EXPERT ROOM

Planning will always take place at the intersection between a number of different disciplines. Planners therefore often encounter challenges in their working day that require clarification and contributions from a number of other experts in other fields. The key to such a process is the development of communication methods that facilitate participation and do not isolate the planner, leaving them with just their own aptitudes and limitations to fall back on. One way of ensuring this type of open discussion is to establish what could be called an 'expert room', where the planner can meet other people who can help him or her solve problems. In this room other people's opinions and thoughts are aired and provided with fresh input from outside, while at the same time the room helps to ensure that a challenge is constantly viewed from new angles. The purpose of such an expert room is to create an arena in which complex interdisciplinary issues can be raised and discussed. The Norwegian Defence Estates Agency has good experience of utilizing professional planning and architectural advice when discussing individual land use plans. This advice can involve contributions made by employing geographic analysis through GIS. The end result should allow for different perspectives suited to the needs of the different participants. Improving the product by employing geographic analysis and presentations is also touched upon. The extent to which the planner wishes to listen to advice is a matter of personal choice. This process ensures that the plan works for various recipients and has been illuminated from a sufficient number of different angles.

The Norwegian Armed Forces and Military Academy employ military staff representing a multidisciplinary expertise which will vary depending on the complexity and difficulty of the operations being planned. During military operational planning, information on one's own forces, enemy forces, and the impact of the terrain on the operation will be essential in order to make good decisions and to fulfil the objectives. This topic is discussed further in Chapter 6 ('Military geography').

Different methods and approaches that are utilized by different professional groups can be conceptualized by the idea of an 'expert room'. In this context GIS naturally fits as one specialist perspective which can provide a common language for analysis and a distinct platform for the exchange and dissemination of knowledge. However, this requires continuous differentiation between the nature of the tool and the opportunities it provides; the focus being on the inputs fed into the system to allow it to generate more refined and accurate representation to handle more complex problems rather than a focus on its technological operations. Staging such challenges or opposition is crucial when it comes to ensuring that planners can question the issues on which they are working.

1.19 The 'expert room' – a meeting place where professional issues can be raised and discussed without the individual's integrity being challenged.

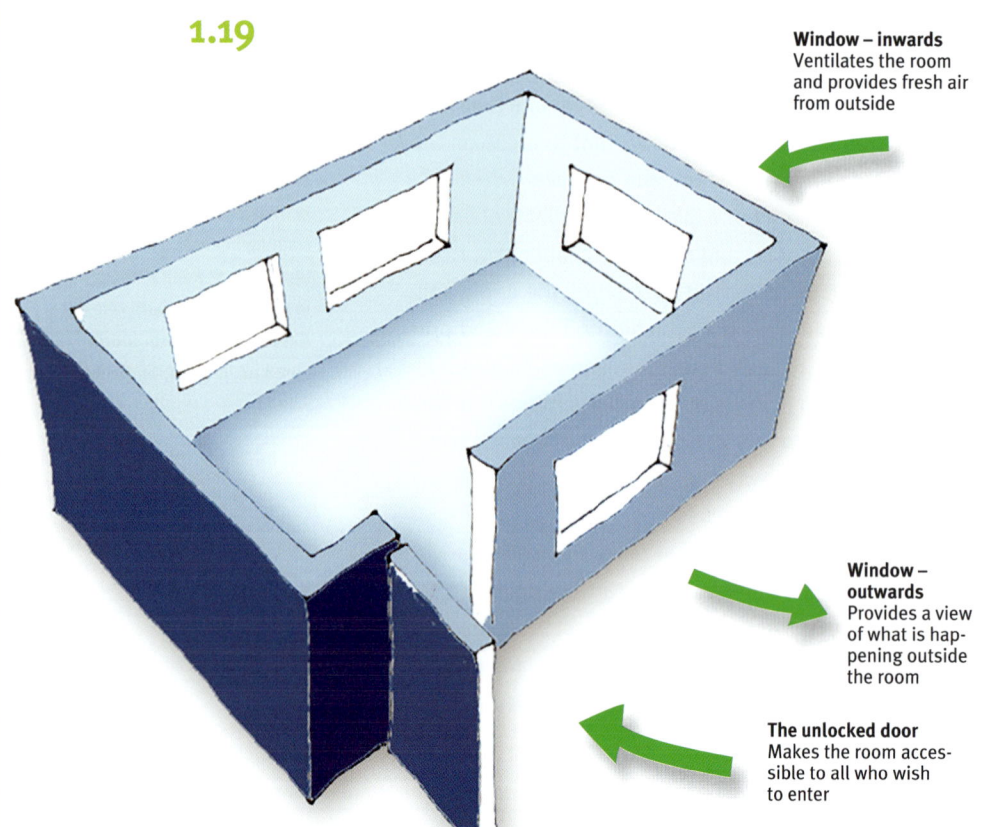

1.19

Window – inwards
Ventilates the room and provides fresh air from outside

Window – outwards
Provides a view of what is happening outside the room

The unlocked door
Makes the room accessible to all who wish to enter

Known goal and method

Unknown goal, known method

Known goal, unknown method

Unknown goal, unknown method

Characteristically, planning tends to have far more actors on the recipients' side than on the planners' or politicians' side. The results of planning affect most people in some way or other. This necessitates an ever greater degree of interdisciplinarity and cooperation between different disciplines, planners, and those affected by the planning. Such interdisciplinarity requires a common language. Traditionally, the lack of ability to communicate has excluded many disciplines and users, and hence excluded a number of actors from participation in planning.

As a geographic language, GIS can help to reverse the tendency for certain voices to be excluded; for example by lowering the threshold for raising questions and coping with challenges, it should be possible to have far more nuanced, accurate, and well-founded planning.

THE TYPE OF PLANNING

Structurally speaking, most of the problems faced by planners can be grouped into four categories, depending on the extent to which the goals and methods are known or are unknown (Obeng's method, 1994).

Painting by numbers

Category 1: If both the goal and the method are known, this can be described as a 'painting by numbers' scenario. Ideally, this is the dream situation for a planner. All the factors in their plan are known and they simply have to use a known method to achieve the correct result. Unfortunately, few situations in planning are like this. In this context one would primarily use a GIS to show results on various map representations.

1.20 Structuring of planning into four main categories depending on the extent to which goals and/or methods are known or unknown.

Filming
Category 2: If the goal is unknown, but the method is known, this is a 'filming' scenario. This is the most common scenario where the goal is uncertain but the method is clearly understood. In such a scenario, it is the most important to ensure a good process and then the goals will become part of the results. Such planning processes require good, strong management as well as strong commitment from the involved parties. The use of GIS in analyses can contribute significantly in this respect. By analysing the impact of the various measures, one can work out the best possible basis for making the right decisions. In principle, the process is about creating the opportunity to ask the right questions, rather than focusing on looking for answers. Once questions have been defined, the answers will become implicit.

Quest
Category 3: When the goal is known and the method is unknown, this is a 'quest' scenario, where the destination is well understood but not the method to achieve this point. This requires a more investigative approach. From a local point of view many of the national goals and success criteria for land use will probably be perceived as such a 'quest'. Many visions and political ambitions can also be said to represent this pursuit of an end product. In this case it is vital to use tools that communicate the vision's point of view and the current challenges. What has to be done in order to realize the vision? Again, the ability and opportunity to ask questions are key criteria for tackling such a situation. One example of the use of a 'quest' is the Norwegian Armed Forces' reduction of its building portfolio from 6 to 3.5 million square metres in the period 2002–2008. In this case the goal was crystal clear and stipulated in parliamentary White Papers and objectives documents. However, achieving the goal depended on good resources and land use planning in the Norwegian Armed Forces.

Walking in the fog
Category 4: When both the goal and the method are unknown, the scenario can best be described as 'walking in the fog'. This is the hardest type of planning. However, if one does not become frustrated by the lack of a framework for the planning and manages to attack the task in a systematic manner, major opportunities can exist in this phase. Nonetheless, such a process requires many different methods for generating opposition and challenges, and one continually has to create situations in which one meets oneself coming the other way and to confront the motives behind the work. In other words, one has to use one's knowledge, will, and common sense to make it through the 'fog'.

Often, the challenge appears to be that solutions to a given task are sought, rather than analyses of the situation and planning how to solve the task through methodical questioning and adopting a reflective approach to the problem. Obeng's method can thus help to understand where one is and how to proceed in order to make progress.

STRATEGY THINKING
The art of warfare
The year is 1863 and America is fighting a war that will come to be known as the bloodiest war in the country's history. The civil war has raged for two years with each side gaining the upper hand only to lose the advantage shortly afterwards. Despite the northern states' technological and economic superiority, the southern states, led by General Robert E. Lee, have gained a slight advantage in the east. General Lee graduated at the top of his class from West Point and served as an engineer before the war broke out. He was known for his high morals, strong sense of leadership, and extremely good tactical evaluations of the terrain. After a few strategically important and impressive victories at Fredericksburg and Chancellorsville, Lee was about to lead his army northwards from Northern Virginia to invade the northern states in order to entice the Army of the Potomac out into the open. This tactic would give the advantage to his
better organized army and thus force a decisive victory.

General Lee moved his army north-west of the Blue Ridge Mountains and used the mountains for protection (see Figure 1.22). The formation and size of his army was then more difficult for the northern states army to detect since it was on the other side of the ridge. In order to make a good evaluation of the terrain, General Lee relied on good intelligence. General James E.B. Stuart led the cavalry division and was Lee's chief collector of intelligence on both the enemy and the terrain. The northern states army, led by newly installed General George G. Meade, was forced to move in the same direction to protect strategically important Washington.

General John Buford, who led Meade's cavalry division, discovered some aspects of Lee's strength at Gettysburg on 30 June 1863, while he (Buford) was on reconnaissance, he gathered vital information on the enemy by climbing up the bell tower of the town's church. He then analysed the visual 'data' in his head and soon realized how he could exploit the terrain to his advantage. Despite the inferior size of his troops he was able to contain major parts of Lee's army. Budeford then had to keep control of the terrain until Meade arrived with reinforcements.

On 1 June 1863, one of the most important battles in American history began – the Battle of Gettysburg. Lee chose to attack Buford's forces, although he had neither received intelligence from Stuart nor gathered his forces. The Union Army of the North, which was

1.21 General Robert E. Lee. Photo: Library of Congress, LC-DIG-cwpb-04402.

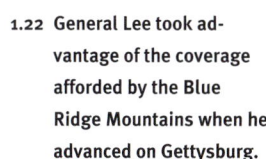

1.23 General George G. Meade. Photo: Library of Congress, LC-DIG-cwpbh-01199.

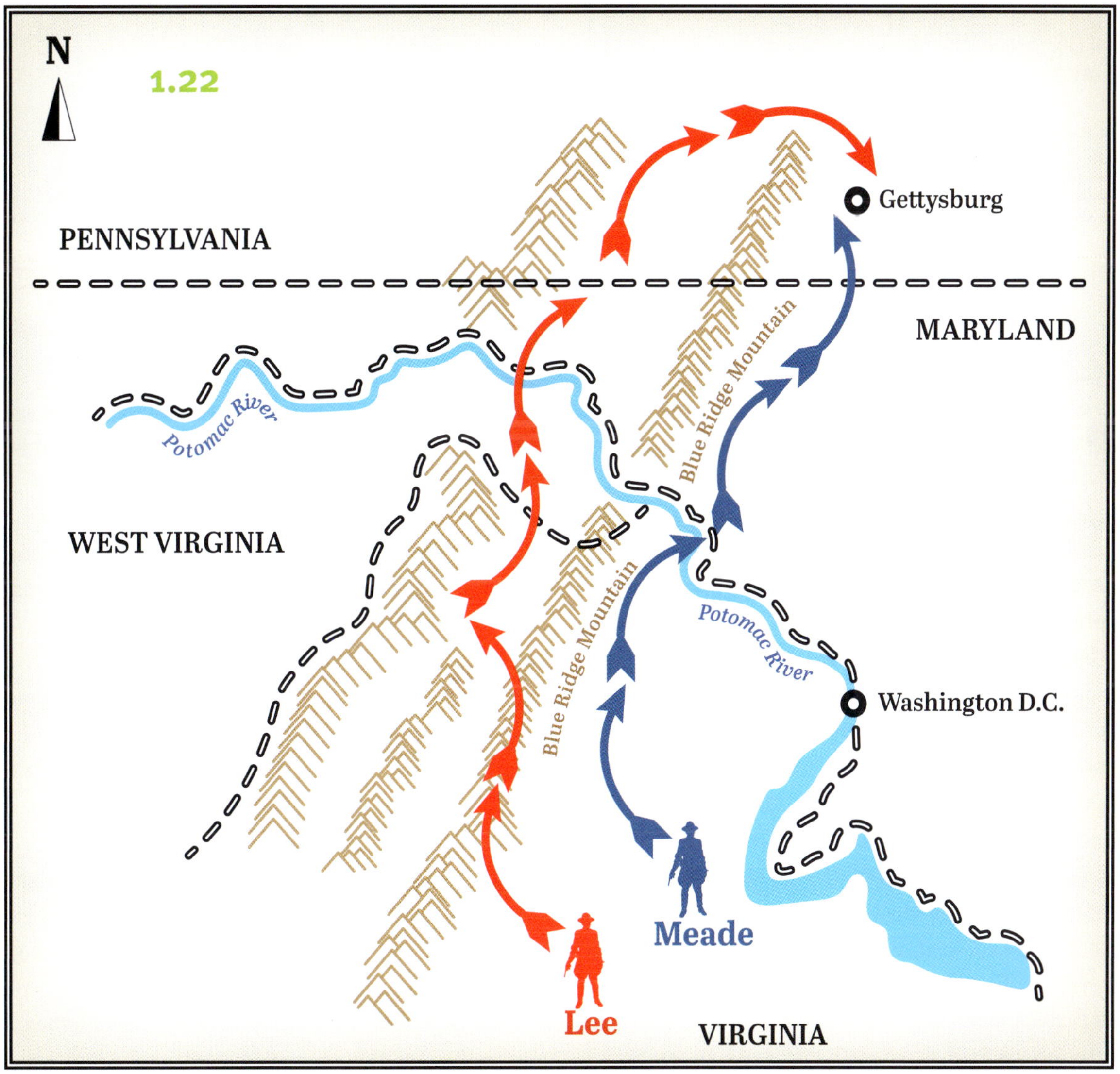

1.22 General Lee took advantage of the coverage afforded by the Blue Ridge Mountains when he advanced on Gettysburg.

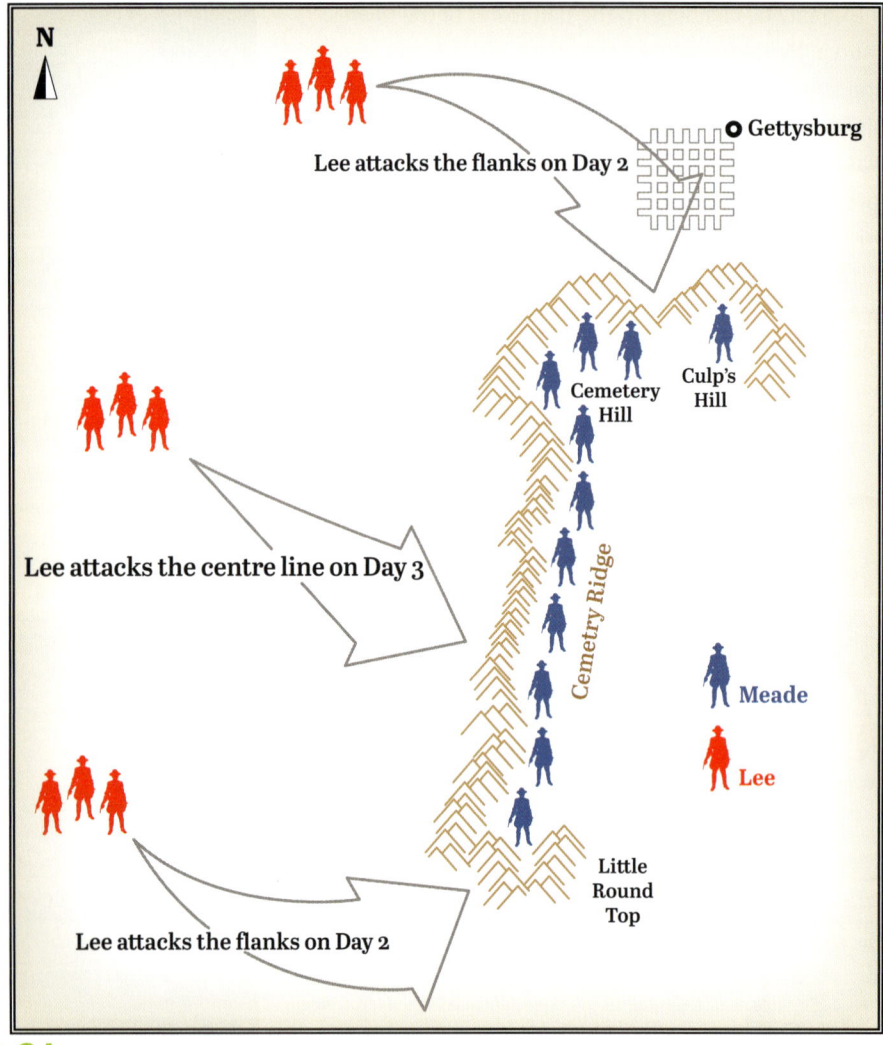

1.24

He had to fight uphill, which was extremely disadvantageous. Both sides suffered heavy losses that day.

On Day 3, Lee decided to concentrate on the centre of the Union line, even though it meant sending a force of 12,500 men out into the open. He believed that Meade would gather his forces on the flanks and therefore be weaker in the middle. The distance to the centre of the line was more than 1 km and the terrain had a slight uphill gradient. Lee's assessment proved fatal and may also have been the turning point in the war. 'Pickett's charge', as the attack was known, ended Lee's chances of winning at Gettysburg. Realizing this defeat, Lee took responsibility by withdrawing the remains of his army to Virginia in order to regroup.

1.25

Strategic planning
A common problem during planning is planners' inability to abstract themselves from the current situation and to anticipate the consequences of the planned measures. In military operations, planning involves first defining the desired outcome and thereafter preparing a plan for how the outcome can be achieved. The lack of foresight regarding the consequences has influenced planning at the global level. Modern planning is largely founded on different forms of impact assessments that are based on legislation. The detailed regulations governing environmental impact assessments, which should also be made applicable to municipalities' plans, are clear indications of Norwegian society's recognition of accurate planning practices. Such practices call for a methodology that avoids inertia. Case studies, scenario construction, and other types of performance methodologies are valuable tools that enable innovation within a comprehensive regime. One such planning method is backcasting (see Figure 1.25), whereby users aim to abstract themselves from a specific current situation and then imagine the situation 10–15 years ahead and look back

1.24 In terms in of the terrain, General Meade had the advantage on Cemetery Ridge, on Days 2 and 3 at Gettysburg.

1.25 'Back-casting' lets you take a step away from the current situation and allows you to take a less constrained look at the future.

even less well organized, was pushed back due to the southern states' numerical superiority. They established a new ridge, Cemetery Ridge, to the south of the city and used this advantage to maintain their strong position until the army had been gathered.

Meade arrived at the battlefield late the same evening and held a council of war with his generals, during which they agreed to maintain their strong position. Lee had not yet heard anything from Stuart, who had made his own assessment and conducted a more extensive reconnaissance than Lee. Nevertheless, Lee chose to attack the heavily fortified flanks (Little Round Top as well as Cemetery Hill and the adjacent Culp's Hill) of the northern states on the following day (Day 2) (see Figure 1.24).

to identify what measures are necessary to bring about such the desired development in the future. Through the systematic use of the time factor in such exercises it should be possible to evaluate differences in opportunities and limitations in the short, medium, and long term. Hence, it should be possible to see whether some short-term solutions would either enable or preclude other solutions in the longer term. Such planning is thus affected by the decisions on measures to be taken along the way, and approaches dynamic and timely planning, yet at the same time long-term choices can form the basis for sustainable development.

With regard to the Battle of Gettysburg, it is interesting to ask the following questions: What would Lee have done if he had had adequate information about the battlefield's terrain? Would he still have chosen to fight there or he would have chosen to withdraw his army to Washington, where he could have chosen a more advantageous battlefield? Could he have exploited the flanks better and surrounded Meade's army? Moreover, could he have done this if the various scenarios based on terrain information had been adequately visualized before the attack strategy had been formulated? Lee had the following strategy options: he could still attack the flanks, he could attack the centre line, or he could withdraw his army towards Washington.

The significance of the terrain alone was not decisive for the outcome of the Battle of Gettysburg, but rather the lack of visibility of the various scenarios that influenced the leaders' strategic decisions was an important factor in Lee's defeat and the turning point for this historically important war.

Restructuring of the Norwegian Armed Forces
Perhaps the most significant objective in restructuring the Norwegian Armed Forces in the 2000s, was the objective to reduce the portfolio of buildings of 6 million m² by almost half (see Figure 1.26). This objective has raised an important question: How would the new Norwegian Armed Forces be housed? In turn, this question highlighted a number of different issues and formed the basis for a new future in which there would be a shift away from earlier ideas of scattered structures towards more concentrated regions or power centres, and so required the use of 'backcast' thinking to try to arrive at the way to achieve least upset while still keeping the desired outcome in view. Once it has been identified that potential processes and initiatives were moving in the right direction, the work on evaluating different strategic choices could begin by comparing them. One important consideration with regard to the restructuring of the Norwegian Armed Forces was how the different choices influenced the local communities. In many respects, the Armed Forces can be considered as a key industry in many regions of the country meaning the loss of jobs and consequent depopulation needed to be factored into the picture. A further important planning factor is flexibility in relation to changes in the security situation worldwide and any threats against Norway.

One example of such 'conceptions' is exemplified by the national plan for the Home Guard. In common with other branches of the armed forces, the Home Guard had a dispersed building structure that was oversized

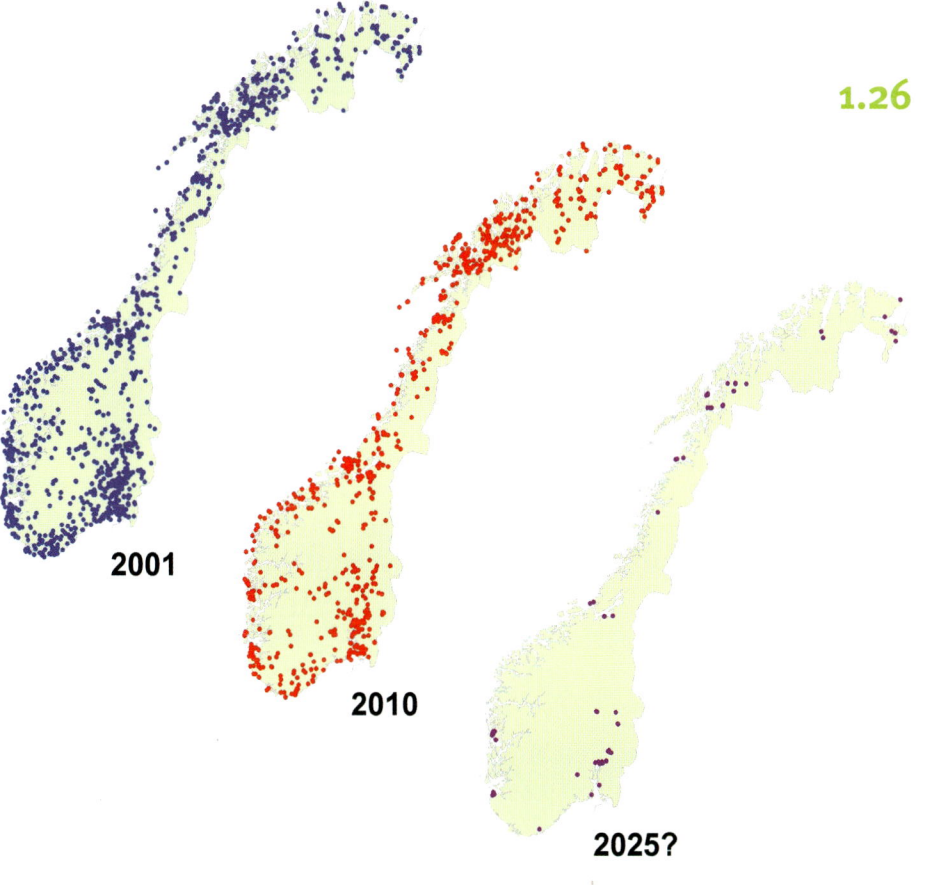

1.26 The Norwegian Armed Forces are moving away from a dispersed presence in almost every local authority to centralisation around some key centres of power.

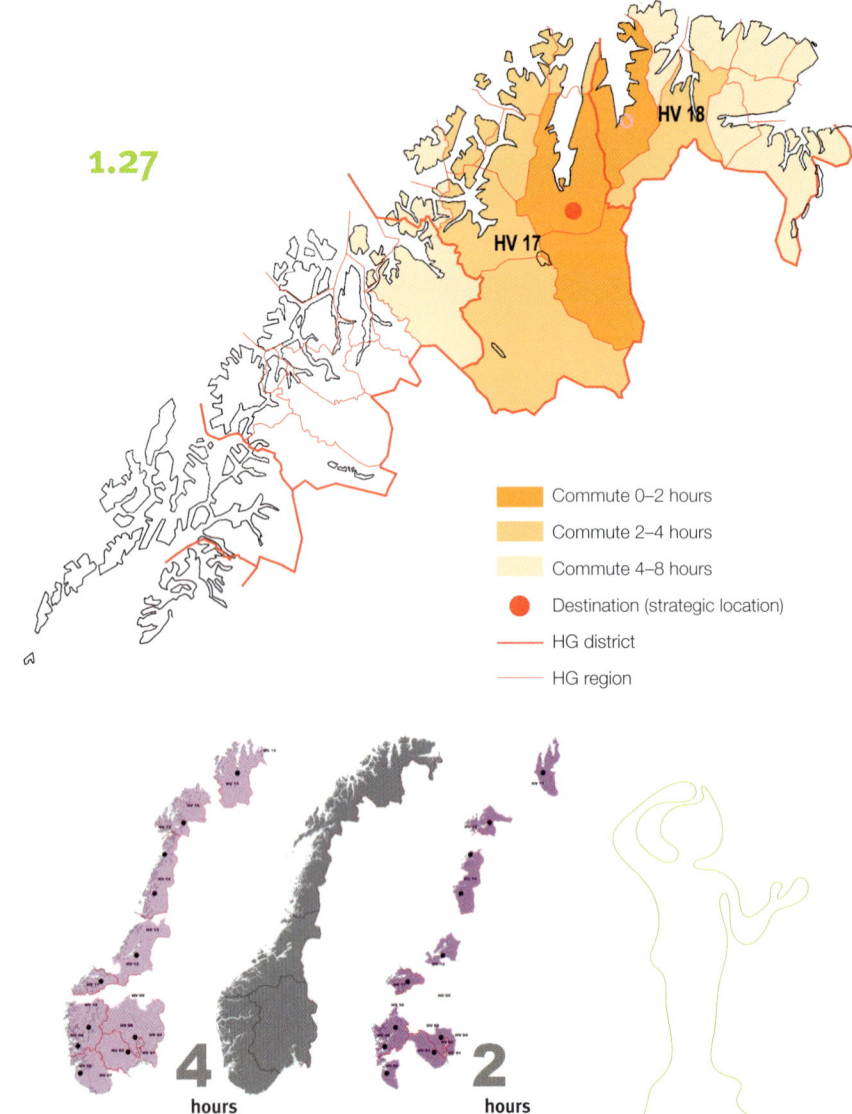

following the new developments. An important fact to consider is that personnel and volunteers of the Home Guard often have to travel long distances from their homes to the regional training centres. When they are called to attend refresher courses at the centres, travel times are an important determining factor.
In order to reduce the total building stock, GIS studies were conducted in the 2000s and they revealed what degree of accessibility could be achieved via the road network and by car transport from all municipalities in the country towards more centralized training centres. The analyses of the personnel's travelling time to such envisaged centres led to the creation of a scenario in which the country was divided into 10 new exercise regions, each with its own fitness centre and with good availability of personnel throughout the country.

From national guidelines to local solutions

Planning is always taking place in Norway at various levels. It can be difficult to discern an understanding of long-term thinking and strategy development from studies of the often pragmatic administration of local problems. The Norwegian Government and Storting (Parliament) take decisions and establish guidelines at a national level. However, solutions are created at a local level and these often bear little resemblance to the national guidelines. A good example of this divergence is studies of the number of dispensations that municipalities granted from their own planning regulations. The criterion that often generates uncertainty about universal plans is the extent to which they can reflect contemporary, local problems. This is often due to the fact that plans are drawn up at different levels and with different time horizons and hence with different prioritizations. However, this lack of correspondence can constitute a fertile feedback process. It is easy to view such scrutiny as a type of expert room. This brings us to the need for communication, language, and shared arenas for discussing the various standpoints with respect to plans.

In this context there are already clear and visible areas for the use of GIS. Organizing collective infor-

1.27–1.28 Commute time analysis by the Norwegian Home Guard illustrating various alternatives for the location of training centres.

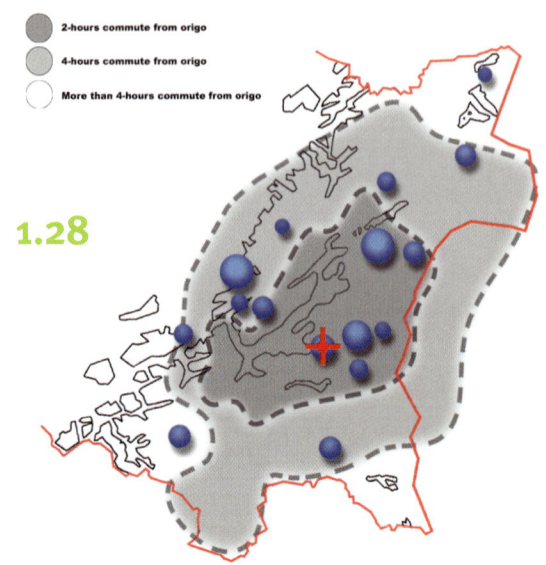

mation distribution solutions, such as *Norge digitalt*, makes information of national significance more easily available to local actors. At the same time, local information becomes available to national actors.

The Norwegian Defence Estates Agency (NDEA) is tasked with fulfilling all of the Norwegian Armed Forces' duties in all types of land use planning, including making the necessary contact with local authorities, county administrations, and other state agencies in this connection. This task was fulfilled by a process in which all municipal and county area development plans and other planned measures were sent to the NDEA for scrutiny. An NDEA planner studied these plans and conferred with professional military resources where necessary. If a conflict of interests existed, the process of addressing the Norwegian Armed Forces' interests was very cumbersome and often entailed having to submit objections to the proposed plan. This was clearly time consuming and expensive, and the use of objections was far too common.

In the case of some prioritized state projects, conflicts of interest have been discussed in the media as well as by the government. One example is the conflict of interests between the Norwegian Armed Forces' radar facilities and new wind power facilities.

For its part, the NDEA wanted to speak with possible initiative-takers in advance. This would have resulted in a more flexible process and the reduced the use of resources and the number of objections, yet at the same time the quality of the work would have increased. In other words, the NDEA wanted to reorganize its planning system in order to create predictability through binding collaboration at all stages of the planning process. To achieve this, it was entirely dependent on the various parties with interests in the land being surveyed and listed somewhere where all local authorities, county administrations, and others could easily get in touch with them.

The White Paper titled *'Norge digitalt' – et felles fundament for verdiskaping* [Norge digitalt – A Common Foundation for Value Creation] (St.meld. nr. 30 (2002–2003)) states: 'All public agencies that have geodata responsibilities or are major users shall con-

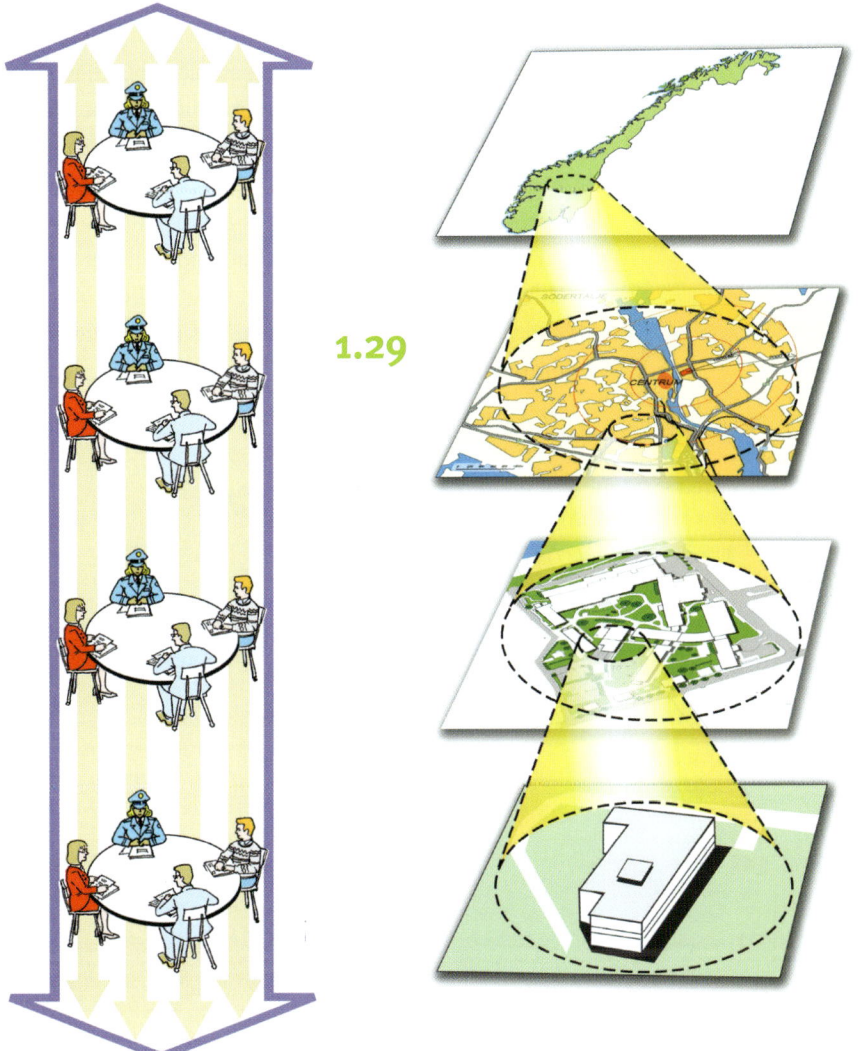

tribute to the establishment, operation and maintenance of Norge digitalt.'

In Norge digitalt it is possible, for example, for the NDEA to define the Norwegian Armed Forces' boundaries and special policy areas in which defence activities require municipal plans to set special conditions for civilian use of the same land. This is a major undertaking and much work remains to be done, but eventually it will streamline administrative procedures, since it enables more efficient, targeted, and appropriate case management through dialogue, with the affected parties' interests clarified in advance.

At the same time as state interests can be made available through Norge digitalt, the local authorities'

1.29 Using new technology makes it possible to see the relationship between national and local levels and vice versa. Figure: Ulf Ranhagen, AB Media.

ongoing prioritizations will, through their own plans, be visible to all as important feedback on national value choices. The use of modern mapping tools allows the retrieval of information (previously unavailable) to be presented in map form, around which interested parties can assess their needs and priotities. This method of providing access can help to lower the threshold for various actors wanting to be involved and to challenge and ask questions about relevant problems.

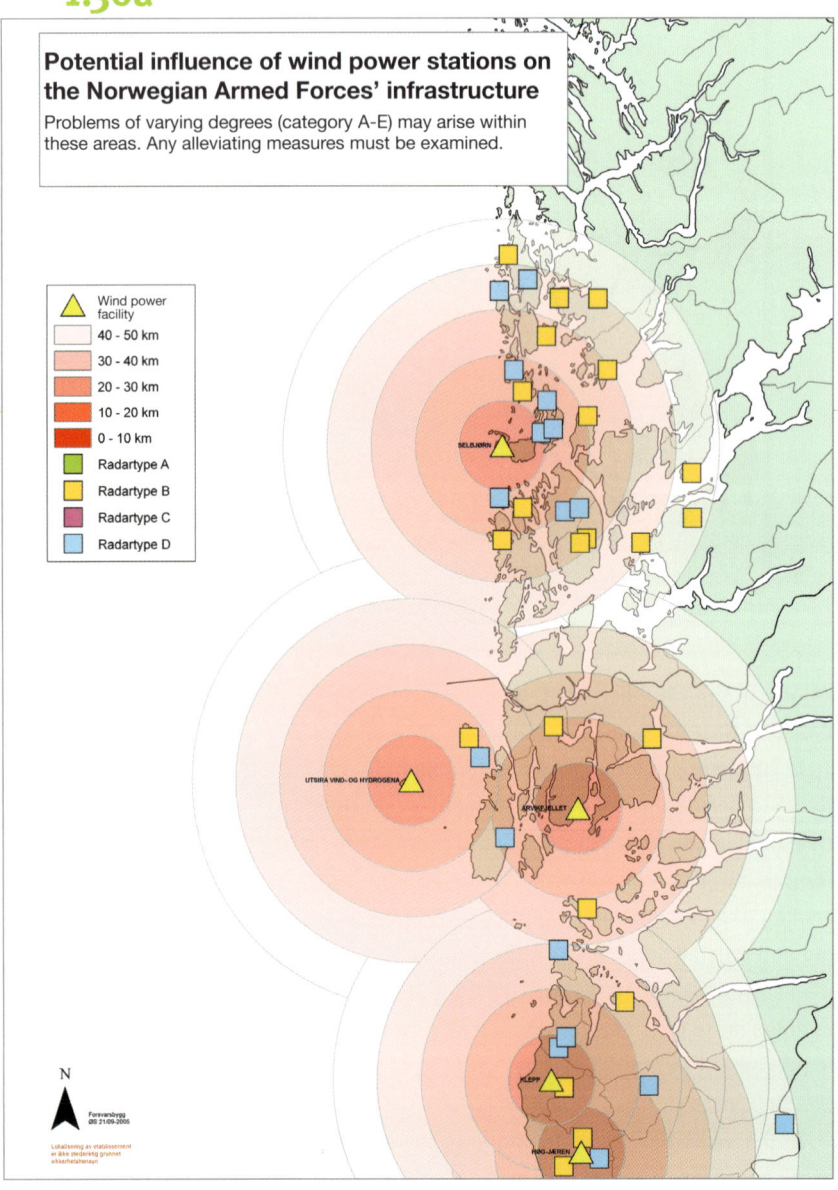

1.30a–c Visualisation of planned wind power stations shows conflicts with existing radar facilities. Source: Gjermund Wallenius (windmill photo) – Torgeir Haugaard/Norwegian Armed Forces Media Centre (radar station photo).

The northern regions

In a speech about the government's Far North policy given in September 2006, the then Norwegian Minister of Defence, Anne Grete Strøm-Erichsen, said 'The world map looks different depending on what we assume is the centre'.

The increased focus of attention on the Barents Sea region is based on different views on how a map of the world appears, from both a geographic and legal perspective. The region has a number of administrative boundaries that have been the subject of considerable disagreement between nations.

Examples of such uncertainty include the 200 nautical mile maritime economic zones that coastal states were able to establish from the late 1970s. The formal basis of these rights can be found in the United Nations' Convention on the Law of the Sea (UNCLOS) of 10 December 1982. The Convention gives coastal states rights over natural resources on the seabed and in the waters above it. This implies that they have exclusive rights to manage natural resources, such as oil, gas, and fishing, within their own economic zone.

Such economic zones can be determined from a buffer analysis, as discussed in more detail in Chapter 4 (Analysis). With the establishment of these zones, each country was granted the authority to control and manage the fishing resources within their zone. As a general rule, foreigners are prohibited from fishing and hunting in Norway's economic zone unless agreements have been made with individual countries, including EU countries. There has been a long, ongoing negotiations about the maritime boundary between Norway and Russia. The Norwegian claim has been based on the application of the median line principle, which is UNCLOS' main principle for determining maritime boundaries. The Russian claim has been based on sector line principle, which follows the geographic meridians or longitudes. In 2010 Norway and Russia signed an agreement on the maritime boundary in the Barents Sea and the Arctic Ocean (see Figure 1.33), and the agreement entered into force on 7 July 2011.

A 200 nautical mile fisheries protection zone has been established around Svalbard and Jan Mayan. However, this protection zone is only recognized by Finland and Canada.

When the Norwegian Coast Guard monitors Norway's maritime waters for any unregulated fishing in the fisheries protection zones, they do this in accordance with the boundaries shown in Figure 1.33. The fact that there is no consensus between Norway and Russia on the boundary between their respective economic zones has led to a number of incidents. Possibly the most widely publicized incident to have resulted from such uncertainty was the 'Elektron incident' in 2005. At that time, Norway and Russia were in dispute over the position of the boundary line.

On 15 October 2005, the Russian fishing vessel Elektron was stopped by the Norwegian Coast Guard when it entered the fisheries protection zone that Norway had established around Svalbard. The Norwegian Coast Guard inspectors discovered several breaches of Norwegian fisheries legislation and therefore placed Elektron's captain under arrest and ordered the trawler to set course for the port of Tromsø. After initally agreeing to sail for Tromsø, Elektron changed course and quickly headed for Russian waters, with two Norwegian Coast Guard inspectors from the Coast Guard ship Tromsø on-board. Despite repeated attempts to force Elektron to change course for the Norwegian port, the Russian captain failed to comply with the Coast Guard's orders. Consequently, the fishing vessel was also in violation of the Norwegian Coastguard Act of 1999. Subsequently, after intense diplomatic discussions between Norwegian and Russian authorities, the two inspectors were eventually handed over to Norway almost one week after the drama had started.

In retrospect, the Norwegian Coast Guard was praised for keeping a cool head and not taking more drastic steps to stop Elektron, but was criticized for failing to enforce Norwegian law more effectivley. Potentially, the incident could have quickly developed into a dramatic crisis with major foreign policy consequences.

Legal proceedings against the Russian company were started in the wake of the incident. In a Russian court, *Elektron*'s captain was acquitted of kidnapping of the two Norwegian Coast Guard inspectors, but was fined RUB 100,000 (around EUR 2,800) for illegal fishing activities.

There was considerable disagreement on the principles concerning *Elektron*'s fishing activities in the protection zone around Svalbard. In an interview, the Head of the Russian Fishermen's Union of the North Gennady Stepakhno, emphasized the Russian perspective as follows: '*Elektron* was in an area that only three countries recognize as Norwegian waters, namely Norway, Finland, and Canada. Russia has never recognized the 200 nautical mile Norwegian protection zone around Svalbard' (Internet newspaper: russland.RU).

In the coming years, the conflicting perceptions of boundaries and rights between states in both the Barents Sea region and the Arctic could have enormous value implications.

1.31 The Norwegian Coast Guard ship, Tromsø. Photo: Jørgen Holst/Norwegian Armed Forces Media Centre.

1.32 A fishing trawler being inspected by the Norwegian Coast Guard. Photo: Morten Karlsen /Norwegian Armed Forces Media Centre.

1.33 Maritime borders in the northern regions. Source: Norwegian Military Geograpic Service.

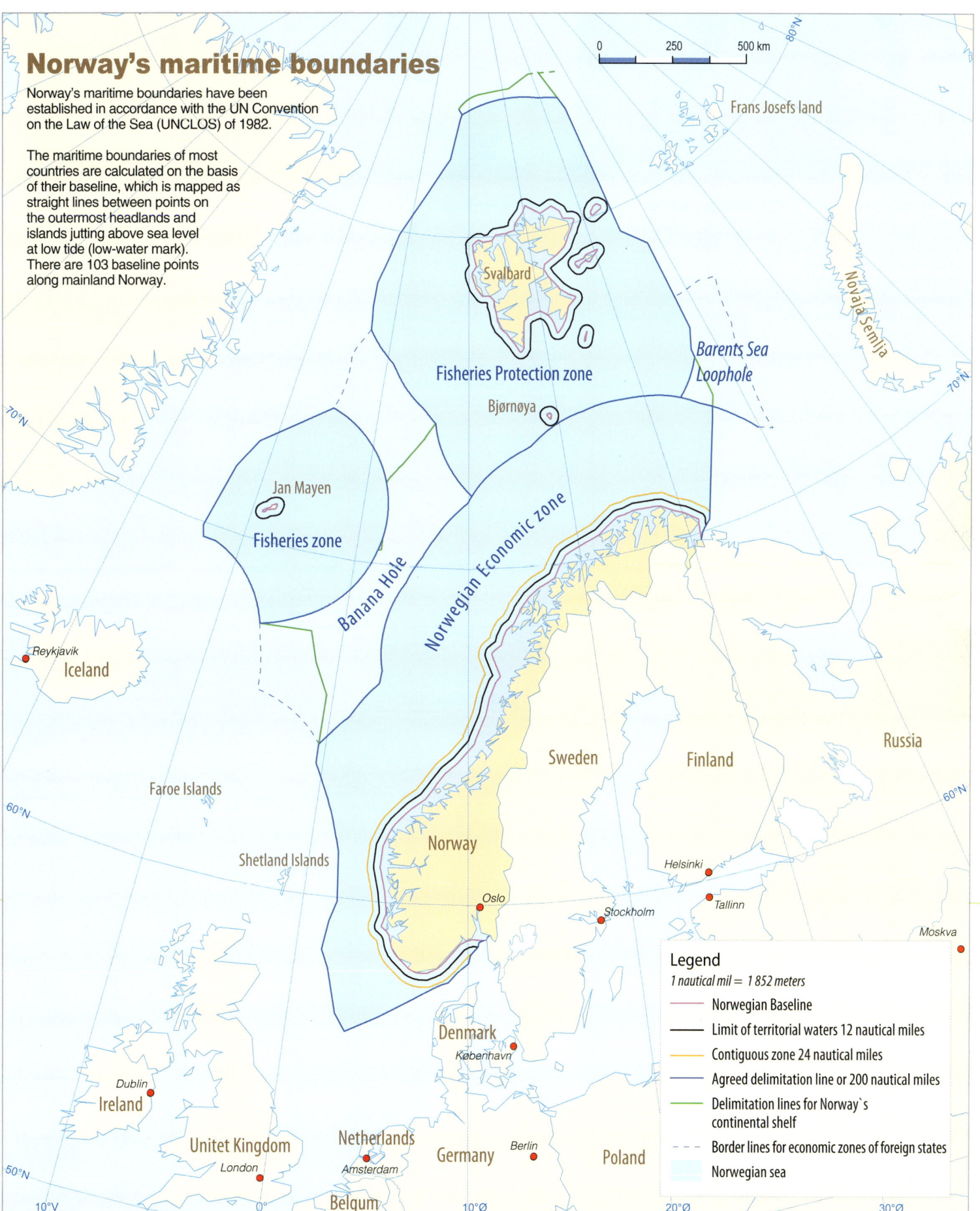

Øystein Sanderud holds a master's degree in geodesy from the Norwegian University of Science and Technology. He works as a senior engineer in GIS and map administration at the Norwegian Defence Estates Agency.

Halvard Bjerke holds a master's degree in geomatics from the Norwegian University of Life Sciences. He is an Assistant Professor of engineering at the Norwegian Military Academy.

Magnus Fjetland graduated from the Norwegian Military Academy in 2011. At the academy he studied Military Geography and got a bachelor degree in civil and military engineering. He holds the rank captain and works as an instructor in engineering at the Norwegian Military Academy. Captain Fjetland has a background from both national and international service in the Norwegian Armed Forces.

The chapter is a revised version of the chapter written by Atle Holten, Per Gunnar Ulveseth and Ingvill Richardsen in the 1st edition.

Systems 2

Makes the earth appear flat	40
A computer model of reality	46
Data collection of geographic data	50
Data quality	60
Sources of errors	63
SOSI	64
Geographic data in databases	64
Web-based services	67
Technological trends	69

Makes the earth appear flat

Chapter 1 (Application) deals with the application and use of GIS. In this respect, knowledge of the earth's shape, gravity field, and changes in the gravity field is vital. Such knowledge is called *geodesy*, a word derived from the Greek word *geodaisia*, which refers to the *division of the earth*. One example of the technical application of geodesy is surveying, which includes the use of coordinates for mapping areas. The purpose of a map is to recreate a two-dimensional surface representing the real world in the most practical manner. For this to be possible, geographic reference systems have to be used as a framework to fix the location of objects. There are several different types of geographic reference systems, but geodetic reference systems are often used to determine precise locations. In order to understand the relationship between the different reference systems, it is essential to have some background knowledge of them.

GEOID

Unlike the surfaces of the oceans and seas, land surfaces are very rough. The large vertical variations between mountains and valleys, as well as several other factors that affect the earth's shape, mean that it is almost impossible to describe the earth's surface by using just one mathematical model. However, the process can be simplified by using an *equipotential surface*, where the potential (the gravity field) is equal over the entire surface. There are a number of models with equipotential surfaces, and the *geoid* is one of them. The geoid represents the mean sea level, but the geoid's potential is still influenced by large mountain masses. In such areas, the geoid can be interpreted as an implied line through the terrain and visualized as an uneven surface, as shown in Figure 2.1. This means that the geoid is the only equipotential surface that is akin to the physical reality on earth. In land and sea topography, the geoid model is used as reference surface and allows for descriptions of gradients and inertial navigation.

ELLIPSOID

A reference surface is needed in order to calculate the position, distance, and direction of the earth's surface. If the earth had a uniform density and an even surface topography, the geoid would be shaped like an *ellipsoid*, which results from an ellipse rotated around its minor axis. This due to the fact that the earth is not perfectly spherical but slightly flattened at the poles, resulting in an elliptical shape, as shown in Figure 2.2. Unfortunately, the reality is not as simple. Due to the topography, the ellipsoid's surface varies relative to the geoid. Since the ellipsoid does not fully correspond to the geoid, multiple ellipsoids have to be calculated in order to provide the best possible reference surface in different areas. The different ellipsoids correspond closely to the geoid and have the highest degree of accuracy in the areas where they are to be used. Ellipsoids, as simplified models of the geoid, form the basis of defining reference systems known as *datum*.

DATUM

Datum – a geodetic reference system – describes the ellipsoid and how it relates to the geoid. Geodetic reference systems include both *vertical* datum and *horizontal* datum, which determine height and position respectively. Vertical datum often refers to the geoid as a reference surface, whereas horizontal datum usually refers to the ellipsoid as a reference surface. Horizontal datum describes the size and location of the ellipsoid

2.1 Geoid. Source: Nicolas Primola / Shutterstock.

(the reference surface) using a *fundamental point*. In the fundamental point – the height of the geoid – is zero, and is the point where the geoid and ellipsoid coincide, as shown in Figure 2.3. There are different types of horizontal datums and they fall into three main categories: *local*, *regional*, and *global* datum.

Local datums are adapted for local conditions and are mainly used within a limited area. Norway's local geodetic datum for 1948 – NGO1948 (Norges Geografiske Oppmålingsdatum for 1948) – was formerly the country's official horizontal datum and the fundamental point was at The Astronomical Observatory in Oslo. NGO1948 was used for economic maps (at scales of 1:5000 – 1:10 000), while European Datum 1950 (ED50) was long used for Norway's main map series at a scale of 1:50 000. Both datums have since been replaced by the European Reference Frame 1989 (EUREF89), which is now Norway's official datum. The ED50 and EUREF89 are regional datums adapted for large areas, in this case Europe. Hence, the fundamental point is located in Europe. The fundamental point for ED50 is in Potsdam, Germany. EUREF89 is based on precise satellite measurements for points across Europe and is really a reflection of the World Geodetic System 1984 (WGS84) as it appeared in 1989. Due to continental drift, various geographical positions in Norway have moved by up to 30 cm, which corresponds to the difference between WGS84 and EUREF89. WGS84 is a global datum calculated using satellite measurements. Unlike local and regional datums, which are based on fundamental points, the centre of the ellipsoid is at the centre at the earth's mass, and therefore it is important to know which datum geographic data relate to. The deviation between EUREF89 and WGS84 is less than 1 m and in many cases this can be regarded as insignificant, but the discrepancy between ED50 and EUREF89 can be up to 200–300 m.

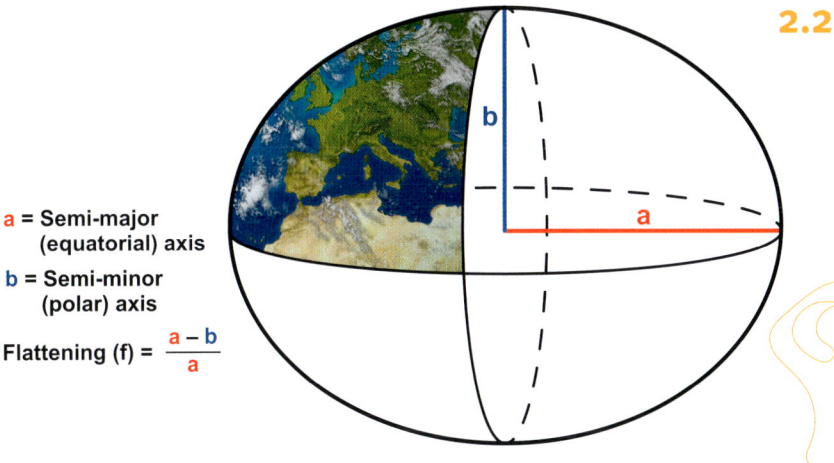

a = Semi-major (equatorial) axis
b = Semi-minor (polar) axis
Flattening $(f) = \dfrac{a - b}{a}$

MAP PROJECTION

In order to prepare three-dimensional maps, the ground surface has to be projected onto a two-dimensional plane. However, since the ellipsoid has curvature in two directions, this does not allow for projection onto a two-dimensional plane and therefore map projection is used instead. *Map projection* can be explained as a three-stage approach. First, terrain points are projected onto a reference surface with two curves, in this case the ellipsoid. Thereafter, the points are projected onto a geometric surface that flattens out into a two-dimensional map. In practice, one could envisage a globe with a light source in the middle. The light from within the globe would cast an image of the globe's surface onto a hypothetical two-dimensional surface.

2.2 Ellipsoid.

2.3 Terms used to specify heights.

Orthometric height (H_0) – height above the geoid
Ellipsoid height (H_e) – height above the ellipsoid
Geoid height ($H_e - H_0 = N$) – difference between the geoid and the ellipsoid

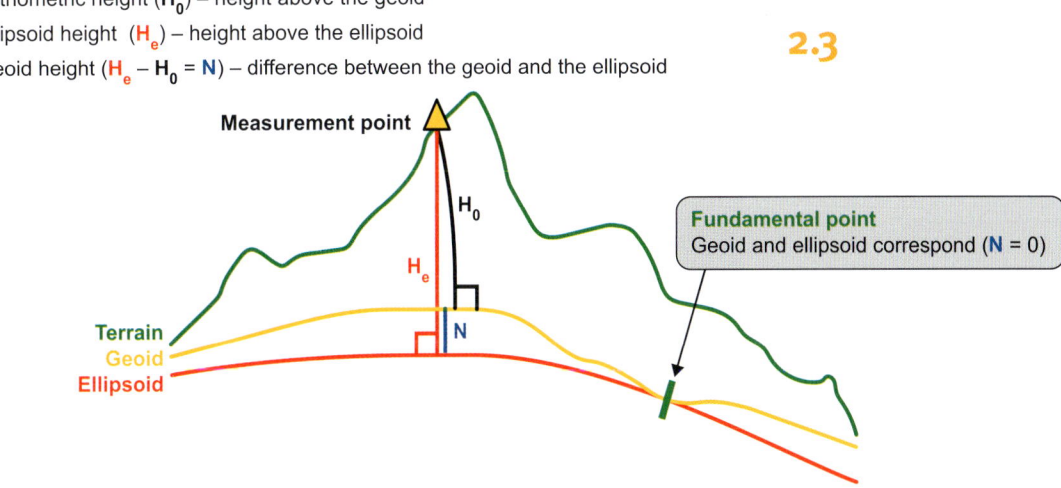

Three different types of surfaces are mainly used in map projection:

- plane projection
- cone projection
- cylinder projection.

Since the ellipsoid is curved in two directions, its properties cannot be retained when an image of its surface is projected onto a two-dimensional surface. As shown in Figure 2.4, the projection results in errors on the mapped surface. Hence, it is necessary to decide whether the projected image should show distances on the surface in the right scale (*conformal projection*), show distances between points correctly (*equidistant projection*), or show surface areas accurately (*equal-area projection*) in relation to the physical reality. It is not possible to project an image of the ellipsoid onto a plane in such a way that all three conditions are satisfied simultaneously. In Norway, transverse (lying) cylinder projection is used for map production, namely Universal Transverse Mercator Projection (UTM). UTM is based on the Gauss-Krüger projection, which maintains perpendicular imaging, such that the original shape of land is preserved in the image.

There are several adaptations of Gauss-Krüger projection, such as Transverse Mercator projection. In a Transverse Mercator projection, the cylinder lies tangential to the surface of the earth in the y-axis, along a north–south meridian (tangential meridian), with the positive direction towards the north. The x-axis follows the equator with a positive direction towards the east. The system can be seen as straight lines in an x- and y-axis system. The other meridians (longitudes) and latitudes are depicted as complex curves. The properties of this conformal projection have led to its widespread use, such as for surveying and the production of large-scale maps. The UTM projection is based on such a Gauss-Krüger projection and is thus a transverse projection with the cylinder in a horizontal position. In the UTM, the globe is divided into 60 zones and each zone is projected using the Gauss-Krüger projection. It involves a significantly higher degree of accuracy compared to projecting the globe in a single operation.

UTM projection has a separate coordinate system, and the adoption of UTM for official map projection has

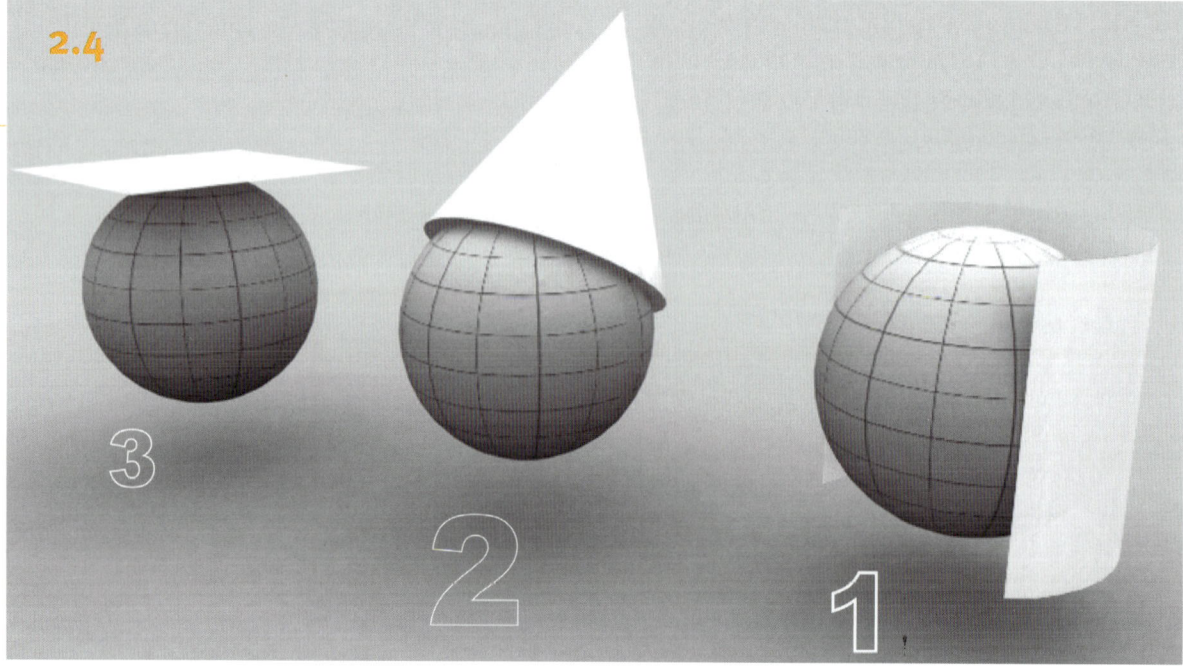

2.4 Three geometric surfaces are mainly used for map projection: 1. cylinder, 2. cone, and 3. plane.

resulted in some undesirable results. Problems arise when there are particularly strict geometric tolerances. UTM projection has a scale factor that indicates that a distance measured in the terrain must be adjusted by up to 400 parts per million (ppm) or 4 cm per 100 m. Thus, in practice, a stretch of road that measures 100 m in the terrain will effectively be 100.04 m in the projection. Although this may be insignificant for most purposes, it creates challenges for projects with strict tolerance requirements. EUREF89 Norwegian Transversal Mercator (NTM) has therefore been introduced as a secondary Norwegian projection. EUREF89 NTM projection uses the same reference surface as UTM. It corrects the scale factor to a great extent, and in practice it has a maximum scale correction of 11.4 ppm for the whole country, from north to south. Data produced by the municipalities and the Norwegian Mapping Authority will continue to be stored and administered in EUREF89 UTM, and the input and output of map data will continue to be administered in EUREF89 UTM.

COORDINATE SYSTEMS

Coordinate systems are used to indicate the position of a terrain point. Perpendicular coordinate systems use x- and y-axes to define a point's location, as well as the z-axis in three-dimensional systems. Two of the most commonly used perpendicular coordinate systems are geographic and Cartesian coordinate systems.

Geographic coordinate system
The concepts of meridians and parallel circles are used when locating a specific point on the earth's surface. The circle on which a point occurs is assigned a meridian reference, which is the point at which a meridian intersects the circle. The circle's position is set first and indicates the point's position on the earth's

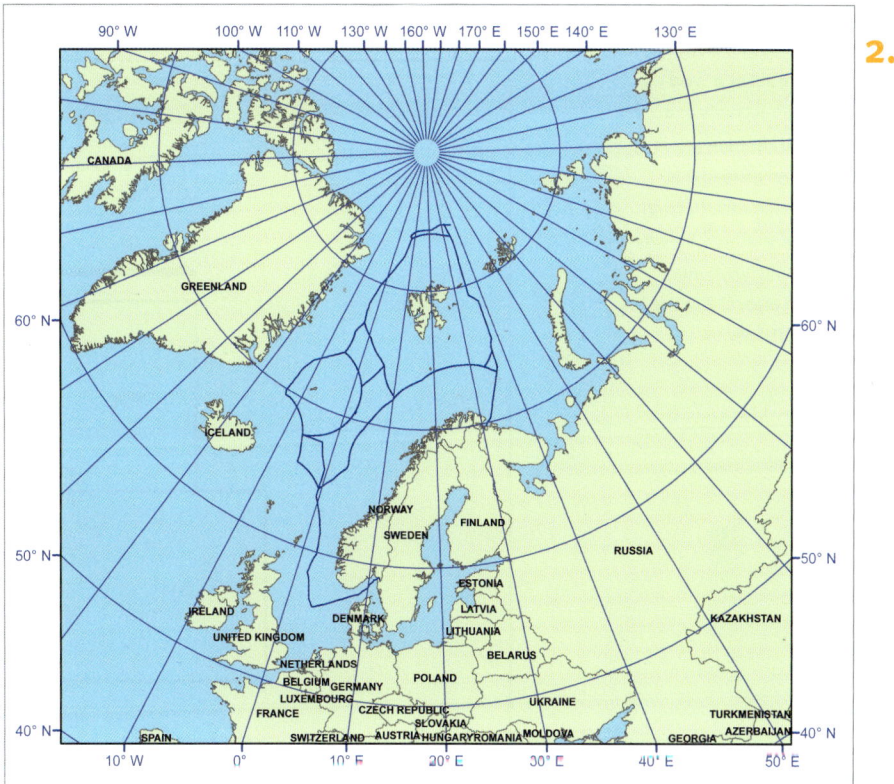

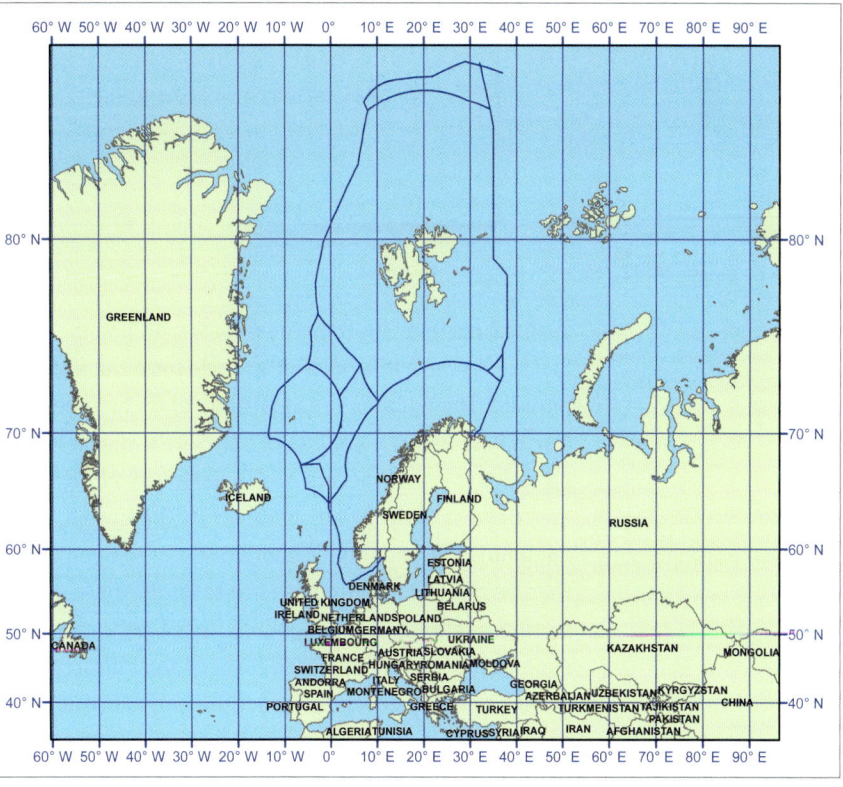

2.5–2.6   Map of the northern regions in two different map projections. The uppermost map is shown as a planar projection, the lowermost one as a cylindrical projection. Presentation of the same region in different projections gives different perceptions of the geographic conditions. See also the section 'The northern regions' in Chapter 1. Map data: ESRI, Data and Maps for ArcGIS (2014) © ESRI; Norway's maritime borders (2015) © Norwegian Mapping Authority.

2.7 Geographic coordinate system with UTM zones for Norway. Intervals are set at 8° and 6°, which corresponds to a UTM zone. UTM zone 32V is a non-uniform zone that is extended to the west to cover the whole of southern Norway and to simplify positioning. Background map: © Norwegian Mapping Authority.

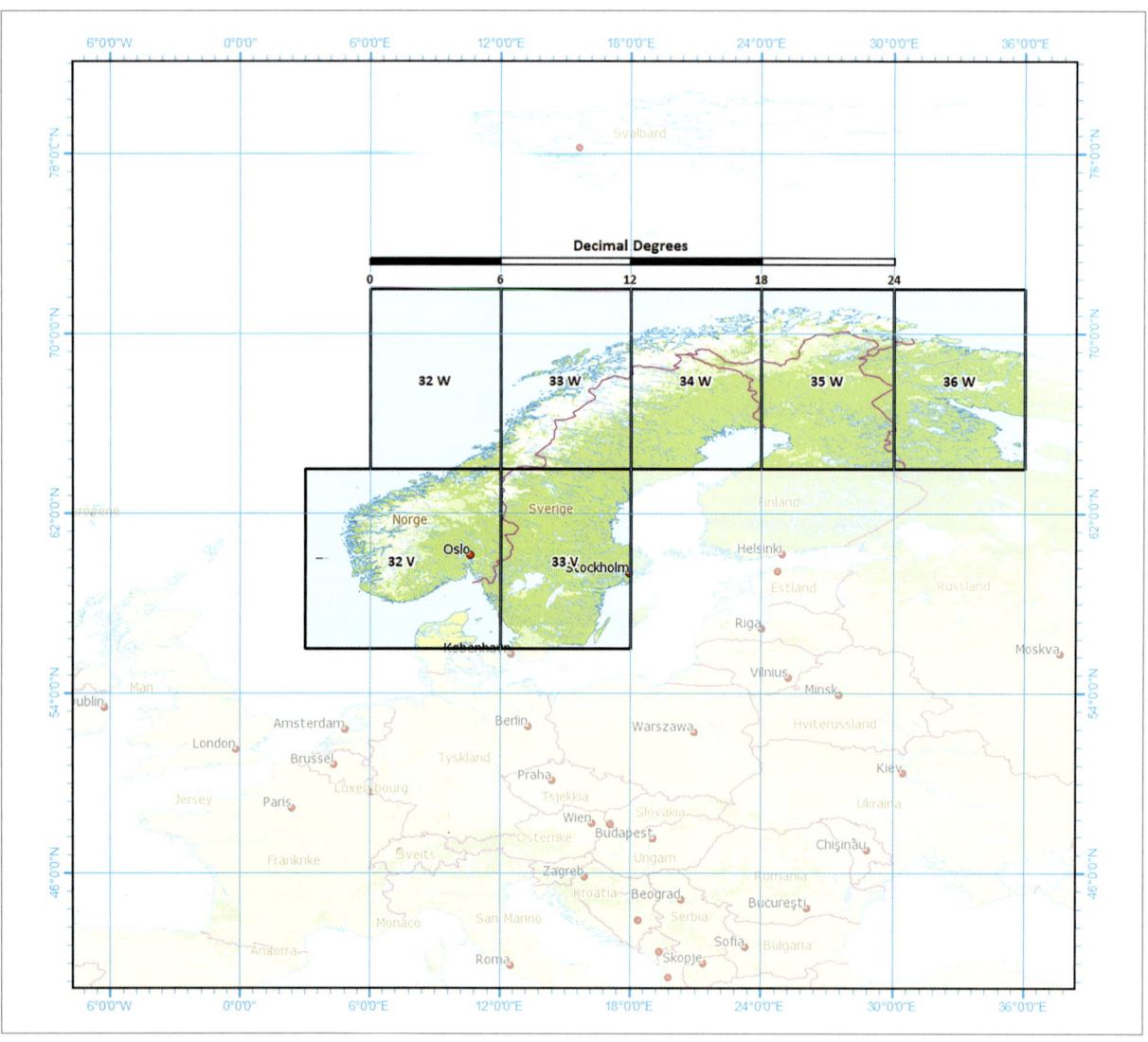

2.7

surface, north or south of the equator. All points located on the same horizontal circle have the same latitude. The Equator lies at latitude of 0°, and the North Pole lies at a latitude of 90° North. The circles on which the point occurs is also assigned a longitude east or west of the prime meridian (at Greenwich, England). The geographic coordinates of the point can then be described with reference to its northern or southern latitude with respect to the equator and at its eastern or western longitude (east or west of Greenwich). For a more accurate description of a location, each degree is subdivided into 60 minutes, and each minute into 60 second.

Cartesian coordinate system
The coordinate axes in a Cartesian coordinate system are perpendicular to each other and consist of an x- and y-axis. The system also has a third dimension known as the z-axis. A point's location is determined by its distance from the coordinate system and is found by reading the values from the two axes or possibly all three exes in a three-dimensional coordinate system. There are many Cartesian coordinate systems, but the one most commonly used in Norwegian cartography is the combined UTM and Military Grid Reference System (MGRS).

The UTM system is a worldwide coordinate system divided into 60 zones in an east–west direction, where

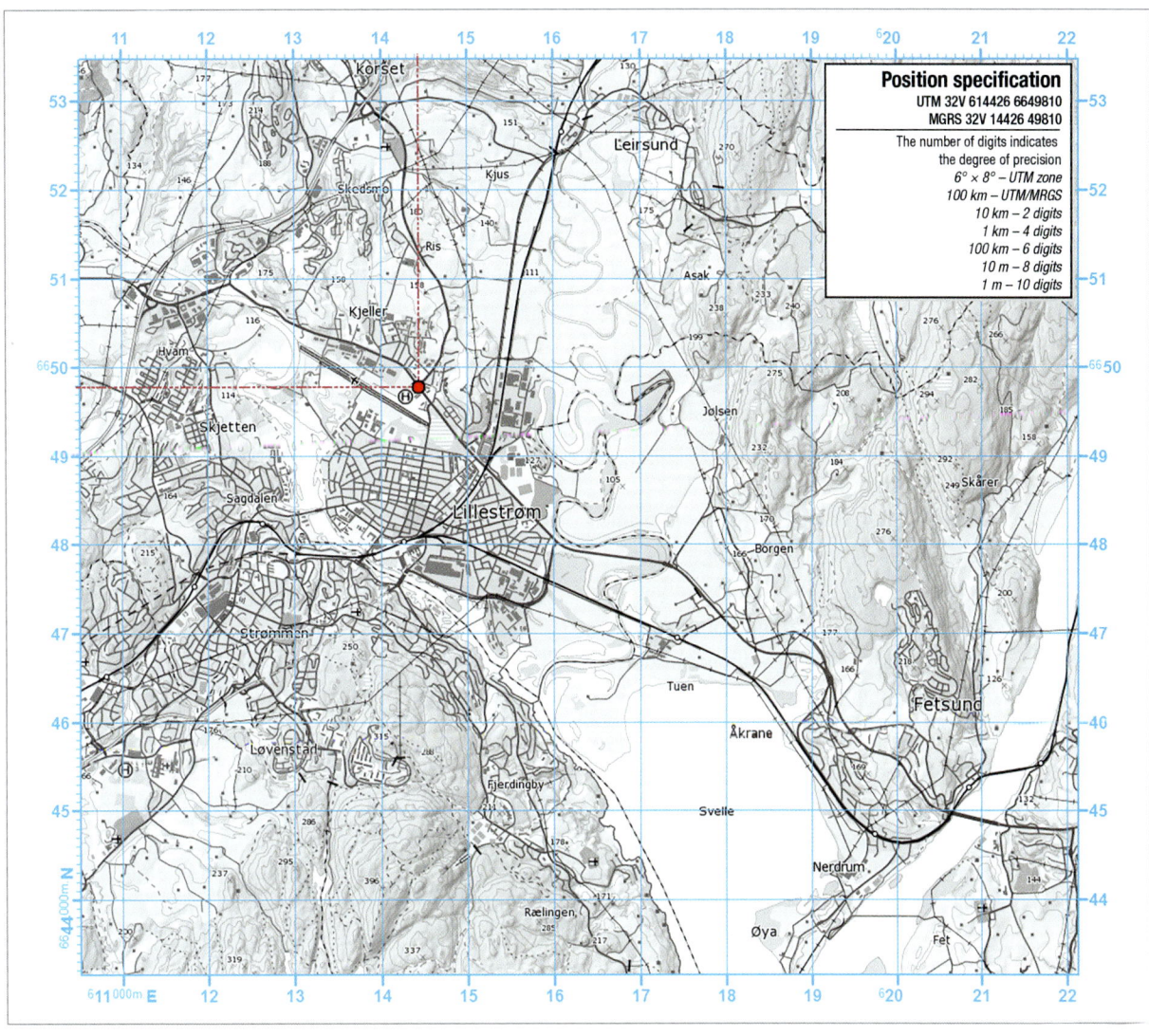

2.8 UTM–MGRS system, showing positions. Background map: © Norwegian Mapping Authority.

each zone corresponds to 6°. Alphabetical letters are used to designate the north–south bands, ranging from the C band in the south to X band in the north. The letters O and I are not used, due to their similarity to the numbers 0 and 1. Each band corresponds to 8°, with the exception of the X band, which has 12° in order to cover most of the land area in the northernmost part of the earth. The A, B, Y, and Z bands, together with associated latitudes are reserved for the North Pole and South Pole, for which there is a separate coordinate system: the Universal Polar Stereographic (UPS) system. As an example of the UTM system, the correct position of the intersection (road junction) between RV22 and RV120 in the county of Akershus (Norway) is: 32V 61442 664981. Figure 2.8 shows how position is marked on a map, but in this case the marked position has an additional two digits, indicating that the level of precision is higher. The same position in the MGRS system is 32V PM 1442 4981, which indicates the considerable difference between the two coordinate systems.

The MGRS system is based on the UTM system, but subdivides each UTM zone into 100 x 100 km grids. These grids as designated a two-letter reference. The MGRS system is thus a simplified version of the UTM system. The MGRS system is used extensively in military contexts, both by the Norwegian Armed Forces and by military service in other nations.

HEIGHTS AND VERTICAL DATA

In order to be able to record heights, it is necessary to define a zero level. Heights are generally expressed as metres above sea level (m a.s.l.) and therefore a smooth sea surface could be used as the zero level. However, sea level is not a static surface and is affected by factors such as tides, barometric pressure, and winds. In addition, large areas of the earth are not covered by water. Therefore, the geoid is used as the potential for vertical datum, in common with horizontal datum. In order to understand vertical datum, it is therefore necessary to know about the vertical relationships between the geoid and the ellipsoid.

Starting from a measuring point on the earth's surface, a vertical line perpendicular to the geoid from the point to be measured is defined. The height of the vertical line from the geoid to the point is called the *orthometric* height and is used for mapping and other purposes. However, satellite-based navigation systems, such as the Global Positioning System (GPS), do not refer to the geoid for height measurements. Instead, height measurements in such systems refer to the ellipsoid and are described as *ellipsoidal* heights. When both the orthometric height and ellipsoidal height are known, the geoid undulations (undulation of the geoid) can be calculated, as shown in Figure 2.9. When a model of geoid undulations is available, global navigation satellite system (GNSS) devices can be used to calculate orthometric height, depending upon which geodetic datum is used by the devices.

At the fundamental point the geoid height is equal to zero. In other words, there is correspondence between the orthometric height and the ellipsoid height. The height of the fundamental point is based on the calculated mean sea level. The Normalnull 1954 (NN1954) height system has been used in Norway ever since it was introduced in 1954. NN1954 is based on calculations of mean sea level at Oslo, Nevlunghavn, Tregde, Stavanger, Bergen, Kjølsdal, and Heimsjø, with the fundamental point in Mandal. NN1954 is gradually being replaced by NN2000, Norway's newest height system. NN2000 is based on a common European theoretical zero level with the fundamental point in Amsterdam.

A computer model of reality

The physical world is complex, and in order for a model of it to be made, it must be simplified. This simplification can be called a *reality model*. A reality model is a common understanding of how one shifts from the physical reality to a model that can be used in a GIS. Such a model is required to simplify and convert real objects into objects that can be stored digitally.

Geographic data comprise *geometry* and associated *properties*. The geometry can be points, lines, or areas, and these represent other phenomena called *entities*. Entities often have different properties called *attributes*. An attribute might be a speed limit on a road, the price of a property, or the age of a tree. The reality model describes the various map objects broken down into entities, attributes, and the relationships between them.

The next step is to create a computer model based on the reality model. A data model is required to store the information digitally. In a data model, the objects are information carriers. The objects in a data model are described according to their type, relationships, geometry, and quality. Within a GIS, there are two ways to present information digitally: the *vector model* and the *raster model*.

2.9 Terms used to specify heights.

Orthometric height (H_0) – height above the geoid
Ellipsoid height (H_e) – height above the ellipsoid
Geoid height ($H_e = H_0 - N$) – difference between the geoid and the ellipsoid

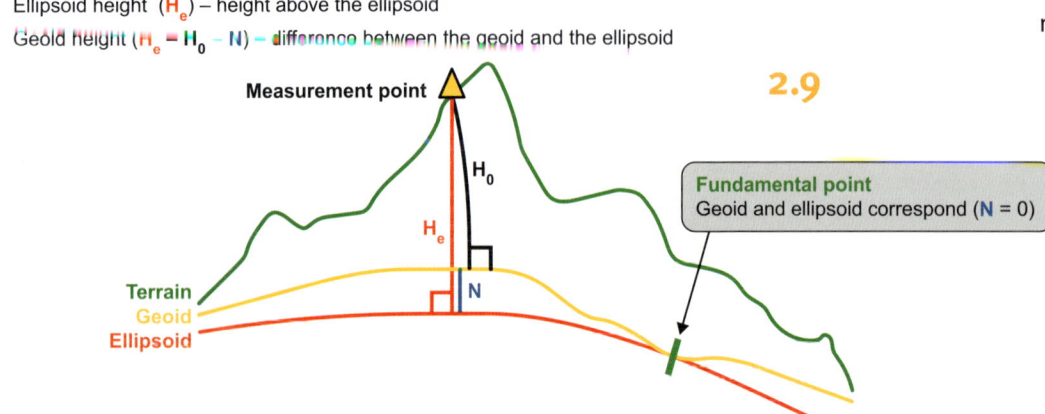

2.10

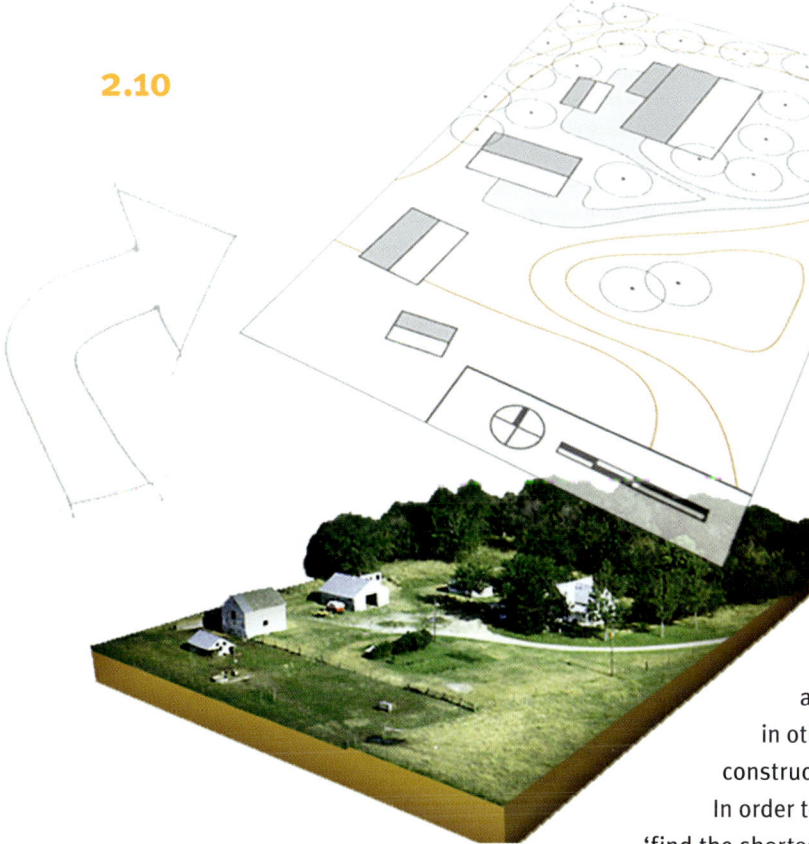

2.10 The geographic reality is modelled with a reality model as a basis. A data model is required to store the information digitally.

THE VECTOR MODEL

The *vector model* is also used in mathematics and is defined as a straight line with a specific direction in a coordinate system. Maps include coordinate systems that are depicted as gridlines with associated numeric values. Known points in the terrain are points with known coordinates that refer to the same coordinate system. In a vector model, the geometry is represented by points, lines, and polygons. A *line* is represented by a line drawn between two individual *points*. An *area*, or *polygon*, is represented by several lines defining a zone. The data in a vector model are called vector data.

Vector data are stored in two different ways: either as *spaghetti data* or in a *topology model*. Spaghetti data can be described as raw data in which the features are not stored with any information about the relationships between them. This can be illustrated by the fact that in the case of two adjacent areas the boundary between them will be recorded twice. This will result in an unnecessarily large amount of storage space being needed and that would make analysis difficult because one would not know how the various features relate to each other. Searching the data would be unnecessarily complicated too. If one uses manual digitizing as the data capture method, the data will appear as spaghetti data. In order to use these map data in connection with analyses, they have to be structured; in other words, a *topology* has to be constructed.

In order to carry out an analysis of the type 'find the shortest transport route between two points' or 'find the distance between two buildings', it is necessary to know something about how the objects relate to one another. Topology is a mathematical expression that describes the relationship between different features. This is done by defining *nodes* and *links*. A node can be an end point where a line ends or where multiple lines cross. Links are lines that start and end at nodes. Attributes can be assigned to nodes and links

2.11 Geometric objects are lines, polygons and points.

2.11

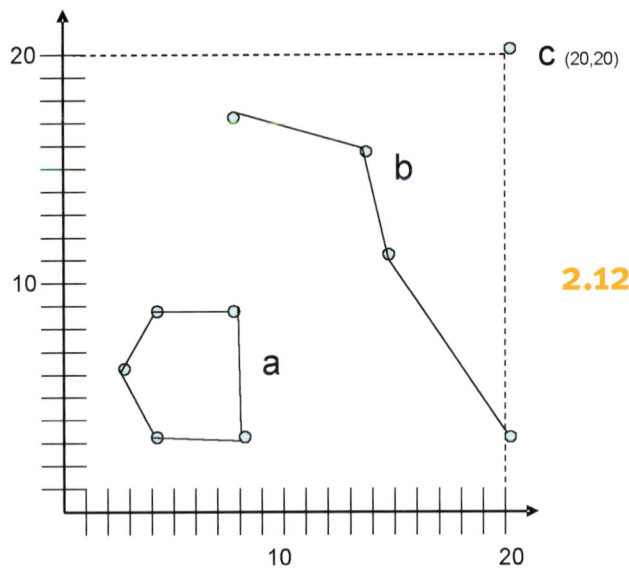

Object	Identity	Attributes	Coordinates
Area	a	m²	(4,3), (2,5), (4,9), (8,9) og (8,3)
Line	b	length	(8,17), (14,16), (15,11) og (20,3)
Point	c	colour	(20,20)

2.12 Various vector data incorporated into a coordinate system.

by assigning each node and link a separate identity with a separate code.

The following are defined in order to describe relationships between geometric objects:

- a polygon table containing data relating to all of the areas (polygons) with data for the links they are built up from
- a node table containing data on which links meet at which nodes
- a link table containing data on which nodes the links start at and end at, and which areas are either side of a link.

Common to these tables is the fact that the various geometric objects (features) are assigned a unique ID code. The geometric objects are thus object-oriented, and this means that the amount of data will be less since lines are not stored twice. Another advantage is that different attributes can be linked to the respective geometric objects.

The different objects are also assigned a thematic code. A thematic code is needed to group or sort objects of the same type. This is done to divide up in the geometric objects into various thematic layers such as roads, infrastructure, buildings, and topography. Once the topology has been defined, various analyses of the data can be performed without having to do geometrical calculations. It is also necessary to organize data using standardized methods such as this one, in order to combine different maps at different scales. For example, a digital property map (DPM) at a scale of 1:5000 can be combined with technical maps at a scale of 1:1000.

The vector model is most frequently used in mapping because it results in a high degree of accuracy and enables different databases containing attributes to be lined to the map objects. The vector model is also relatively simple to update.

RASTER MODEL

A *raster* is based on a reality model, where a grid is laid over an image of the terrain, and each grid square is simplified to contain a given value. One square is called a pixel and each *pixel* is thus allocated a *pixel value*. The pixel value can describe properties such as colour, terrain elevation, gradient, land use, temperature, or vegetation classes. In a *georeferenced* model, each pixel in the grid will link the model to the physical reality. The data in such a raster model are often called *raster data*.

The grid size or *pixel size* in raster models vary according to the required degree of *resolution*. Large cells results in lower resolution, whereas small cells result in high resolution, as shown in Figure 2.13. Generally, the grid sizes within a raster map are the same size, but their sizes can also vary.

As with a vector model, one can construct different thematic layers in a raster model. The difference is that in a vector model a feature is coded with a thematic code, whereas in a raster model the same pixels have to be coded several times if one wants to link more than one attribute to each pixel. This can be illustrated by the fact that in a vector model a building can be assigned a number of attributes, such

2.13 Orthophotos of the university campus at Ås with different resolutions. In the uppermost photo the pixel size is 1 x 1 m², whereas in the lowermost one it is 10 x 10 m². The smaller the pixel size, the higher the resolution and the more details will be visible. Source: © Norwegian Mapping Authority.

as size, type, and utility, whereas in a raster model every pixel covering the building has to be coded with the building's size. Thereafter, a new thematic layer has to be defined in order to assign a code for the building type. Hence, raster data often require a lot of space, although they can be compressed for storage purposes.

Raster models have a number of advantages: data collection is fast and simple, and the data can be organized in a simple data structure; spatial analyses can be performed; and the data for a given area can easily be linked for analytical purposes. The disadvantage of raster models are: network analyses cannot be carried out; the degree of accuracy is poor; and the amount of data is very large. Today, raster models are used to link images to maps, such as through the use of satellite imagery in GIS.

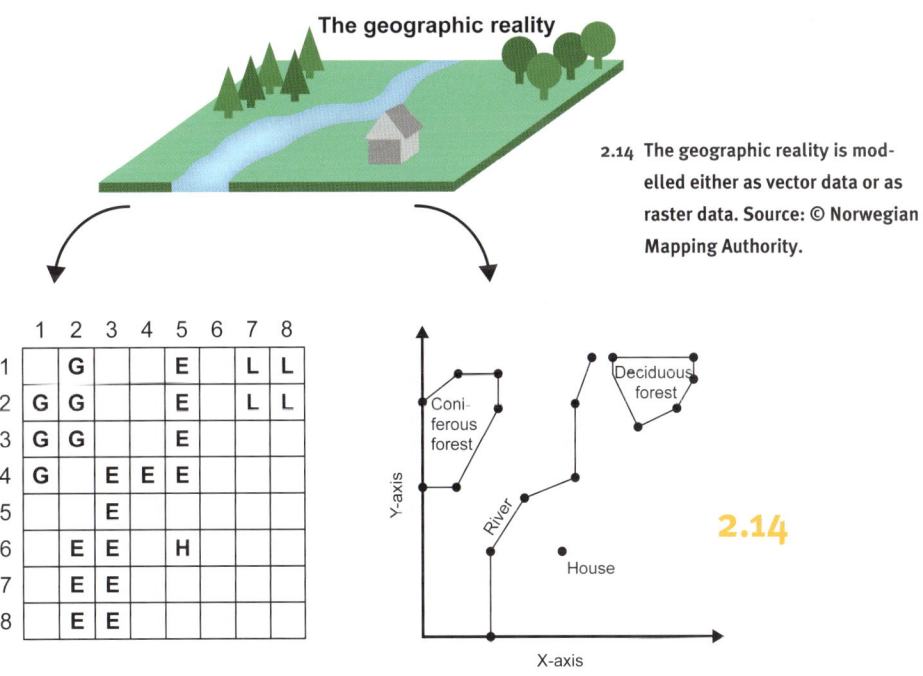

2.14 The geographic reality is modelled either as vector data or as raster data. Source: © Norwegian Mapping Authority.

Data collection of geographic data

The basis of a geographic information system is the actual data to be processed. Without geographic data, no analyses can be performed. There are various methods for collecting data and this section describes some of the common ones. Data collection methods can be divided into three main types: *primary data collection*, *secondary data collection*, and *data transfer*. In primary data collection, data are gathered directly and often they can be data that one has generated oneself. Examples of primary data collection are mobile image that can be localized or a distance measured with a laser distance finder. Secondary data collection involves data derived from other data, such as schools or churches highlighted in an aerial photo.

The most common way to acquire geographic data is to use already existing data (i.e. data transfer). Large amounts of data are made available by others through various forms of cooperation. Alternatively, data can be purchased from various data providers. The Norge digitalt initiative is a result of collaboration between many Norwegian actors who administer geographic information. Norge digitalt and some of the data that are available online are described in more detail in Chapter 3 (Geographic data).

Several methods for data acquisition involve the use of *remote sensing*, whereby a *sensor* collects data about an object without being in physical contact with the object itself. Examples of remote sensing can be taking pictures with a camera or the use of lasers by the police to measure the speeds of vehicles. These are two different types of sensors: *passive sensors* (cameras) and *active sensors* (lasers). A passive sensor detects energy emitted from an object, whether in the form of radiation directly from the object or as reflected energy from other sources. By contrast, an active sensor emits its own energy and then measures the reflected radiation from objects. Whereas a camera records reflected energy from sunlight, a laser emits its own energy in the form of laser beams that are reflected from, for example, a car, and then recorded.

The chosen method of data collection depends on many factors. Some of the questions that should be asked before starting extensive collection of geographic data are: What data should be collected? For what purpose will the data be used? Are there any requirements for accuracy? What technical aids are available? Are there existing data? Which data format are the existing data in? When were the data produced? The methods described in this section are:

- surveying with a total station
- surveying with GNSS
- sea measurements
- aerial photography
- satellite photography
- radar measurement
- laser scanning
- digitization
- local knowledge from users.

SURVEYING WITH A TOTAL STATION

Surveying is the measurement of the earth's surface and altitude using sophisticated instruments. Chapter 1 (Application) includes a description of how people have always made representations of the world, and surveying has been central in this regard.

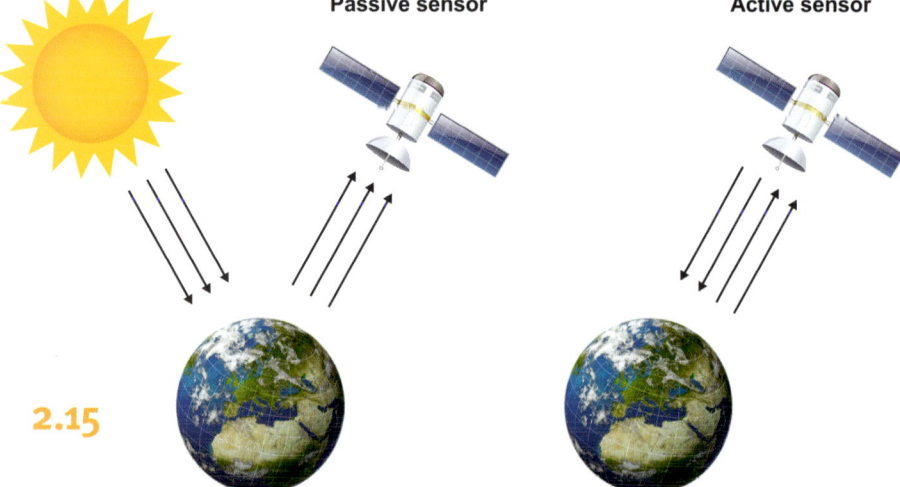

2.15 A distinction is made between passive and active sensors in remote sensing. Source: Lightspring/Shutterstock.

In combination including mathematics, physics, astronomy, and technology, surveying has developed into a large specialist field with many opportunities for data capture.

Traditionally, surveying has been done using a *theodolite*, an instrument that uses a telescopic sight to measure the angles between points in both horizontal and vertical planes. In common with other disciplines, modern technology has meant that surveying methods have evolved too. The introduction of electronic rangefinders in combination with computer technology has meant that point determinations can be done faster, yet still as accurately. An instrument that combines a theodolite, a rangefinder, and an integrated computer is called a *total station*.

A total station can be used to determine the location of points in the terrain by setting the instrument over a *benchmark*. A benchmark is a point where the coordinates (degrees north and east) and height are known. From this permanent mark, one also needs an unobstructed view of one or more other known points, in order to orient the total station and to reveal any defects in the selected benchmark. If more than one benchmark can be accessed, it is customary to set up the total station at a random position that affords a good overview of whatever is to be surveyed. This method of determining the new alignment point is called *resection* (also known as free stationing).

Previously, it was common to have an operator to operate the total station and an assistant who moved around in the terrain with a mirror or *prism* mounted on a pole, known as a *prism pole* or reflector pole. The assistant stood at the measuring points, while the operator measured the distance to the prism by sending out a signal from the total station that was then reflected back to the instrument by the prism. The on-board computer could then calculate the position of the prism pole directly.

Newer models of total stations can track the prism automatically, which makes it possible for surveys to be conducted by one person, who operates the station from a field controller mounted on the prism pole. In addition, in newer models data are stored in a *digital data collector* that can transfer the data to a computer or elsewhere for further processing.

Surveying with total station is used in cases where there are high demands for precise location coordinates. A level of accuracy down to 1 mm should be expected, given that the benchmark has comparable accuracy and the method of surveying is good. Examples of objects that require highly precise location coordinates are boundary markers and house corners.

The establishment of satellite systems for navigation and positioning has meant that total stations are used to a lesser extent than before. However, one important factor that favours the total station is that it does not depend on satellite coverage. Nevertheless, total stations still have to be used in tunnels or other areas with poor satellite coverage.

SURVEYING WITH GNSS

Global Navigation Satellite System (GNSS) is a generic term for satellite systems for navigation and positioning with global coverage. The first systems were developed for military purposes, but today the same systems are used in civilian life. Both the American GPS and the Russian GLONASS are global systems. China is in the process of expanding its regional system into a global system known as COMPASS, while Europe is working to establish its GALILEO system. GALILEO will be under civilian control, unlike GPS and GLONASS, which will remain under military control.

All GNSS consist mainly of three segments. GPS satellites and their location in space form the *space segment*. To keep track of where these satellites are located, there is a *control segment* that monitors and communicates with them. The third segment comprises *users*, who are recipients of GNSS data and users of GNSS.

2.16 A total station – a theodolite with on-board software for electronic distance-measuring.

2.17 All GNSS consist of three segments. Source: abstractdesignlabs/Shutterstock; Tatyana Prokofieva/Shutterstock; Nucleartist/Shutterstock.

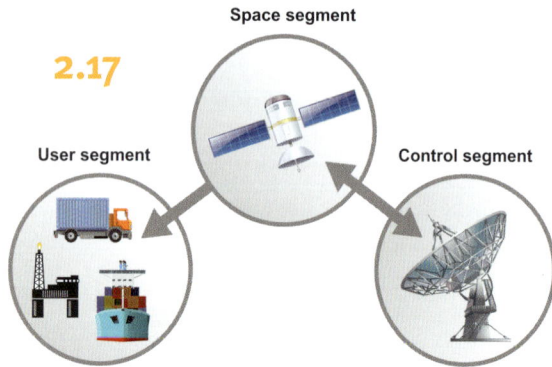

Surveying with GNSS is based on measuring distances between satellites and one or more receivers on the ground. The satellites' orbits are known and therefore the position of satellites can always be considered known points. Each satellite continually sends unique signals that are received by the GNSS receivers on earth. The receivers record the time when the signal was sent, and since the signal travels at the speed of light, the distance between the receiver and satellite can be calculated. This is done for all of the satellites that are visible to the receiver and in this way the position of the receiver can be determined.

One of the common methods of surveying with GNSS is the use of *Real Time Kinematic* (RTK), also called real-time measurement. RTK is a variant of what is known as *differential GNSS*, which is based on receiving corrections from a recipient at a known point.

In principle, the system consists of a *base station* that is located at a known point, a *rover* that processes the measured data, and a *communication channel* that transmits corrections from the base station to the rover.

Within *traditional RTK*, one can configure one's own base station. A base station has a satellite dish, a GNSS receiver, and a communication device, which is usually a radio with a radio antenna. Alternatively, a *network RTK* can be used instead of a base station. Then, one has to subscribe to services that send out corrections via mobile network or the Internet from established base stations. A number of subscription services have been established in Norway under the auspices of the Norwegian Mapping Authority, of which *DPOS* and *CPOS* are two of the most commonly used services and can provide measurements to within an accuracy of 10 cm and 1 cm respectively.

The rover usually consists of a portable antenna, a bag with a GNSS receiver, and a communication device, whether a radio, mobile connected to a mobile network, or the Internet, depending on the type of RTK used. In addition, the rover will have a digital data collector linked to a receiver for processing correction data and measurements relating to the task in hand.

The base station calculates the coordinate for it's position on the basis of measurements of the dis-

2.18 Both traditional RTK and network RTK are used extensively in surveying. Traditional RTK (left) creates its own base, whereas Network RTK (right) uses already established bases.

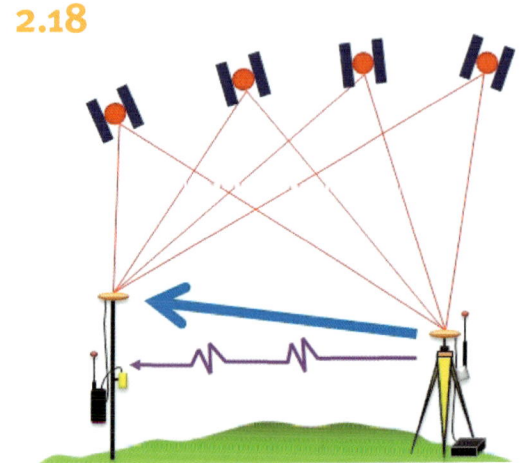
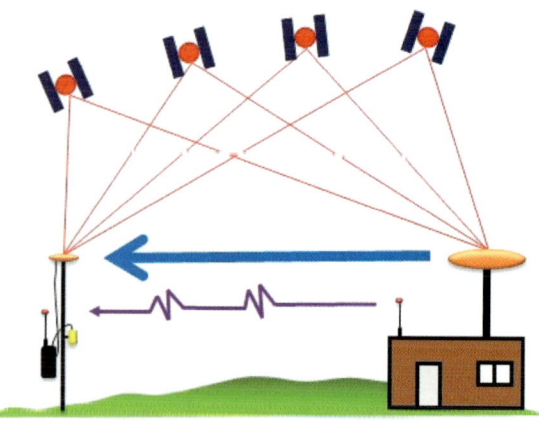

tance to the GNSS satellites, and then compares this calculated position with the known coordinate of the same position to calculate the correction values. It is assumed that these corrections will be the same within a range of c.10–15 km from the base station. The base station then sends the calculated corrections to the rover, which in turn corrects its measured position and presents the corrected coordinates in the digital data collector.

Both traditional RTK and network RTK provide coordinates with an accuracy to the nearest 1 cm in plan view and 2 cm in height. This is dependent upon the position of the base being correct, the equipment being in good order, and that there is network coverage. In some cases, buildings, trees, or high mountains can block or interfere with satellite, radio, and mobile signals and this may complicate measurements and reduce accuracy.

A simpler form of surveying with GNSS is the use of *handheld GNSS*. These devices have become increasingly more common in recent years, and are now integrated into most mobile phones and new cars. Their level of accuracy is not as good as RTK, since correction values are not normally used. The accuracy of handheld equipment varies, but on average it is the range of 5–15 m both in plan and elevation, depending upon the satellite coverage.

Handheld GNSS with map data are a useful tool in many contexts. For example, emergency vehicles can be guided to the right address, hunters with dogs are able to locate their dogs, and images taken with mobile phones can be linked to the places where they were taken. In the transport industry, cars can be equipped with devices that send their position to a centre that can track their current location and then optimize their route.

MEASURING AT SEA

For a coastal and ocean nation such as Norway, knowledge of what lies below sea level is vital. Traditionally, *bathymetry* (ocean topography) has been used to create charts, for which the main focus is on safe navigation. This is still perhaps the most important application of bathymetry.

Depths can be measured by various methods. The simplest but also the most time-consuming method is to use a weighted line to measure the depth of the place at a given location. This method was used extensively until the development of *sonar* (echo sounding), which can do the same by sending out an acoustic signal and then measuring how long it takes before the echo is heard. Then, the speed of sound in water can be use to determine the depth. Special types of sonar can be used to collect information about the characteristics of the seabed in addition to the varying depths. Gradually, sonar was developed into *multibeam sonar*. In contrast to sonar, which emits a sound signal towards the bottom of the sea, a multibeam sonar emits many

2.19

2.19 Traditional RTK with a base station and rover unit.

2.20 Multibeam echo sounder.

simultaneous audio signals to the seabed in a given sector, such as long a vessel's course. Together with the vessel's position and orientation, when the signal is sent and received, tens of thousands of points can be measured in a matter of seconds. By contrast, measurements taken at great depths with a weighted line could take the best part of a day, which means there have been huge improvements in the technology.

The development of multibeam sonar and the possibility of collecting high-resolution data relating to the bottom of the sea have led to a wide range of new applications. For example, high-resolution data are becoming increasingly important for fields such as oil and gas, fisheries, aquaculture, maritime transport, research, and the Royal Norwegian Navy.

Data collection using sonar is still very time-consuming and resource-intensive, and this is the main reason why up-to-date measurements are missing for large parts of the Norwegian coast. Advances in remote sensing have also reached the maritime domain, and laser scanning tests from aircraft have shown promising results for measurements in areas with shallow water (as described below).

AERIAL PHOTOGRAPHY

In 1858, French photographer Félix Nadar photographed Paris from a hot air balloon (Biography.com 2015). This is considered the first attempt to take a photograph from the air. Gradually, aircraft replaced the use of balloons, and aerial photography has long been one of the most common forms of data collection within GIS.

Traditionally, two types of aerial photographs have been used. A *vertical photograph* is one taken from above, vertically above the ground surface. This type of aerial photography is most often used in connection with surveys of large areas. An *oblique photograph* is one taken with the camera tilted towards the ground, so that world is seen from a bird's-eye perspective. The question of whether to use vertical or oblique photographs depends on how the geographic information is to be presented.

In a GIS, all geographic data should have a geographic position. This also applies to aerial photographs. *Photogrammetry* refers to the use of photography to ascertain measurements between places and objects on the ground. Photogrammetry can be described simply as measuring on pictures, and has long been the most common source of data capture. In addition to aircraft, there are currently several other platforms that take images for mapping purposes. Photogrammetry is also used for photographs taken by either a satellite or a camera mounted on a car if the images are to be linked to positions.

In order to collect data using aerial photography, one first needs to calculate the position of the airplane or more precisely the camera itself at the moment of exposure. Aircraft have inbuilt GNSS receivers and an inertial navigation system (INS) that keep track of the

2.21 Mapping with the use of aerial photography.

2 – SYSTEMS 55

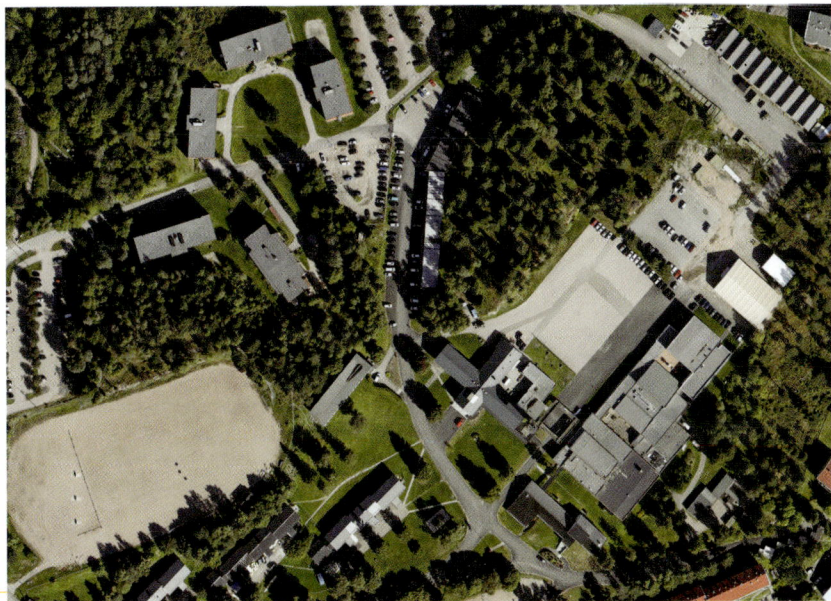

2.22

position and orientation of the aircraft or the camera.

In addition to the position and orientation of the aircraft, some points on the ground need to be highlighted. These points should be marked with reflective signal discs in the terrain before the flight, so that they will be easily identifiable in the aerial photos and can be used to indicate the orientation of the images. Prior to producing the map, the aerial photos and airstrips have to be orientated to one another and oriented in relation to the terrain. This process is called *aerotriangulation*. When the aircraft's position is known and the aerial photographs have been oriented relative to each other and to the terrain, it will be possible to calculate the coordinates of all points in the terrain.

The aerial photographs can be registered to *orthophotos*, which are images with the same geometrical properties as a map. An orthophoto is related to a coordinate system. An aerial photograph is taken with the camera pointing straight down towards the ground, in a central projection of the terrain. This causes hills and mountain ranges to be projected at a skewed angle relative to a map. Orthophotos are therefore adjusted so that the objects in the images correspond to their places shown on the map, which is known as *orthogonal projection*. Thus, orthophotos can be used in place of conventional maps or in combination with maps and thematic information. The creation of orthophotos is dependent on the existence of a good height model in combination with aerial photos. For more information about digital elevation models, see Chapter 4 (Analysis).

The use of drones (remote-controlled, unmanned aircraft) is becoming more prevalent. A carrier drone can be flown over the area in order to map and collect geographic data. In addition to taking images, both filming and *laser scanning* can be done using unmanned aircraft.

2.22 Two types of aerial photographs showing Linderud military base. Upper: a vertical photo; Lower: an oblique photo.

2.23 Drone operator from the 1st Cavalry Squadron's Patrol Troop in the Armoured Battalion, Norway. Photo: Ole-Sverre Haugli, Norwegian Army.

2.23

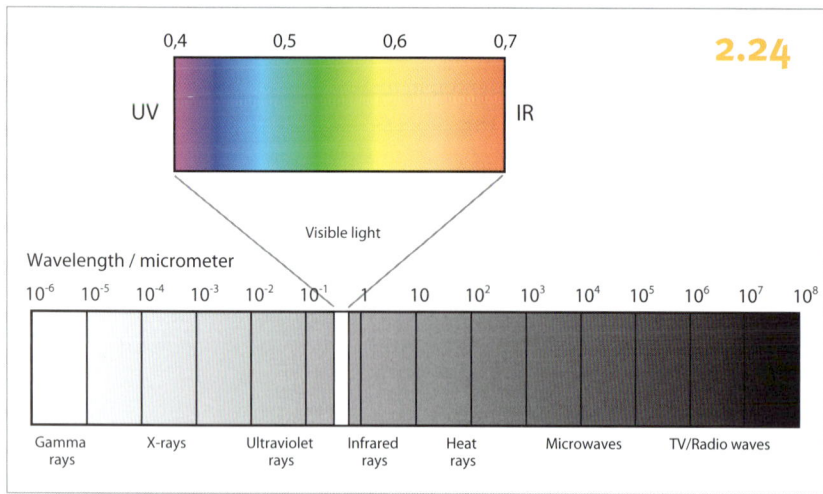

2.24 The electromagnetic spectrum.

2.25 Radiation reflection and absorption.

2.26 The first official image of Oslo taken from the Copernicus Sentinel-2A satellite. Photo: Copernicus Sentinel data 2015.

SATELLITE PHOTOGRAPHY

The Norwegian Space Centre defines all activities relating to data collection via satellite as *earth observation* (Norsk Romsenter, 2015). Satellites have been used to take photographs of the earth's surface for a number of years. They have advanced camera equipment and satellite imagery is used more often than before because the quality of the images is constantly being improved. Commercial satellite images currently have a pixel size as small as 30 x 30 cm, and can therefore be used instead of aerial photographs or in addition to them. They cover large areas from a record height of 600–900 km, and are relatively inexpensive to purchase.

In order to understand the use of satellite data, it is necessary to have some understanding of light, radiation, instruments, and data access.

Electromagnetic radiation is energy transferred by electromagnetic waves through empty space or a permeable medium. The wavelengths can vary widely, from gamma rays (c.10^{-6} microns) to long wave radio frequency energy (c.10^8 microns). The wave lengths between these two extremes constitute the *electromagnetic spectrum* (see Figure 2.24).

Roughly in the middle of the electromagnetic spectrum, from 0.4 to 0.7 microns, there is a field in which the human eye perceives radiation as light. Just under 0.4 microns, there is the ultraviolet radiation and above 0.7 microns there is infrared radiation. At these two wavelengths, nerve cells in the human eye register 0.4 microns as violet and 0.7 microns as red. The colour of an object will thus depend on the surface's ability to absorb and reflect electromagnetic radiation within the range of visible light. A white surface will reflect all radiation, whereas a black surface will absorb all radiation. A yellow surface will reflect visible radiation in the wavelength that the eye perceives as yellow.

2.27 Sentinel-1 is a radar satellite in the European Copernicus programme for monitoring the environment and climate. Source: ESA/ATG medialab.

Satellites are passive sensors. Hence, the radiation that hits a satellite can be direct sunlight, radiation reflected from the atmosphere, or radiation emitted or reflected from the earth. The reflected radiation comes from the sea, land, clouds, or atmosphere. Some of the radiation from the ground is absorbed by clouds and some is absorbed by the atmosphere, as shown in Figure 2.25.

RADAR MEASUREMENT

Radar (radio detection and ranging) is a remote sensing technique that uses radio waves to determine the distance to one or more objects. As a measuring device, radar can be used from different platforms. In the context of GIS, radar measurements are mainly taken from airplanes and satellites. One type of radar that is much used is synthetic aperture radar (SAR) which is an active microwave radar (i.e. an active sensor).

The spatial resolution in a radar image is dependent on the size of the antenna of the radar. For SAR, the radar is mounted on a platform that can move. Thus, one can combine signals from different places as the radar's position changes. The combinations will produce the same effect as if the signals had come from a long antenna. SAR can therefore be used to achieve high resolution, typically 1 m for images taken from satellites and down to 10 cm for images taken from aircraft. SAR is used, for example, to study the waves, wind, ocean currents, sea ice, emissions, and pollutants. One advantage of using SAR instead of other techniques, such as images with visible light, is that the microwaves penetrate clouds and images can be taken at night.

SAR was adopted in 1950, and in common with many other types of sensors, military services were the first to adopt the technique for monitoring. SAR has been used increasingly more for civilian purposes too. It has also been developed over the years. By using phase shift in the signal, differences in height can be used to form three-dimensional images with very high level of accuracy. In this way, very small movements in the earth's surface can be recorded.

LASER SCANNING

Laser scanning or light detection and ranging (LiDAR) is a remote sensing technique that uses light to determine the distance to one or more objects. LiDAR has been used extensively in atmospheric studies and meteorology, and laser devices are used by the police for checking traffic speeds. The technology is also being used more frequently for surveying and mapping the earth. The lunar surface was mapped using LiDAR as early as 1971, during the Apollo 15 mission.

2.28 The Norwegian police force used ground-based laser scanning to document the aftermath of the widespread fire Laerdal in 2014. Source: Kripos.

2.29 An attempt to demonstrate accuracy by projecting a trajectory from a point cloud data collected from laser scanning. Source: Kripos.

The term laser scanning is often used, since usually a laser continuously emits laser light in the form of pulses over a surface. The laser registers the time from when the pulse is emitted to when it is received, and since the speed of light is known, it is possible to calculate the distance from the laser to the object that reflected the pulse. This type of sensor is therefore an active sensor.

LiDAR can be used from different platforms. The laser can be mounted on a tripod in a room, on a car driving along a street, or attached to an airplane, helicopter, or a drone. A distinction is often made between ground-based and airborne laser scanning. Common to all situations is that one needs to know where the meter is located and oriented in space at any given time. In the same way as for aerial photography, GNSS and INS are respectively used to position and orient the laser. All points can be measured by combining the laser beam's position and orientation with the distance to the point to be measured. Following a laser scan, one is left with information in the form of a georeferenced point cloud.

Laser scanning from aircraft is widely used as a quick way to collect elevation data over large areas. Millions of laser pulses hit the ground, even in dense forests, and this makes it possible to obtain very realistic elevation models, as shown in Figure 2.30.

In addition to height data, an intensity value is registered by a LiDAR device. An intensity value provides an indication of how much of the emitted laser pulse is reflected. This property can be used to gain a good idea of what the laser has swept across. For example, snow reflects a lot of laser light, whereas asphalt reflects comparatively little laser light. Compact and loose soils reflect light slightly differently, and this quality can be used to identify where excavations might have been carried out.

LiDAR can also be used for measuring depths in relatively shallow water. The use of LiDAR in water presents a number of challenges. The light used in conventional LiDAR has a wavelength (red light) that is easily absorbed by water. This means that this light is not reflected back. For this reason, either blue or green light is used when measurements are taken in water. A further challenge is that when a beam of the light pass from the air and into the water, the beam bends. Although it is known how light is broken at the water surface, calculating this means that one must know where the water surface is relative to the laser beam. This could be done by using laser with both infrared light and green light. Then, there would be a reflection from the infrared laser from the water surface, and one would know the angle at which the green laser hit the water surface. Put simply, if one disregards the waves on the water surface, this method in conjunction with knowledge of how light is refracted can be used to determine how the green laser light was bent. One could then calculate the point from which the green laser beam was reflected. Yet another problem of using LiDAR in water is that shallow waters often have a lot of particles from rivers or waves breaking on beaches, or they may contain algae, which makes it difficult to for the light to reach the bottom and be reflected. Alternatively, the light may be so diffuse that a false echo from the bottom is given. Together, all of these challenges mean that using LiDAR for depth measurements will cost more than using it to map land surfaces, although the method will usually be much safer than sending boats into shallow water conditions.

DIGITALIZATION

Although digital data are becoming more prominent, analogue products still exist. In order to include these in a GIS, *digitization* is used as data capture method. Raster data and vector data are respectively established by *rasterization* and *vectorization*.

Rasterization is usually done by conducting a *scan* of an analogue product, such as a purchase contract or a paper map. The result is a raster that can be processed digitally. For the scan result to be used in a GIS, it must be corrected geometrically and georeferenced so that it is properly positioned in relation to other geographic information. Scanning is a simple and fast process, but the amount of data is large. The method provides a high degree of stability and accuracy, and raster are therefore often used as basemaps for other geographic data. When rasterization is done with a scanner, one will end up with an image that has a lot of information. In some cases, only selected objects such as roads, buildings, manholes, or protected areas will be of interest. Then, the amount of scanned data will be unnecessarily large and vectorization will be a more appropriate method.

Earlier, vector data were established by *manual digitizing*. An analogue map was placed on a digitizing table, and using a recording unit, the objects were selected and given specific locations relative to the coordinates on the digitizing table. The method was simple to perform, but was dependent on correct interpretation by the person performing digitization. There was considerable room for misinterpretation.

Manual digitizing is still done, but the digitizing table has been replaced by a digital display. Since most geographic data have become digital, vector data are increasingly established from raster data. Analogue maps are no longer used as the basis for digitization. Interpretations of the source and marking selected relevant objects are still done, but the source is primarily digital images. Today, software is available

2.30 Laser scanning can provide excellent elevation models. Left: part of an elevation model from the Norwegian Mapping Authority's database. The model has a ground resolution of 10 x 10 m. Right: an elevation model based on laser scanning in which the ground resolution is 1 x 1 m. Source: © Norwegian Mapping Authority, Geovekst.

2.31 A scanner consists of a laser head that reads the original map by registering variations in colour values and converting them to raster data.

2.32 Mobile client (app) with digital map base. Photo: Torgeir Haugaard, Norwegian Armed Forces.

that makes this process easier by deriving the vector data from a raster automatically. This places great demands on inspection to identify errors in the automatic vector process.

Vector data can be converted to raster during image processing. This form of rasterization is sometimes used in GIS in connection with the preparation of data for further analysis (see Chapter 4 (Analysis)).

LOCAL KNOWLEDGE FROM USERS

It can be assumed that feedback from the users of maps has been common since maps first appeared. When errors have been detected, they have been corrected in new editions of the maps. GIS and the Internet now provide even better opportunities for the public to provide feedback on available map data. At the same time, actors can contribute voluntarily to the creation of large databases of geographic information based on knowledge held by private individuals. One example is *OpenStreetMap*, which is run by a community of mappers who use aerial photos and GPS devices to collect information about roads, railways, rails, and other features worldwide. The data are shared with anyone as long as they formally acknowledge OpenStreetMap and other contributors. Another example is the Norwegian Mapping Authority's solution *Rett i kartet* ('Update the map') where the public can report errors and gaps in the authority's data.

Data quality

Different purposes require data of different quality. For example, in order to find an even gradient for a drainage pipe, equipment is needed that provides precision to the nearest centimetre when precision to the nearest metre is not good enough. Meanwhile, the demands for high data quality often mean high costs, and it is therefore important to consider carefully how the data will be used. Information about the origin of data can give an indication of how the data should be used.

Depending on how geographic data are to be used, there will be different requirements for its level of accuracy. The data quality must be appropriate for the specific application. Data quality is also often a question of cost; the higher the level of accuracy, the more it will cost to establish data. It is therefore important to consider the quality of data needed when working with GIS. Data quality will change during a GIS analysis and will generally be worse afterwards. When using GIS, there is a need to be aware of the errors found in the data and the errors incurred during the work, so that ultimately the accuracy of the work can be assessed.

Metadata are data about other data. They describe how the data have been collected, how old they are, their accuracy, who collected the data, what purpose they were collected for, the type of information contained in the data, and so forth. Metadata are important information when searching for suitable data for analytical purposes. Data quality depends on many different quality factors, some of which are:

- localization accuracy
- attribute accuracy
- topological accuracy
- completeness
- updating.

ACCURACY

Accuracy is generally indicated in the form of a standard deviation. This describes where a given position is in relation to the exact location in the terrain. Accuracy can have both a systematic and a random component. As long as one works with data from the same source, systematic errors will be of less importance, but it will be more difficult to ensure accuracy when data are compiled from multiple sources.

Precision is a consideration too. *Precision* is not a measure of the accuracy of a position, but rather a measure of how well a measurement can be repeated. The term accuracy always refers to a measured value deviation from its true value. Take, for example, the case of two shooters. Their task is to hit the centre of a target. Suppose they have ten shots each, and Shooter 1 shoots ten 'nines', all of them grouped to the lower left of the centre, and Shooter 2 shoots eight 'nines' and two 'tens', with all the 'nines' scattered around the target. Shooter 1 was less accurate but more precise than Shooter 2.

LOCALIZATION ACCURACY

Localization Accuracy describes how a geographic position is given in relation to the true value of the terrain, and is generally expressed as a standard deviation. The standard deviation describes how there is a degree of probability that the true value lies within the specified value ± standard deviation. There is usually a link between localization accuracy and map scale. Large-scale maps have greater localization accuracy than small-scale maps. A map of Norway does not need to show roads with the same degree of accuracy as a map of Bergen city centre.

Coarse localization accuracy on a map will not be a problem if the map is at the scale for which it is adapted, such as an accuracy of ±500 m for natural areas shown for the whole of Europe. However, with a GIS it will be possible to zoom in farther than the scale at which the data were adapted for use. Given the increasingly large amounts of readily available data, care should be taken when using coarse data, especially in the case of relatively new users who are not particularly familiar with the data and their quality.

ATTRIBUTE ACCURACY

The accuracy of an object's attributes – *attribute accuracy* – is as important as the object's localization accuracy. It is of little help to have accurate positions for the data if the data have incorrect attributes. Such attribute errors can be a point designated as the fire station, when in reality it represents the library and the fire station is actually located in a very different

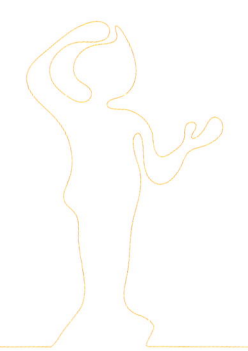

 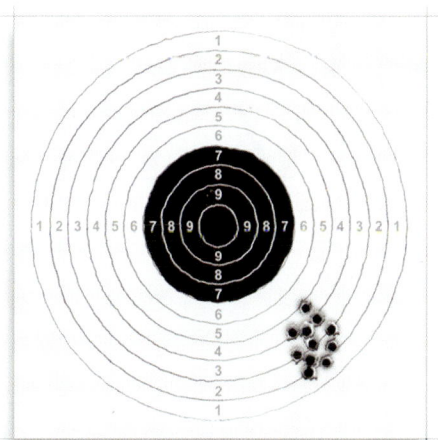

2.33 The shooter on the left achieved the best accuracy; the shooter on the right achieved the best precision.

2.33

Data quality

place in town. If the map is used to calculate response times, attribute errors could have consequences for the fire department that is assigned to deal with an incident instead of fire stations that are located closer to the incident.

TOPOLOGICAL ACCURACY

Topological accuracy refers to the accuracy of interrelated data. Are there errors in the form of crossed lines or surface overlap, or are lines connected properly? As an example, roads are represented by lines. When roads are represented by two lines that are not connected, this could lead to large errors when calculating the shortest route between two points.

COMPLETENESS

Data set *completeness* means how thoroughly an object type is registered, such as whether all buildings in an area are registered. Completeness can apply to both geometry and attributes. One can have 100% completeness in buildings' locations, but only 50% of these might have been given a building type. In this case, the completeness property 'construction type' would only be 50%.

A poor level of completeness can have serious consequences in certain contexts. For example, GIS is used to calculate who should be notified about construction work on a property. If the property database is incomplete, it may result in a neighbour not being notified, which in turn might mean that the building application has to be postponed and that would incur increased costs for the developer. Another example might be the use of information about underground cables, such as how old they are and what kind of material they are made of, in order to assess the condition of the cable network in an area. If the cables in an area are only partially recorded, such as only the most recently laid cables, the assessment would only be based on those with registrations, thus giving a false image. Similarly, it will be important for a decision-maker to gain information about what percentage of the cables are not included in the analysis and on this basis to assess whether, regardless of their condition appearing to be good, there should be

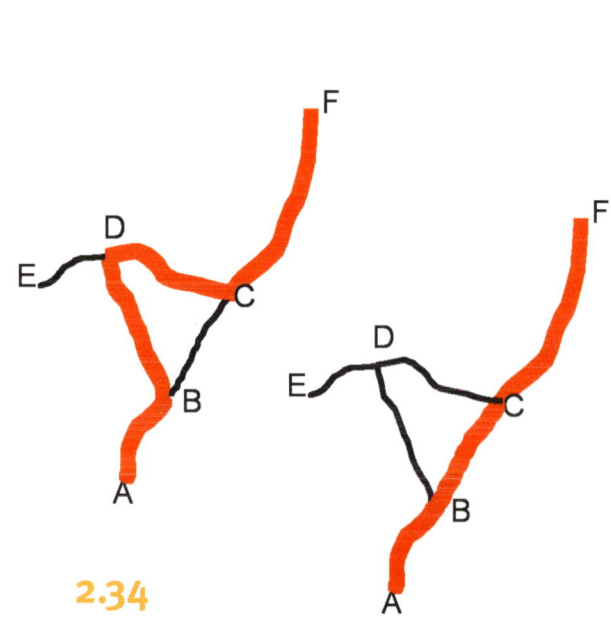

2.34 In this example an error will occur when calculating the fastest route between A and F. There is a topological error at point B in figure a, meaning that that the connection between B and C is broken. The shortest route will therefore be via point D. In figure b the error has been corrected and the shortest route is now along the road between B and C.

2.35 The topological error if it was transferred to the real world.

further investigations into the cables for which there is no information.

UPDATES

Timeliness or *updating* describes whether data are current. It is not usually possible to update old data. A data set that is not regularly updated may contain errors due to changes to existing objects not being taken care of or there may be new objects that have not been included in the data set. The frequency with which different data are updated will vary according to the data type. For example, the railway network in Norway changes relatively little from year to year and therefore it may be sufficient to update the data sets every five years or fewer. By contrast, buildings in a municipality are constantly changing and this means it may be necessary to update the data sets each month. An update date is normally the date when data were last updated. Often, data include information about when they were first registered. One example of the consequences of using not updated data could be when a logistics company has not updated a route calculation program with information that the Oslofjord tunnel is closed to heavy transport. By calculating the time from, for example, Drammen to Drøbak, a delivery time of just 40 minutes will be calculated, yet in real terms it will take twice as long because a detour will have to be made through Oslo.

Sources of errors

Data errors occur in all phases of data processing. If an error occurs at any phase, it may be very difficult to identify the same error at a later phase.

ERRORS REGARDLESS OF GIS PROCESSING

Errors may occur regardless of data being processed in a GIS, such as errors that occur during data entry. Typical data entry errors result from the instruments used in data collection such as GNSS, rangefinders, and aerial cameras. In addition, errors occur during processing of the data, such as errors in calculations and errors in the instruments used to design maps. Errors due the appearance of new objects in the terrain after the data have been produced may also occur. Moreover, errors can occur if the data collection is done using erroneous methods, such as bad point density when measuring the terrain.

INCORRECT GIS TREATMENT

Errors may occur during data entry in GIS, such as inaccuracies in digitization, conversion errors, or typing errors.

Furthermore, errors can occur when data are saved on a computer, due to insufficient numerical precision, such as number of decimal places for stored coordinates.

When processing the data, new errors may occur as a result of generalization and thinning data, interpolation, conversion to other formats, merging data, or developing topologies. Such errors can also result from the use of improper methods for analyses. The presentation of data may also contribute to errors in the data if the printers are not sufficiently accurate or if the paper becomes stretched in hot areas.

FUZZY DATA SETS

Fuzzy data are data with blurred boundaries. As an example, there is no clear boundary in transition from marsh to woodland and therefore if a number of people were asked to construct a map showing the boundary there would be different answers, many of which would be correct. It is possible to perform analyses that take into account such smooth transitions, but these are relatively complicated. In most cases, it is therefore only necessary to know that such blurred lines exist in the data sets being used

2.36 The location of Leknes (marked with a red pint) differs with different map applications. Source: www.BillionPhotos.com/Shutterstock.

and to take this into account when making assessments of the results.

ERROR PROPAGATION

Errors found in data will propagate further in the processing of those data. New errors will add to existing errors and therefore the results of a GIS analysis always have less accuracy than the original data. The propagation of errors is governed by statistical laws.

SOSI

Coordinated plan for spatial information

SOSI is a Norwegian standard within geographic data. This standard was introduced in 1987 and today it is a standard for describing geometry, topology, coordinates, data quality, metadata, and area constraints. SOSI is not only a syntax for exchanging geographic data, but also an object directory for the description of geographic objects and their properties.

SOSI is used as a raw data format, a storage format, and an interchange format. There is ongoing internationalization of geospatial standards in CEN (the European Committee for Standardization) and the ISO (International Organization for Standardization) and future versions of SOSI will probably be in accordance with these international standards.

The SOSI standard describes how objects and properties are given an unambiguous geographic connection with a demand for precision in place attachment. There are two main principles:

1. direct localization using coordinates georeferenced to a geographic coordinate system
2. 'indirect localization' by reference to known, limited areas without coordinates, also called the label method because a city name can be used as identification in the field one is working with.

The following points describe unambiguous data relating to the coordinates for localizing a map:

1. map projection
2. geodetic horizontal datum
3. axis and/or zone
4. north value (positive direction northwards) in metres
5. east value (positive direction eastward) in metres
6. height (positive direction along vertical line) in metres
7. geodetic vertical datum (vertical reference level).

Points 1 to 5 apply to horizontal coordinates. If height is applied, also a geodetic vertical datum is used (see points 6 and 7).

The objects in SOSI object catalogues can be realized according to SOSI syntax, or other formats such as Geography Markup Language (GML). GML is a standard for exchanging geographic data defined by the Open Geospatial Consortium (OGC). GML will eventually be able to take over the role SOSI's exchange format within Norway's national geographic infrastructure. For example, already the object list of 'Cord Data 4.5' is only realized as GML and not as SOSI.

Geographic data in databases

In order to store maps and attribute data, one needs a database. A database is a collection of related facts that naturally belong together. These facts reflect aspects of the real world and are established to meet specific needs. A telephone directory is an example of a database. The directory itself is not the database, but rather the data stored on the various pages.

A traditional database consists of fields, records, and files. A *field* is a single item of factual data, whereas a *record* is a complete set of fields, and a file is a collection of records. In a telephone directory example, the directory will be the file that contains a list of records, and 'name', 'address, and 'telephone number' will be the fields in each record.

In digital terms, a database is usually made up of numerous *tables*. A table is a collection of closely related data, such as facts about people, places, contracts, or property. A table is made up of rows and columns. A *row* describes facts about a specific object, a person, a place, or a contract, and so forth. A *column* contains facts of the same type about many different objects, such as people's ages, postcodes, or rents stipulated in contracts. Tables can be linked together via a relationship in which one column is linked to the same column repeated in another table containing other data.

RELATIONAL DATABASES

Many different database models have been developed to date, such as the network model, the relational model, and the object model, to mention just a few. Today, the vast majority of database systems use the relational model.

In order for geographic information to be retrieved quickly and efficiently, the data have to be stored and structured in a rational manner. In the past, efficient storage of geographic data was challenging, due the amount of data and the many different data types. Previously, geographic information was stored as binary files, but today relational databases are increasingly used. Three different types of database solutions exist: the hierarchical database model, in which data are stored in several levels with a *one-to-many relationship*; the network model, in which data have a *many-to-many relationship*; and the relational model, in which all objects can be related to each other in one-to-one relationship. The relational model is the most widely used today and the trend is towards object-oriented relational databases.

Object-oriented relational databases provide a number of advantages, since a number of people can work with the same data at the same time and, not least, these databases permit the use of object-oriented data models. These models are efficient when describing spatial objects in geographic data models. Such objects (features) have attributes and behaviours that allow the creation of more accurate models of the real world.

A relational database consists of *objects* and the relationships between those objects. Database objects are representations of physical features in the real world, such as a person, a place, or a project. In other words, objects are clearly definable in the real world. Objects are defined by their attributes, such as the name, age, gender, and/or national identification number in the case of a person. Features of the same type are collated in tables, in which different attributes that describe an individual object are stored in a single row, whereas attributes of the same type that describe one aspect of many objects are stored in a single column.

Further, *relationships* can be defined between the objects in the database. A relationship reflects the link between two objects and is needed to link the same objects in the database. A relationship between two objects could be exemplified by an employee who has a relationship with the department in which he or she works or by an employee who participates in several projects. There are three types of relationships: one-to-one, one-to-many, and many-to-many relationships.

A *one-to-one relationship* means that an object of one type only has a relationship with *one* object of another type. If, for example, we have two objects – 'Sarpsborg' and 'Østfold County' – and a one-to-one relationship between with them called 'County capital', this would mean that 'Sarpsborg' could only be the 'County capital' of one county, 'Østfold County', and that 'Østfold County' could only have one 'County capital', 'Sarpsborg'.

2.37 Networked computers, servers, and databases. Photo: SCOTTCHAN/Shutterstock.

Geographic data in databases

Property

| Property ID |
| Municipality |
| Gnr (Cadastral unit number) |
| Bnr (Property unit number) |
| Fnr (Leasehold number) |
| Snr (Condominium unit number) |
| Area |

1 — Association — m

Building

| Building ID |
| Area |
| Building type |
| Building status |

2.38

2.38 One-to-many relationship.

A *one-to-many relationship* means that an object of one type has relationships with many objects of another type. For example, the objects 'building' and 'real estate' could have a one-to-many relationship called 'belongs to'. In this example, 'real estate' would be related to a collection of 'buildings' situated on the real estate object. This means that one 'real estate' could have more than one 'building' object, but each 'building' object could only belong to one 'real estate'. Unlike a one-to-one relationship, the relationship would only be between 'building' and 'real estate'. Hence, one could not find out from the 'real estate' object alone which physical buildings belonged to the physical real estate because the 'real estate' object would have to store a reference to all 'building' objects related to the 'real estate' object.

Many-to-many relationships between objects are more difficult to represent in databases because specific object relationships (tables) have to be established in order to store the relationships. The objects 'property' and 'property title holder' can have the relationship 'property title conditions'. In reality, this means that a property can have a number of property title holders and that a property title holder can hold titles relating to a number of properties. For example, Kari and Ola Hansen each own 50% of

2.39 Many-to-many relationship.

Property

| Property ID |
| Municipality |
| Gnr (Cadastral unit number) |
| Bnr (Property unit number) |
| Fnr (Leasehold number) |
| Snr (Condominium unit number) |
| Area |

m — **Legal conditions** — m

Share
Legal conditions

Title deeds holder

| Person no./ |
| Organization no |
| First name |
| Surname |

2.39

properties 93/1 and 97/2 as well as 25% of property 93/3, whereas Hans Hansen owns 50% of property 93/3 and leases 97/1/12. This information cannot be represented only by the two objects 'property' and 'property title holder', since the objects would have to have references to an unknown number of objects. Alternatively, the objects could be repeated several times in one of the tables. Instead, a new object is established to define the relationships, such as the object 'property title conditions', which contains references to both the landed property and the property title holder. The object 'property title conditions' can also contain more information, such as a description of the type of the 'property title conditions' and how those conditions are divided. Figures 2.39 and 2.40 show how the example would appear in tables in a relational database.

In order to implement the above-described relationships, the objects in the database would need to have attributes that define the relationships. Such attributes are called keys. Objects have two types of keys: *primary keys* and *foreign keys*.

Primary keys are attributes of an object that uniquely identify that object. A primary key can consist of a number of attributes or just one attribute. A national identification number is an example of a primary key that uniquely identifies an object, namely a person. In the case of the object 'property', mentioned above, 'property' could be uniquely identified by, for example, the attributes 'municipality', 'cadastral unit number', 'property unit number', 'lease number', and 'condominium unit number', and these five fields would constitute the primary key for the object 'property'. Often, unique system-generated numbers (e.g. 'property number') are used for objects that do not have obvious primary keys.

A foreign key is an attribute that defines the real-world relationship. A foreign key is an attribute of one object that is the primary key of another object. For example, the object 'building' could have the attribute 'property number', which is a reference to the property to which the building belongs. In this case, 'the property number' would be a foreign key in the building table and a primary key in the property table.

2.40 The relationship between property and property title holders depicted using keys.

In short, in order to refer to a specific row in another table, one has to refer to the primary key in the first table. Thereafter, a column is created that relates to the same primary key that becomes the foreign key in the first table. Thus, 'national identification number' would be a primary key in the 'property title holder' table and a foreign key in the 'property title conditions table.

One example of a relevant relational database for the GIS environment is a land register known as the Matrikkel, which is Norway's official database with records of all landed properties, addresses, and buildings. In reality, the Matrikkel comprises three different databases that are connected together by keys, such that a 'Real estate' in the landed property database is connected to its respective 'Address' in the address database and its respective 'Building' in the building database. This makes it possible to retrieve information from the overall database, based only on owner's name, municipality, cadastral unit number, and property unit number, address, or other similar data.

Web-based services

An increasing number of major geodata users are realizing that it is impractical to hold large quantities of map data that they do not update themselves. Therefore, they want, as far as possible, to retrieve this data directly from the data provider over the Internet. This will ensure that they are using up-to-date data at any given time, and save them storage space. However, this presents a number of challenges given the fact that geographic information often involves large quantities of data, and good broadband access is therefore a must.

WMS

Various services have been developed for exchanging geodata over the Internet. The most widely used is an ISO standard called WMS (Web Map Service), which is divided into four main elements:

- GetCapabilities
  WMS works as follows: the user (the client) first sends a query to the server about what the service is offering. The server responds with a XML file telling you about the various formats, reference systems and map sections that are available, as well as the actual path from which the map can be retrieved.
- GetMap
  Based on the response from the server, a new query is sent to retrieve a map. This order includes, among other things, the geographic section, desired layers and desired format. The map section is delivered as an image file, usually in PNG format.
- GetFeatureInfo
  Gives us properties of an object on the chart, in the form of a text file or XML file. Unlike the two previous methods, a WMS service is not needed to support this method, but often it is appropriate that it should be possible to retrieve information about objects in maps, in the same was as a number of WMS services provide opportunities.

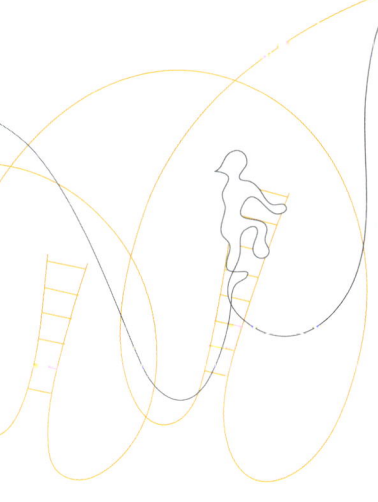

Web-based services

2.41

WMTS will often be faster, since the images have already been preprocessed. WMTS is well suited for the distribution of mapping data that do not change often, such as background maps (e.g. the topographic maps of Norway).

WEB FEATURE SERVICE
WMS and WMTS only give a graphical representation of the data in the form of a map image, not the actual data. Hence, in order to use the actual data rather than a map image, it is necessary to use another standard called the Web Feature Service (WFS). The use of a WFS makes it possible to distribute spatial data as vector data in GML format.

WEB COVERAGE SERVICE
Web Coverage Service (WCS) is used, for example, to distribute raster data, such as temperature over time, snow, or altitude models.

GEOPROCESSING
As with any data, WFS services, map images, and WMS services available via the Internet, geoprocessing services can be offered too. These services are designed to solve specific tasks or to perform specific analyses. For example, a geoprocessing service might receive a coordinate and from this position it will perform an analysis to find all kindergartens within a 15-minute travel zone. The service could then return a list of kindergartens with addresses or a map showing selected kindergartens and the travel times to them.

WEB PORTALS AND MAP CATALOGUES
While WMS services are relatively simple to implement, WFS services are significantly more demanding. However, the growing need for a common database with information means these servers are growing in number and popularity. In order to make map data available to large numbers of users, the Internet is the best solution when the distribution of physical storage media costs too much and is problematic in terms of implementation in various applications.

2.41 Web map services provide quick and easy access to large volumes of geospatial data. Source: agsandrew/ Shutterstock.

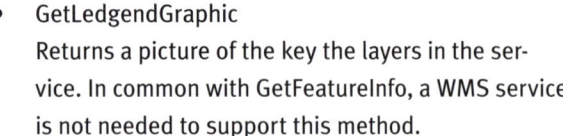

- GetLedgendGraphic
  Returns a picture of the key the layers in the service. In common with GetFeatureInfo, a WMS service is not needed to support this method.

WEB MAP TILE SERVICE
Web Map Tile Service (WMTS) can in some ways be compared with WMS, but unlike WMS in which the service receives a large picture, WMTS distributes already finished small cached image tiles. That means

2.42 Web map services provide quick and easy access to large volumes of geospatial data. Source: agsandrew/Shutterstock.

A number of portals offer WMS services and large amounts of data are currently available via these services. Compared to the WMS services, the WFS services are less widespread but nevertheless essential if data are to be used for analyses. It is thus likely that more and more data will become available via WFS services.

As increasingly more geographic information becomes available through download and review services, it may be necessary to have good solutions to identify the most appropriate data and services. Therefore, metadata directories have been created where users can search for data. One such solution is Geonorge's map catalogue of data available through Norge digitalt.

Technological trends

The 21st century has so far seen high levels of innovation in the development and use of new technological solutions. Information and communications technology (ICT) analysis companies make annual analysis of these technology trends, and at the start of the century the company Gartner Consulting (provider of analytical services) highlighted four 'megatrends' for the coming decade (Gartner Consulting, 2010):

- industrialization of ICT
- increased scope of shared services
- increased level of social interaction and information sharing online
- unlimited mobility.

These technology trends have since been broken down and concretized for shorter planning periods to show how they affect corporate strategic planning. The trends are of great importance for how geographic data and information systems are used. Hence, below, they are examined more closely on the basis of how they affect and enable better utilization of geographic information systems.

INDUSTRIALIZATION OF ICT

With the industrialization of ICT, it is conceivable that initially off-the-shelf infrastructure and standard solutions will be used. Costly customized solutions for individual companies will thus be avoidable. This will increase competition and push up the prices of ICT services. Consequently, open standards will become increasingly important because they will reduce the risk of unsound investments and binding times to providers. In other words, this concerns the use of established technologies in a smart way that saves time and money in daily work.

3D printing is part of this industrialization that is expected to increase. This market is growing as a result of 3D printers becoming cheaper and applications becoming more widely known. This means 3D printing will become a viable and cost-effective production method for physical models.

New solutions for payment for various services via smartphones and computer chips are creating new possibilities for how companies can charge for their services. Eventually, it is expected that the advanced algorithms will make systems more intelligent and informative. In this way, they will increasingly understand their surroundings and seemingly act independently.

INCREASED SCOPE OF SERVICES SHARED BY MORE

The use of cloud technologies and services makes it possible to share, disseminate, and apply all available information efficiently. In both professional and private arenas, there is a plethora of providers of cloud services. These common interactive spaces make it possible to receive, share, and manipulate documents, reports, and information in a much more efficient way than before. The advantage of the cloud solution is that they make information and applications available anytime and anywhere on various technological platforms. Shared services will also be cheaper because the costs will be shared and the services will become more streamlined.

Simple tools, often called apps, adapted for mobile phones and tablets make it possible to use and update geographic data in a new way. Geographic data are thus real-time information that are readily accessible to users via the Internet. The scope of the services shared by many users is increasing and in a business context, the following questions need to be asked:

- What does my business need from geographic data? How do we ensure that this is real-time information that will be available to our employees when they need it?
- A central question is whether the company itself should update the geographic data or whether it is important that we have the expertise to apply it and only have access to information? What represents the most efficient use of money?

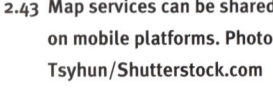
2.43 Map services can be shared on mobile platforms. Photo: Tsyhun/Shutterstock.com

2.43

- How can we cost effectively analyse structured and unstructured data within and outside the company in such a way that we can use the data to increase our insights and value creation?
- How can we ensure a good understanding of what are 'trade secrets' and what is open information in our master data?

In this regard it is conceivable that users must have a proven relationship with what is shared with others via cloud solutions. As stated above, there will be major challenges in terms of security and privacy with the increasing use of cloud solutions. Much can therefore indicate that most businesses will at some point come to the realization that it is not possible to have solutions that are 100% secure. This is expected to lead to an increased use of risk analysis and risk mitigation measures. Hence, to a greater extent, that will determine what constitutes an acceptable level of risk for initiated measures.

UNLIMITED MOBILITY

The developments in ICT means that also work and leisure are becoming more and more integrated. It could be said that society is becoming increasingly more flexible. Today's mobile platforms such as tablet, computers, and smartphones mean that e-mails can be sent and received while simultaneously streaming a film or watching television. This creates expectations in society that e-mails or postings on social media will be responded to quickly. Similarly, geographic information is used all the time on mobile platforms. Eventually, industrialization will reach the stage where GIS analyses that were previously both time and resource-demanding to carry out will be performed on simple mobile platforms. This will mean that the recipient of such simple GIS analyses will be required to evaluate and understanding the contents against the quality of the data on which the analyses are based.

INCREASED DEGREE OF SOCIAL INTERACTION AND INFORMATION SHARING ON THE INTERNET

One of the current trends is that the use of e-mail is losing more and more ground to the communication of real-time information via social media. Also, the gap between the professional and private arenas is becoming increasingly smaller. The openness of society is therefore expected to increase and create favourable conditions for learning, innovation, and development. This means that businesses will have to be better at using social media for collaboration and information sharing. Much information is fresh and it is often important that it is communicated quickly, without too many intermediaries. The focus will therefore be on the needs for information and how to service them in different contexts and on different platforms.

Per Gunnar Ulveseth holds a master's degree in mapping technologies from the Norwegian College of Agriculture (now the Norwegian University of Life Sciences). He works as a senior engineer in GIS and mapping at the Norwegian Defence Estates Agency.

The chapter is a revised version of the chapter written by Arvid Lillethun, Edith Bø-Nygaard, Jens-Petter Seppola and Nils Ivar Nilsen in the 1st edition.

Geographic data 3

What is geographic information? *74*
Geographic infrastructure *75*
Basic geodata *77*
Thematic data *88*
Land use planning *123*
Civil protection and emergency preparedness *127*
Statistical data *133*

What is geographic information?

The use of a GIS is highly dependent on access to relevant and up to date map data. This chapter focuses on what constitute geographic data and how the data can be systematized.

Geographic or place-based information is information about objects or conditions that can be placed in a geographic space. This can involve either general information that describes specific physical objects, such those found on topographic maps, or data that provide detailed information about specific themes, such as a bedrock map. The information can be presented on a surface (e.g. a map) in such a way that the distances between the depicted objects correspond to the distances between the represented physical objects in real life. Alternatively, the information can relate to data that have a geographic location, but where the presentation of the data does not show the geographic relationships in the information,

3.1 Topographic map at a scale of 1:100,000 of part of Troms. Published by the Norwegian Military Geographic Service.

3 – GEOGRAPHIC DATA

3.2 Geonorge is the national website for map data and other location information in Norway. Geonorge is the hub of Norway's geographic infrastructure.
See www.geonorge.no

such as a table listing comparative dates for municipalities.

Geographic data are collected at various levels. Much information is collected at the local level such as municipalities, while other data are collected by specialist agencies at national level, such as the Norwegian Mapping Authority and the Norwegian Environment Agency, or by private firms such as power companies that need an overview of their distribution networks. We all collect geographic data, in many contexts.

The following section presents Norge digitalt, followed by a number of examples of types of geographic information that are available to the general user. Since there are numerous providers of geographic data, the descriptions only cover a sample.

Geographic infrastructure

Norge digitalt

Norge digitalt represents a contractual cooperation between public authorities at national and regional level that have responsibility for obtaining geographic data and/or are major users of such data. The partners are national agencies, county and municipality authorities, and other major geodata users. The purpose of Norge digitalt is to ensure proper support for access to basic data and thematic geodata through online services and download solutions. Norge digitalt thus forms the geographic infrastructure in Norway.

The development of Norge digitalt is rooted in geodata law and related regulations. Under the Geodata Act of 2010 (Lov om infrastruktur for geografisk informasjon), all public agencies with responsibility for geodata or that are major users of such data should contribute to the establishment, operation, and maintenance of the country's infrastructure for geographic information.

Geographic infrastructure

The Ministry of Local Government and Modernisation has overall responsibility for Norge digitalt, and the Norwegian Mapping Authority coordinates and supervises the use of national geodata. The mapping authority has primary responsibility for managing and coordinating the website regarding administrative and technical matters.

Standardization of geographic data
There are a number of standards and guidelines relating to geographic data, and the Norwegian Mapping Authority provides an overview of most of them on its web pages. Currently, the largest and most important standard for geographic data in Norway is the *SOSI-Standard*. In addition to being Norway's official data format for the exchange of geographic data, SOSI (Systematic Organization of Spatial Information) includes a comprehensive standard with its own database in a general features catalogue. The features catalogue contains standardized definitions of geographic features, properties, and tag values, together with explanations. It is a very important source of information for those who wanting to view geographic data or possibly to specify their own geographic data. See kartverket.no/sosi for more information. The Norwegian Mapping Authority also lists a number of other standards relating to map production, surveying, and geodesy, as well as other industry standards.

Structuring and metadata for geographic data sets
In order for geographic data sets to be used in various mobile and web-based solutions, it is important that they are uniformly structured when they are produced or updated. This is also important if the data are to be used by others. The Norwegian Mapping Authority and Norge digitalt have established requirements for specifications and information relating to geographic data sets.

SOSI data specifications. *SOSI data specifications* should be established for all data sets, and are specifications and methodologies for establishing basic geodata and thematic geodata. The specifications are based on the SOSI features catalogue and they describe in detail how the data should be established within defined subject fields. In addition to their technical content, the specifications include descriptions of the use of the data and links to technical guides. The specifications are provided to Norge digitalt by national specialist data providers, both for their own data and for geodata that will be established by other users. The Norwegian Mapping Authority provides technical guides for the preparation of specifications. The specifications for a wide range of thematic data sets can be accessed via the website Geonorge.no.

Metadata. *Metadata* should be established for all data sets,to provide some insights into what the data sets contain, how they are formed, the contact person, and brief information on their accuracy and the frequency of updates. For Norge digitalt, there is a requirement that separate fact sheets should be provided for metadata, together with rules for the presentation of all data sets. These are also made available in Geonorge's map catalogue.

3.4

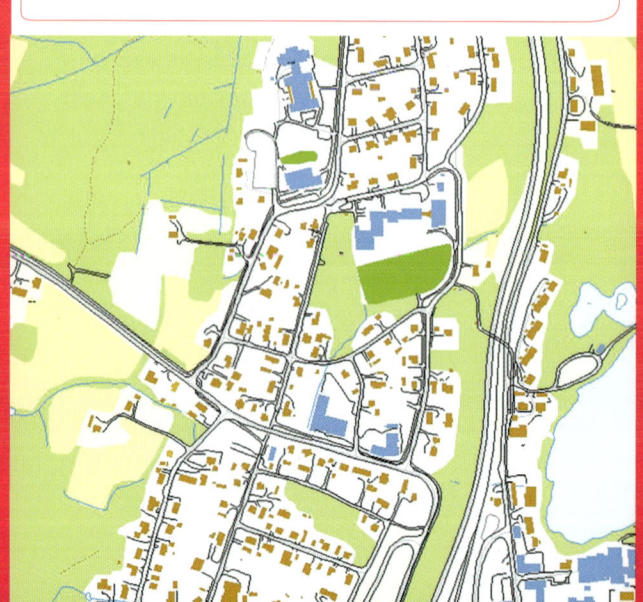

INSPIRE

Infrastructure for Spatial Information in Europe (INSPIRE) is a pan-European initiative for cooperation over the sharing of geographic data. INSPIRE was established in accordance with the EU's INSPIRE Directive of 2007, which describes how the various member countries in Europe should organize their geographic infrastructure, the types of data that should be made available, which service types the countries should offer, and how data should be harmonized across national borders. Under the directive, INSPIRE should ensure access to geographic information administered at national level, including information relating to nature, transport, buildings, populations, and environmental conditions. Norway has adopted the INSPIRE initiative and mainly conforms through Norge digitalt (Kartverket, 2014).

Basic geodata

Basic geodata include various types of basic data such as topographic data, property information, orthophotos, height and depth data, administrative boundaries, place names, and geodetic points. Topographic surveys have a century-long tradition, and have formed the basis for more recent geodata. The Norwegian Mapping Authority and local authorities are among the main actors involved in the establishment of basic geodata, and various schemes have been introduced for co-financing the maintenance of the data. The Geovekst initiative represents a partnership between municipalities, road authorities, the Norwegian Mapping Authority, agriculture, energy producers, and Telenor (telecommunications company) as well as some other national agencies that fund basic geodata relating to land.

Basic geodata are essential for businesses, property management, the construction industry, and the management of environmental resources, for which the lack of quality data could affect various sectors of society. Through Norge digitalt, agreements have been made with a wide range of public authorities to ensure there is adequate long-term financing for the provision of important basic data. Simultaneously, provisions have been made to ensure that such data are accessible to all users of Norge digitalt.

3.3 For simplicity, map data can be divided into basic spatial geodata and thematic geodata.

3.4 Maps produced from basic geodata range from technical maps at a scale of 1:1000 to overview maps at a scale of 1:250 000.

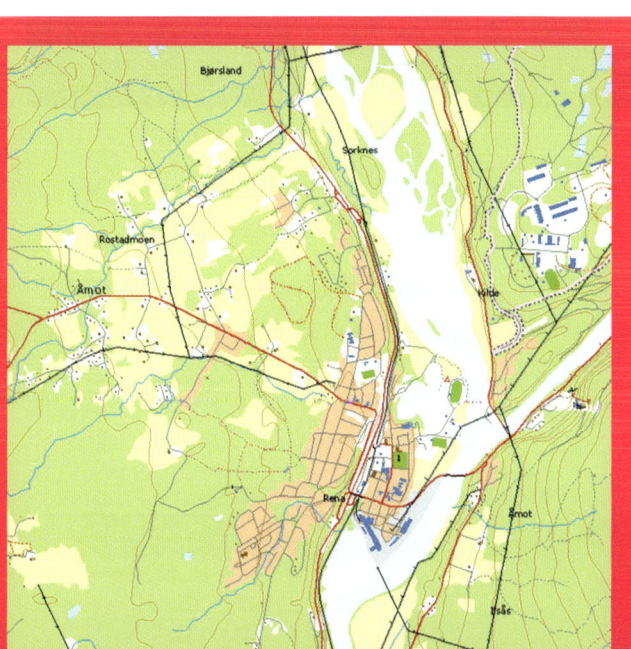

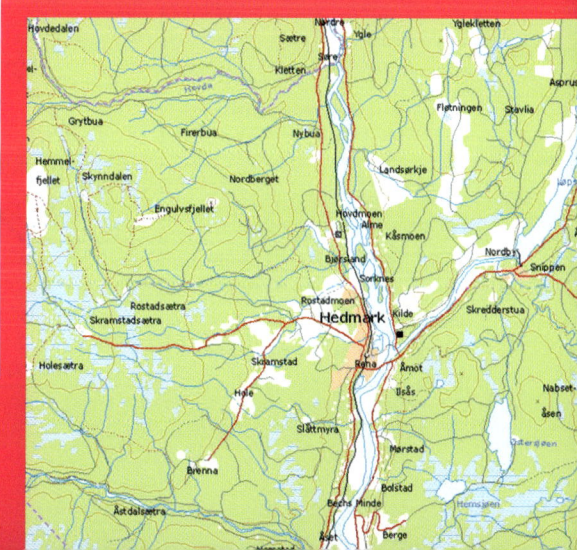

FKB DATA

The most detailed basemap data are collated in the *Felles Kartdatabase (FKB)*, a national map database. FKB data are grouped according to different data fields, such as building, construction, road, and airport data, and the divisions are specified in more detail in the SOSI standard. FKB data are classed according to four degrees of accuracy, where FKB-A are the most accurate data and FKB-D data are the least accurate.

FKB-A data are suitable for detailed design and 3D modelling, and are mainly used for maps of densely populated areas. FKB-B data are suitable for base maps used in the preparation of development plans. They can also be used for 3D modelling, but are slightly less accurate and rather less detailed than FKB-A data. All 3D modelling is dependent upon the availability of detailed elevation data.

The level of detail in FKB-C data corresponds to maps at a scale of 1:5000 and the data are used for maps of areas outside built-up areas. FKB-D data have a level of detail that corresponds to maps at a scale of 1:50 000. The work of compiling FKB data (A, B and C) is financed through Geovekst projects.

FKB data have varying degrees of accuracy and content for different types of areas. Figure 3.5 shows how FKB-A data can be used for maps of urban and densely populated areas, FKB-B data for maps of areas designated as residential and industrial areas, FKB-C data for maps of farmland, and FKB-D data for maps of mountain and forest areas that lack any significant economic activity. Ideally, there should be no geographic overlap between the FKB standards. The FKB data sets are established seamlessly. Hence, it is important to be aware that the level of detail and quality varies according to the type of area.

In order to increase access to and the use of established FKB data, a set of standardized data products has been specified: *FKB data products*. The products are obtained by generalization of the original FKB data and are specially adapted to the needs of different users.

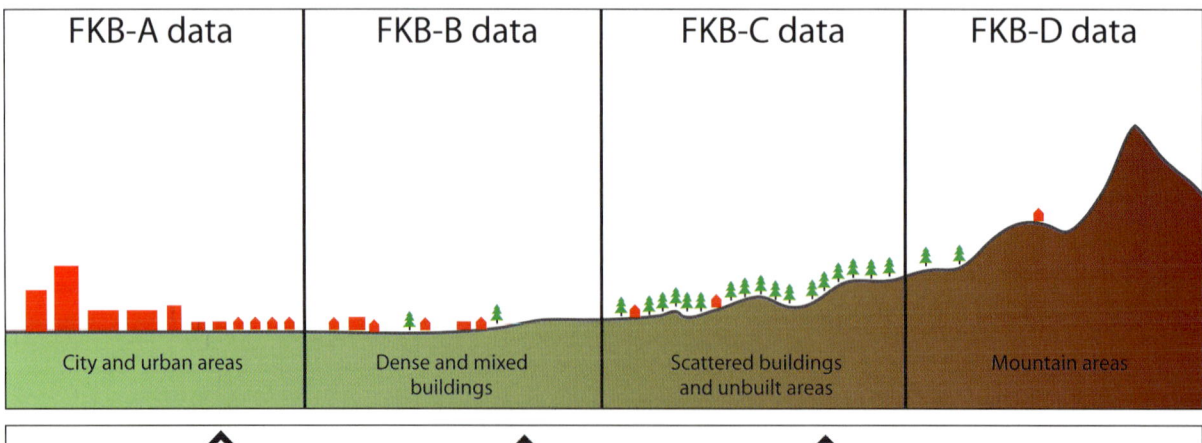

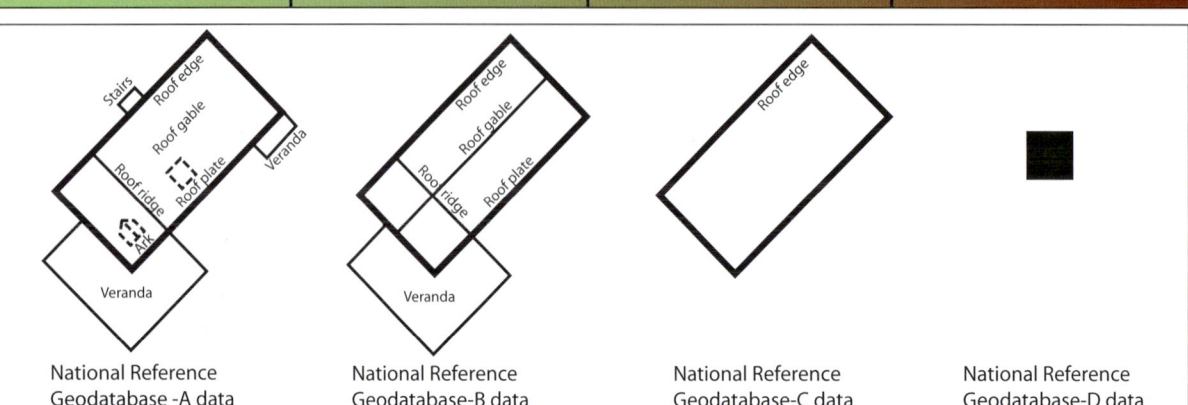

3.5 The National Reference Geodatabase (FKB) is categorised into four map classes. The level of detail and spatial positioning is most accurate in FKB-A data and least accurate in FKB-D data.

The following FKB data products are available: FKB data, cadastral maps, N5 map data, N20 map data, N20 buildings, and N5 raster. Orthophotos are available too.

PLACE NAMES

The Norwegian Place Names Act of 1990 (Lov om stadnamn (stadnamnlova)) stipulates that the spelling of place names should be based on traditional local pronunciation and applicable spelling principles. Under the Place Names Act, the same name in the same place should be adopted as the official form of that name. A municipality can make decisions about the spelling forms of official addresses and the names of villages, hamlets, municipal streets, roads, market places, boroughs, housing estates, facilities, and so forth. The Norwegian Mapping Authority's Mapping and Cadastre Division can make decisions about spelling forms for all national contexts and for the names of rural locations and traditional estate and farm names. Once the spelling of a place name is has been adopted in accordance with the Place Names Act, it has to be used by all official bodies. The official form of a place name is registered in the Sentral stedsnavnregister (SSR), which is the central register of place names maintained by the Norwegian Mapping Authority. The register contains more than 950,000 place names. The SSR is a useful tool for map production and administration of the law on place names.

ADMINISTRATIVE UNITS DATABASE (ABAS)

The Norwegian Mapping Authority provides a separate data set for the demarcation of administrative units. The units are nations, counties, and municipalities, as well as the boundaries that define them. Geographic data relating to these administrative levels are used in many contexts, such as to show the area covered by a municipality, to present statistics, to refine other data, or to show an area that an agency or organization has responsibility for, including its boundaries.

ROAD DATA

The availability of accurate and up-to-date data on road networks is very important for efficient transport planning and reliable in-car navigation systems. The data set FKB-veg, which is part of FKB, includes the geometric design of roads, such as the edge strips and surfaces. These data are primarily photo-

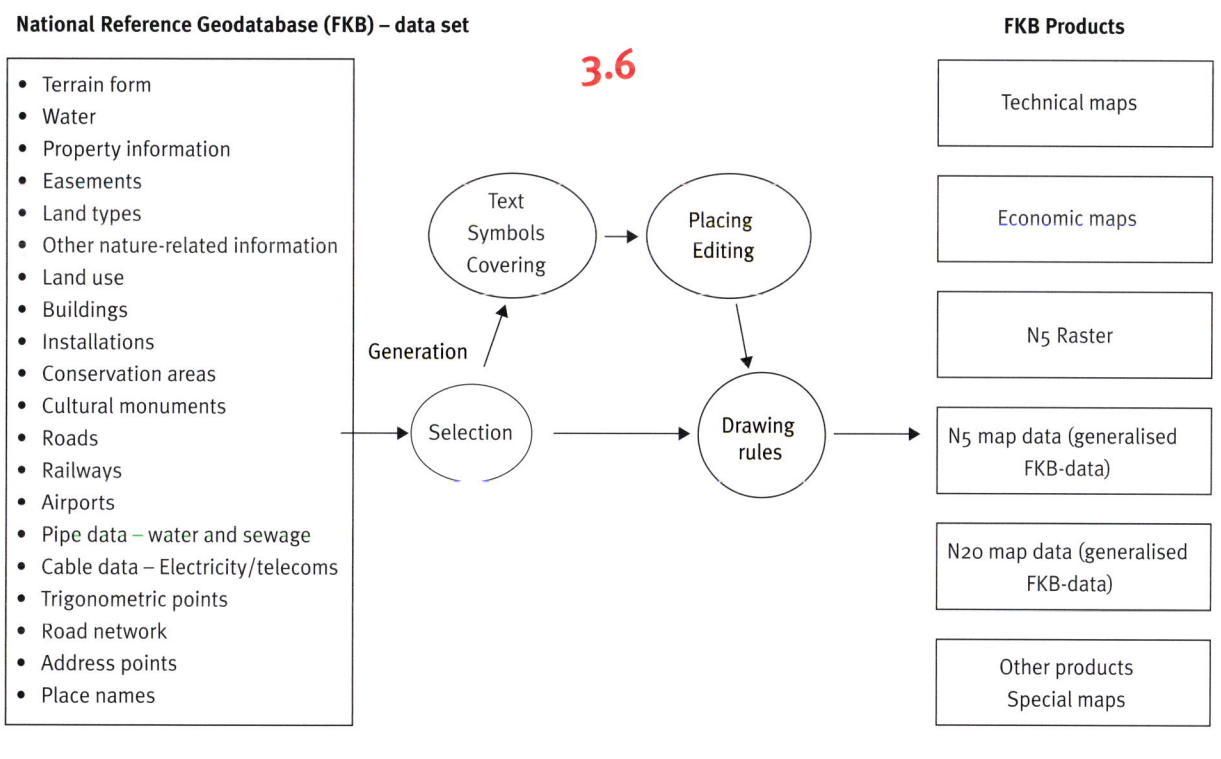

3.6 Schematic diagram showing how independent data sets can be combined into various map products. Source: Product specification for the National Reference Geodatabase (FKB).

grammetric data and are important for an up-to-date map database. They are useful for land use planning and area calculations, but on their own they are not sufficient for navigation purposes or transportation planning. Instead, the road data have to be stored in a network topology.

The digital road network is administrated through the National Road DataBase (*Nasjonal vegdatabank*, NVDB), which contains data on all drivable roads, footpaths, and cycle paths, stored in a bus network. The road lines are stored together with their relevant properties such as speed limits and road widths, and the NVDB is thus a very good source of data for navigation, transport planning, or fleet management. The NVDB has a separate data set called ELVEG, which contains data on all drivable roads and their properties, and is used for navigation and transportation planning (Stoeng et al., 2014). For more information on topologies and network analysis, see Chapter 4.

The data relating to European, national, and county roads are continuously updated by the Norwegian Public Roads Administration (NPRA), whereas data for municipal roads, private roads, and forest roads are administered by the Norwegian Mapping Authority following periodic updates to the NVDB reported by the municipalities. The data set used for roads administration by the municipalities is the FKB-Vegnett (FKB-Road network). The municipalities also have an additional data set called the FKB-TraktorvegSti, which contains data for other routes, such as trails, tractor routes, and pedestrian routes. There are also many different data sets containing data on, for example, hiking and walking routes or cross-country ski trails.

ELEVATION DATA

Normal topographic maps have contour lines that provide information about differing elevations in the terrain. Contour lines are *isolines* drawn between points at the

3.7 The data held in the National Road DataBase includes data about accidents. The data is available via: www.vegvesen.no/vegkart. Source: The Norwegian Public Roads Administration, National Road DataBase.

same height above sea level. The number of contour lines varies according to the level of detail in the map base. Detailed elevation data can be found in the national map database (FKB). Elevation data can also be used to form a *terrain model*, which enables 3D processing of the data (see Chapter 4 for a more thorough description). Elevation data in the form of *elevation layers* provide a basis for various types of visualization such as contour lines, in which each elevation interval forms a surface with colour symbolization or *shading* (mountain shadows). Shading has long been used as a technique to show terrain in relief on topographic maps.

LiDAR data
Light detections and ranging (LiDAR) data are used to produce good elevation models. Lasers can be used to generate point data sets with a high point density, where each point comprises a coordinate and an elevation value. In addition, the data are provided with an intensity value. Based on the intensity value, data can be analysed and categorized, and thus provide the user with very good information about the terrain shape, vegetation, and buildings and constructions in the area represented. LiDAR data are usually established as part of Geovekst projects or in connection with construction projects. Efforts are being made to establish a central management solution for laser data.

National elevation model
Currently, a project is underway to establish a nationwide detailed elevation model for Norway. The model should cover the entire Norwegian mainland, including the skerries and larger coastal islands, as well as a test project in the coastal zone down to 5 m below sea level. Several national agencies form part of this project, which will probably be a five-year national programme. The collection of elevation data involves a combination of laser scanning, image matching, and the use of laser data.

There is also a need for detailed elevation data in a number of important local-level projects, and it is expected that the national elevation model will be highly advantageous in this regard. A White Paper on climate change (Meld. St. 33 (2012–2013)) underlines the importance of municipalities having a detailed map basis for regulato-

ry risk and vulnerability analyses, and a White Paper on floods and landslides (Meld. St. 15 (2011–2012)) highlights the importance of detailed elevation data in order to implement good analyses of floods and landslides.

In an emergency preparedness context, it is important to know, for example, how a river behaves during flooding or how climate change can influence the landscape. A detailed elevation model will provide better knowledge of forest resources and will be a useful tool for preventing landslides and erosion. In addition, it is important to have easily accessible and detailed elevation data for planning and industrial development (Kartverket, 2014).

ORTHOPHOTOS
An orthophoto is an aerial photo with the same geometric properties as a map and is linked to a map coordinate system, which means it can be used in conjunction with map data. The use of orthophotos has become very popular. Most people find it easier to orient themselves using an aerial photo than a map of the same area because all of the features in the terrain can be recognized immediately, without the need for a map key and scale. Orthophotos can be used on their own or in conjunction with map data or thematic data for land administration, land use planning applications, and other purposes. Orthophotos are produced by municipalities,

3.8 Elevation data for the Ryfylke district presented with both shading and elevation layer colours.

Basic geodata

the Norwegian Public Roads Administration, the Norwegian Mapping Authority, the Norwegian Institute of Bioeconomy Research, and other actors.

Orthophotos are made available either through Geovekst projects or through the national programme for repeated aerial photograph coverage of Norway (Nasjonalt program for omløpsfotografering). When orthophotos are produced as part of Geovekst projects, the flying should ideally be done in the spring, before leaves appear on the trees, in order to generate detailed orthophotos with either a pixel size of c.10 cm or c.20 cm (i.e. resolution) that is suitable for map production. By contrast, aerial photography for the national coverage programme tends to be done slightly later in the year and the photos have a slightly lower resolution, with a pixel size of 25 cm. The programme is an overall administration and distribution solution for orthophotos and satellite images in Norway. The data can be accessed via the website norgeibilder.no and are also distributed via online services.

Oblique photos
Oblique aerial photos are photos taken at an angle. These data are particularly popular when they relate to cities and towns. Overhead imagery provides an intuitive visual experience of an area and therefore oblique aerial photography has many potential applications. For example, oblique photos can be used for planning, for particular addresses, or for building permits. They can also be used for visual inspection, to check cabins and houses in a coastal zone or in remote areas. Oblique photos are less often produced for Geovekst projects or for Norge digitalt and therefore no separate national administrative system exists for these data.

CABLES AND PIPELINES

In Norway there are significant numbers of cables and pipes below ground level. A good overview of where they are located is therefore essential for society to function in everyday life. Often we do not appreciate the importance of data relating to the location of cables until supplies from services are disrupted. If we are without power or if drains become blocked, we may soon realize how important it is for these data to be maintained and for the cable and pipeline owners to fix the problems quickly.

Cable and pipeline networks are operated by both public and private actors, which have their own databases and maps. There are many systems available on the market to help various actors to keep track of their cables and pipes. All users are likely to agree on the importance of having control over their cables and pipes and the use of a relevant database, but there are still huge differences in the number and quality of the various bases. Currently, very many metres of cables and pipes in the ground have not yet registered in any databases, and this creates major challenges when the ground is to be excavated. In many cases, excavation

3.9 Ortophotos are aerial photos that have been processed so that they are in the same scale as a map.

work will be associated with great risks: digging in the vicinity of a high voltage cable is extremely dangerous. Hence, it would be difficult, if not impossible, for a large municipality to function without knowing the geographic location of cables and pipes.

Service provision can be broadly divided into the following types:

- water and drainage systems
- electricity grids
- telecommunications
- energy (district heating)
- gas lines
- POL (petroleum oil lubricant) lines.

With good, comprehensive infrastructure databases, it will be considerably easier for information about planned excavations to be made available to all involved actors. The data must be reliable and readily accessible to the actors. It can be difficult to find a balance between the need for to hide sensitive or classified data and the need to make the data available for excavation work.

There are major differences in the quality of underground cables and pipes. For example, in some parts places in Norway the water networks are in such poor condition that they can pose risks in terms of leaks and water quality. The renewal of such pipelines and cables would incur huge costs and therefore it is very important to have knowledge not only about their location but also about their properties. Properties such as the construction materials, dimensions, and the year when laid can provide an indication of the quality of a cable network and can help a manager to take the right action.

It is important that managers set clear requirements for surveys of cable and data, so that they can accurately document their geographic location and properties, as this knowledge is needed in order to operate the distribution systems effectively.

PROPERTY

The development of ownership rights

Land resources are important but limited in all societies. Disagreements over the right to use these resources can cause major conflicts between people and create the need for the allocation, exercising, and transfer of the rights. Information about who owns the rights to land at any given time is thus important.

In places where hunting, fishing, and grazing have traditionally been the most important economic activities, the land use rights have generally belonged to groups (e.g. tribes and families), whereas in places where farming involved agriculture, those who tilled the soils were regarded as having individual rights. Such rights could be handed down from generation to generation before they were transformed into what are now known as individual ownership rights. As the intensity of cultivation increased, so too did the need for overviews of the related rights.

Information about ownership rights is important with respect to the economy since the rights concern transactions of real estate. One of the functions of a property register is to reduce transaction costs. In his book titled *The Mystery of Capital* (2000), Hernando De Soto (a professor at the University of Lima, Peru) concludes that a modern market economy would be impossible without a formal system for the establishment

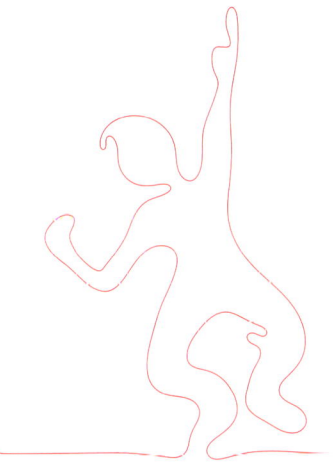

3.10 Cables and pipelines is a complex subject.
Photo: Øyvind Mauseth.

and registration of properties. It is only when an owner gains formal ownership of land that a bank will be able to use the property as collateral and allow the owner to take out bank loans and make investments. Thus, a credible property registration system is a necessary prerequisite for economic development in a society.

The registration units
In order to achieve a well functioning property register there is need for a system based on uniquely defined units of land. In Norway, cadastral unit numbers (*gårdsnummer*, Gnr), property unit numbers (*bruksnummer*, Bnr), leasehold numbers (*festenummer*, Fnr), and condominium unit numbers (*seksjonsnummer*, Snr) are used.

Each municipality in Norway is subdivided into cadastral unit numbers. These are stable units that have remained largely unchanged since the 1600s. Over time, an estate that formed the original unit will have been subdivided into several property units.

All cadastral units have been assigned a property unit number. Hence, every cadastral number will comprise one or more subordinate property unit numbers. Unlike the cadastral number, the number of subordinate properties can vary; the number will increase when land is subdivided and will decrease when land is amalgamated.

A property can have a number of property unit numbers, and each property unit number can consist of a number of parcels of land. A unique designation for a property in a municipality will also requires a municipality number (*kommunenummer*, Knr), in addition to the cadastral number and property unit number.

The holding number is characterized as the legal land unit and transactions involving purchases, sales, mortgages, inheritance, and so forth are all linked to this unit. A leasehold number is assigned when land is leased. Leasehold land is rented often with a contract period lasting 50–99 years.

When a property is divided into separate dwellings (e.g. tower blocks and horizontally divided housing) each unit is assigned a condominium unit number.

Ownership rights
Ownership rights relate to different types of property, such as hunting rights, grazing rights, and fishing rights. An individual, limited right to access property is called an easement (e.g. a right of way).

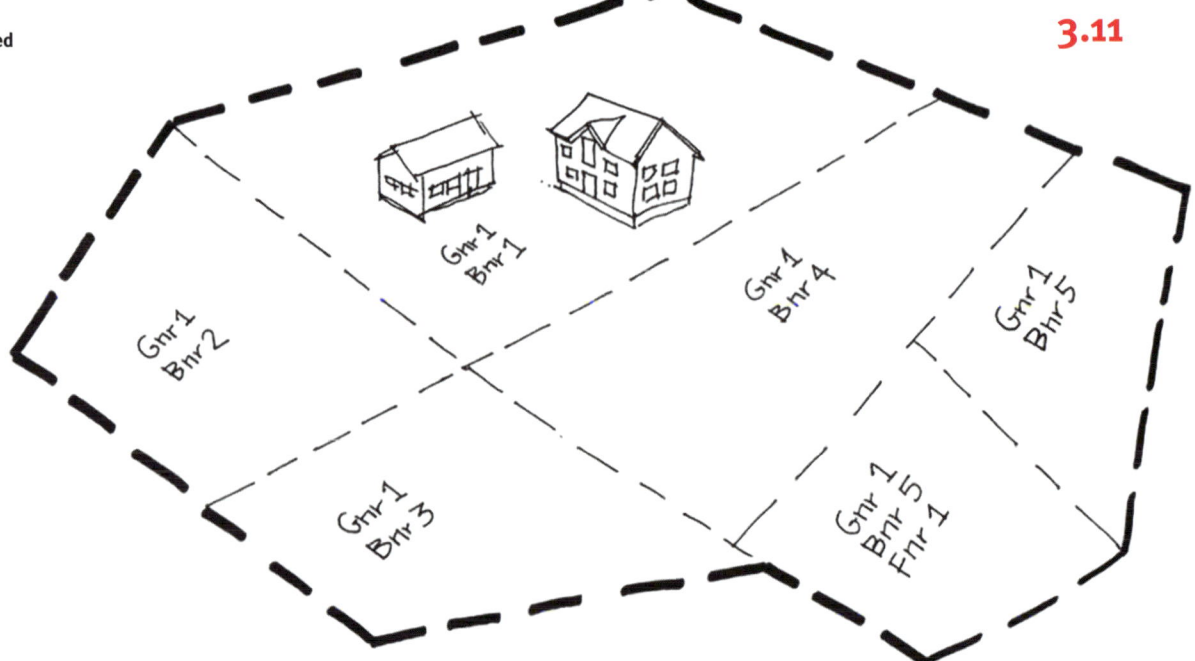

3.11 A cadastral number is subdivided into multiple holding numbers.
Source: Bjørn Bergesen.

Jointly owned common land is property that is owned by other landed properties. If one of the landed properties is sold, then its share of the jointly owned common land is sold too. The landed property owners each own a notional share of the jointly owned common land. Jointly owned common land cannot be sold.

Historically, many forest and mountain regions have not been subject to private ownership rights. The people living in rural districts are entitled to utilize the common land resources. In the case of state-owned common land, the state owns the formal ownership rights, whereas rural common lands are jointly owned by the farmers who hold the rights to them.

Property formation

A landed property can be formed in various ways:
- *subdivision* – e.g. from farming property to housing plots, second homes used for leisure, and commercial buildings
- *amalgamation* – amalgamation of multiple property unit numbers with one and the same owner
- *site leasehold* – the leasing of land for housing, second homes used for leisure or for other purposes
- *land consolidation* – the conversion of properties in the event of land exchange (mutual transfer of properties by deed of exchange) and exchanges of land and rights; the purpose of land consolidation is to create more functional and practical properties
- *sectioning* – an owner-occupied section represents an ownership share in built property with an associated exclusive right to use one or more apartments or other user units in the property
- *compulsory purchase* – this arises when ownership rights to real estate or to a building are expropriated by force for public purposes, as described in Paragraph 2 of the Norwegian Expropriation Act of 1959 (Lov om oreigning av fast eigedom).

Property registration

Property registration is a basic function in Norwegian society. The registration of real estate is a prerequisite to ensure that property can be sold or used as collateral. Prior to the start of the 21st century, properties were registered on digital property maps (Digitalt EindomsKartverk, DEK), and property information was registered in the GAB register – a register of all landed properties, addresses, and building (Grunneiendommer, Adresser og Bygninger).

With a background in the law on property registration of 2005 (Lov om eigedomsregistrering), geographic and property information is now collected and administered in a unified system called the Land Registry (*Matrikkelen*). The Land Registry is the official property register and includes information on properties' geometry, building point, and address point in a common database.

The Land Registry (property), the National Registry (population), and the Central Coordinating Register for Legal Entities (CCR) form the official national registers in Norway.

A recent revision of the property registration law has been the introduction of the concept of *fixed property*, which makes it possible to allocate special cadastral numbers and property unit numbers to properties below or above the ground level, such as tunnels and multistorey car parks.

Property (cadastral unit)

The Land Registry contains records of information on, for example, *Land Registry numbers, legal provisions for use, property boundaries, cultural sites, and soil contamination.*

Buildings

The Land Registry contains records for all buildings that are registered on a property. Currently, there are more than 4 million registered buildings in Norway. The registry also includes information on: *the numbers of buildings, building types, industrial group, built-up areas, number of floors, service units, type of water supply and heating, contact person, and reference pair of coordinates for the buildings' geographic location.*

Addresses

An official address can be either a road address or a Land Registry address (Gnr, Bnr, and possibly also an Fnr or Snr) as well as any house number. A road address consists of a road name and a house number. A house number can also include a letter. The Land

Registry address is derived from the cadastral unit number, property unit number, and any lease number, as well as any lower order numbers used to identify an address. Currently, there is an ongoing effort to introduce road addresses for all properties in the country. Road addresses are crucial for the emergency services to be able to find the right property.

Matrikkelbrev
A *matrikkelbrev* is a certified copy from the Land Registry of all records relating to a registered property for a given date. It includes a property map and a list of coordinates, as well as a description of how the border points are marked.

Probate Registry
Information about who has ownership rights or more limited rights relating to real estate is recorded in the Cadastre. The Cadastre contains a record of the purchases, sales, encumbraces and mortgages on real property. Now these documents are officially entered in the Land Registry. Previously, the records were kept in a book – *Grunnboken* – but today they are maintained in an electronic database.

The Cadastre contains information about the title-holder of a property (i.e. whoever owns the property). In addition, it has information about the dates when ownership of the property was exchanged, the purchase price, and information about encumbrances which can be either easements or monetary encumbrances).

The Cadastre ensures that all rights related to real estate belong to whoever holds the title deeds officially registered for that property, not to anyone who is not the registered title deeds holder. This particularly applies to third parties, such as banks, purchasers, and contract handling parties.

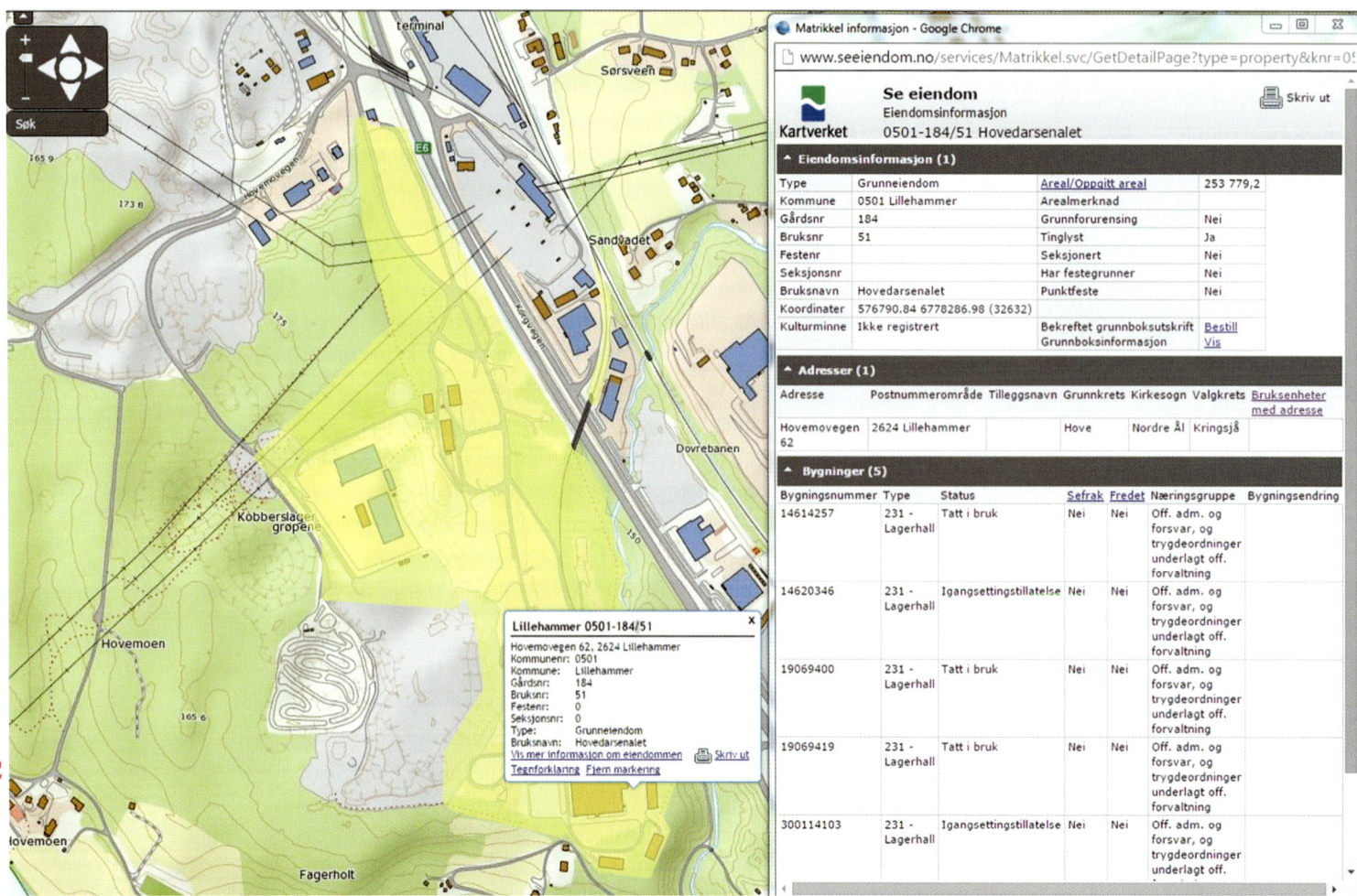

3.12 'Se eiendom.' ('See property') is the Norwegian Mapping Authority's webpage for accessing information from the Land Registry and the Cadastre. Source: seeiendom.no

3 – GEOGRAPHIC DATA

3.13 A reliable property information system is a necessary condition for the secure transfer of real estate. Photo: Michal Knitl, ScanStockPhoto.

The Cadastre was previously kept according to the laws relating to Cadastre applied in the law courts, but today the responsibility rests with the Norwegian Mapping Authority.

Access to property information
Information from Grunnboken can be ordered directly from the Norwegian Mapping Authority or from individual service providers. The Land Registry is the main source for information on the distribution of property maps and well as information on buildings and addresses. The Norwegian Mapping Authority offers free access to information held in the Land Registry through the mapping service seeiendom.no. In addition, several software suppliers offer their own solutions for obtaining information from the Land Registry.

OPEN ACCESS TO MAP DATA
A primary national goal is that map data should be freely accessible to users, as far as possible. As part of this principle, the Norwegian Mapping Authority has released large quantities of geospatial data. In autumn 2013, all map data series from N50 to N5000 were made available for anyone to download free of charge. In addition, terrain models, place name data, some maritime data, data sets relating to administrative boundaries, and the national register of obstacles to aviation has been made available. Developers can use the data freely in their own products. All of the above-mentioned data are available in the Norwegian Mapping Authority's own map-viewer tools, *seeiendom.no* and *norgeskart.no*. When the Norwegian Mapping Authority made these data available, data from the National Road DataBase (Nasjonal vegdatabank)

were released too. Since 2013, increasingly more public data sets have become either cheaper or freely available, and this is likely to continue in the coming years. However, the data in the most detailed database – the national map database (FKB) – still cannot be accessed free of charge.

Thematic data

Thematic geodata are administered by various agencies. The principle followed in Norge digitalt is that any part that establishes thematic data should be responsible for financing, operating, and maintaining the quality of their data series. State agencies are working to standardize and document their geographic data and make them available through Norge digitalt. The following section starts with a description of the *public map data*, and then describes some of the data sets in detail.

PUBLIC MAP DATABASE

The public map database contains public and quality-assured geographic data that are intended to facilitate planning and building work. The background to the public map database is the Planning and Building Act of 2008 (Plan- og bygningsloven) and the official guidelines to regulations regarding maps, spatial information, land use objectives and digital planning registers (short name: Mapping and planning regulations (Kart- og planforskriften)) of 2009, which states that municipalities have to ensure that their data are up to date in the public map database. The data should provide an overview of administrative and physical conditions in the municipalities, and cover coastal areas up to 1 nautical mile beyond the baseline. The information should be made readily available for use in planning and building applications and for other private and public purposes (Kartverket, 2015, Det offentlige kartgrunnlaget). Previously, the basemap was considered public basemap, originally in the form of an map series relating to the economy (Økonomisk kartverk, ØK) but later also as national map database data (FKB data). Over the years there has been an increasing need for thematic data to be included in the public map data too. Municipalities therefore cooperate with the State to organize access to a selection of basic data and thematic data. The lists of data sets accessible to the public are updated regularly.

3 – GEOGRAPHIC DATA 89

3.14

3.14 From Jølster in Sogn. The Norwegian landscape was formed by the forces of the ice during the last ice age. Photo: Gustav P. Jensen /The Norwegian Armed Forces Media Centre.

Thematic data

GEOLOGICAL DATA

Geology is the study of the earth's structure and history, and describes in simple terms the forces and processes that affect the earth. An understanding of the geological relationships and conditions is important in all geographic contexts that have an impact of the ground or elsewhere. Norway has extensive industry related to the resources below ground level, such as minerals, oil, and gas, and therefore knowledge of the geology in an area is of great social and economic importance.

The Geological Survey of Norway (NGU) delivers a range of data related to geology.

Landforms and soils

Norway has a spectacular topography and a unique landscape, with mountains, valleys, and fjords that few other places can match. Many geological processes have helped to shape the country. Mountains and mountain plateaus, and valleys and fjords are the result of geological folding, isostatic uplift, and weathering and erosion. The landforms are a result of the earth's internal and external forces. Upbuilding forces are due to movements in the interior of the earth and are manifested as volcanic activity, faults, folds, and tectonic plate movements.

The external erosional forces, represented by weathering, landslides, running water, ice, and oceans, also help to create the landforms. Alpine and glacial mountain forms have been created by younger landforms with sharp edges and steep mountains, and debris fans made up of weathered materials. Where glaciers have shaped the landscape, there are corries (also known as cwms or cirques), U-shaped valleys, and hanging valleys. The mountain plateaus are vast expanses with several higher mountain peaks and rounded peaks that reflect the former landforms in the old land surfaces.

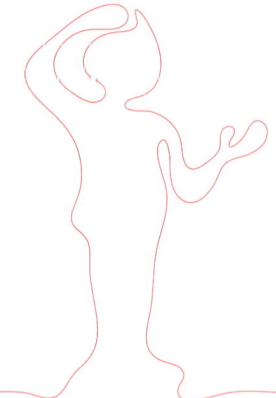

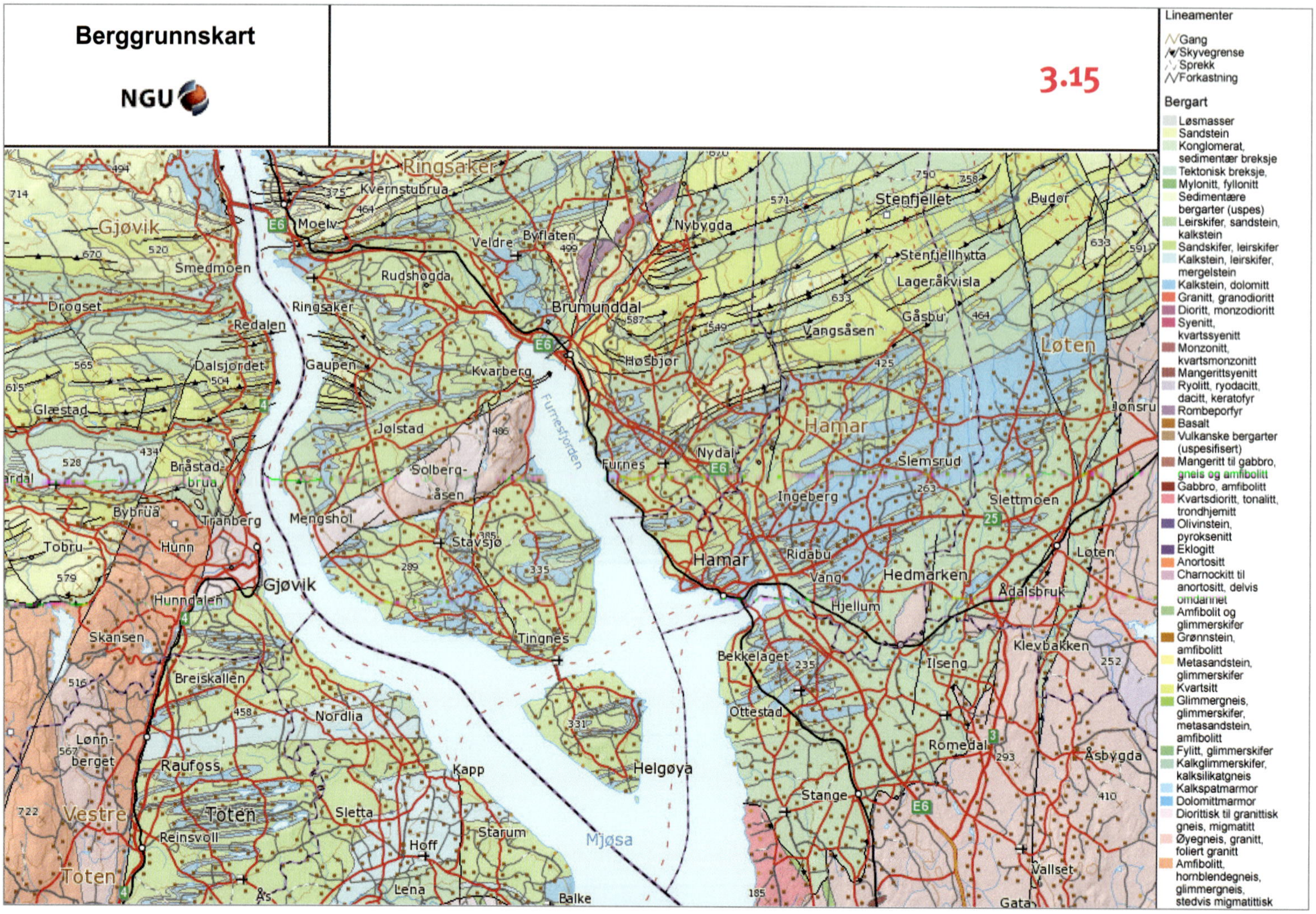

3.15 Bedrock geology data from the Geological Survey of Norway.

3.16 Bedrock in daylight on Spitsbergen. Photo: Ingvill Richardsen.

3.16

Geological deposits are valuable natural resources that constitute the basis for plant and animal life, and are also important for agriculture and settlement. Data relating to such deposits are necessary in order to make best use of the land, and can also be used to make avalanche maps and emergency preparedness maps.

Bedrock data
Bedrock data are classified according to when and how the rocks were formed. Rocks dating from the earliest era in the earth's history – the Precambrian era – form the bedrock in Norway. The bedrock consists of hard, nutrient-poor rocks, and there are three major areas where bedrock is exposed today. In southern Norway the bedrock formations are divided by the Oslo Rift (Oslo Graben).

The Geological Survey of Norway's data on bedrock provide an overview of the prevalence of various rocks in Norway. The data are mainly based on the nationwide, printed, bedrock geological map series at scales of 1:50 000 and 1:250 000. Data relating to major fractures and faults form a separate data set for bedrock structures.

Mineral resources
Mineral resources are raw materials used in everyday life and are essential for the survival of modern societies. This is well illustrated by that fact that, on average, each person in Norway consumed c.13 tons of mineral raw materials in Norway in 2014 (NGU og Direktoratet for mineralforvaltning, 2015). Mineral resources are generally divided into building materials, natural stone, industrial minerals, metallic ores, and energy minerals. The Geological Survey of Norway (NGU) provides a separate data set of mineral resources that gives an overview of recorded deposits and extraction sites. By combining the registered data with map data relating to bedrock, it is possible to indicate potential areas for new extractions of resources. NGU is responsible for administering the extensive map database for deposits and extraction sites for gravel, crushed stone, and sand.

Groundwater
Groundwater is found below ground level, where it fills pores and cracks in geological deposits soil and in the mountains. Groundwater is a very important resource and

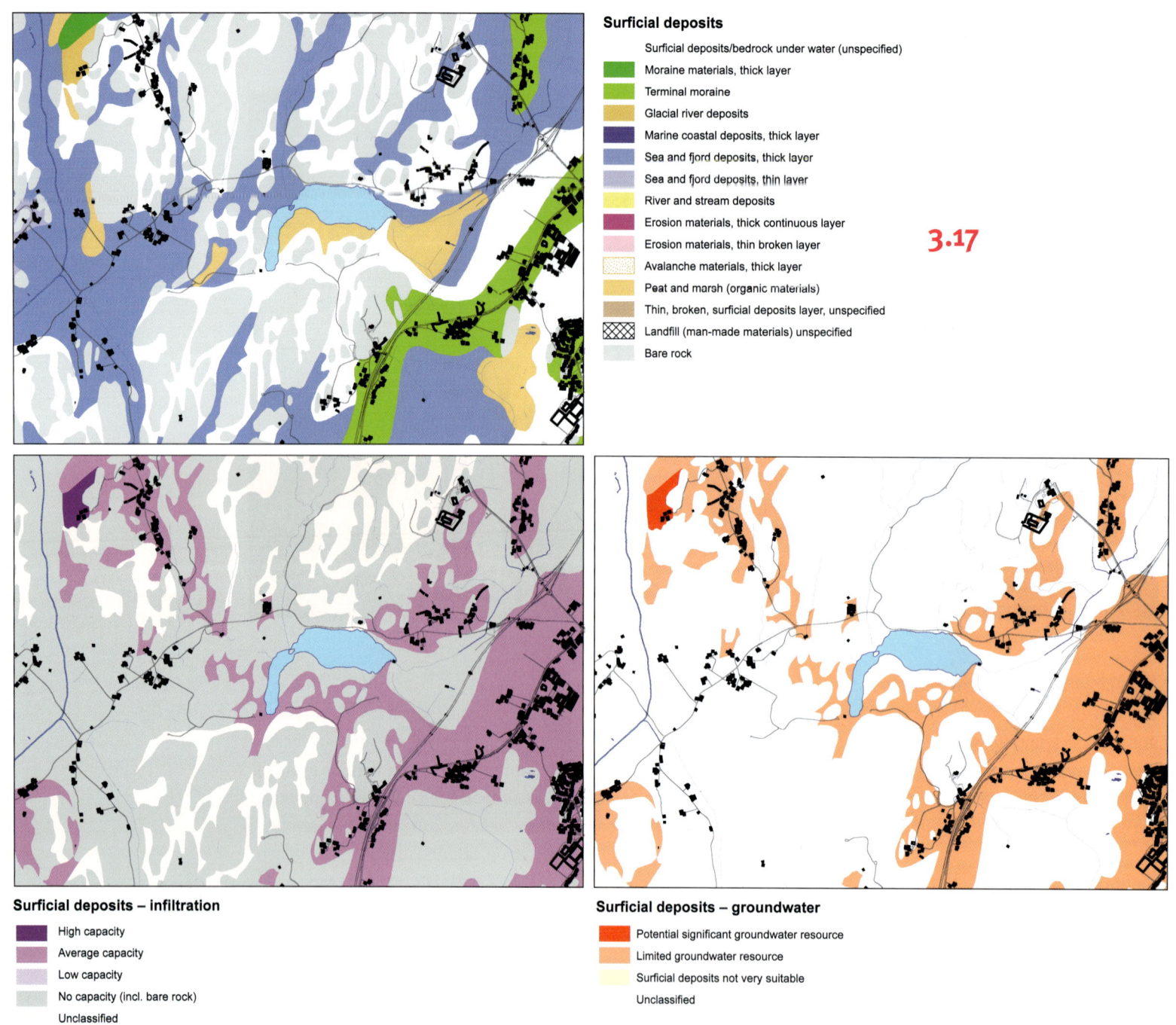

3.17 Maps of surficial deposits, such as morainic material or other loose masses, provide base data for derived maps of infiltration properties and groundwater potential. Source: The Geological Survey of Norway.

is primarily used for water supplies worldwide. Norway is fortunate to have large supplies of surface water, and only c.15% of the drinking water is groundwater. By contrast, the corresponding percentage for other Nordic countries, such as Denmark and Iceland, is over 95%. In addition to water supplies, groundwater is used for energy production and irrigating agricultural land (NGU, 2015).

The Geological Survey of Norway is responsible for the National Groundwater Database, *GRANADA*. The database provides geographic information on groundwater and geothermal wells, and is especially designed for groundwater management in accordance with the EU Water Framework Directive. The purpose of this directive is to monitor the chemical and quantitative properties of groundwater, in order to discourage the release of pollutants and to protect groundwater deposits (NGU, 2015). The use of groundwater is regulated by the Water Resources Act of 2000 (Vannressursloven).

LANDSCAPE

Norway's landscape is very varied and ranges from flat, open farming country or dark forests to bare mountains and mountain plateaus. There is also great variation along the coast, from the outermost windswept skerries to the deepest fjords surrounded by high mountains. Various forces have had an impact on the landscape and different natural factors influence its formation; the landscape is governed by the large, geologically produced landforms. In addition, human activities have had an impact on the various landscape formations. The practice of grazing farm animals, burning heather, and felling has resulted in open landscapes in the mountains and along the coast. The reference system for landscape (Nasjonalt referansesystem for landskap) in Norway divides the country into 45 landscape regions based on significant similarities in the landscape. The Norwegian Institute of Bioeconomy Research (NIBIO) provides its own data sets showing landscape regions at either national level or county level.

Valuable cultural landscapes

More and more cultural landscapes in Norway are likely to be reclaimed or change in character in the future. Farms and villages are gradually being abandoned, and what were once important local, regional, or national areas are now at risk of being forgotten or lost. The Norwegian Environment Agency's data set Verdifulle kulturlandskap (Valuable cultural landscapes) contains data on high-priority cultural for use in administration related to biological and cultural heritage. The data are recorded on the basis of a method for mapping described by the former Directorate for Nature Management (1994). The data set is used in efforts to maintain a living landscape and to take care of cultural monuments and other cultural remains, such as traditional methods of cultivation and biodiversity that are dependent on cultural influences.

BIODIVERSITY AND NATURE CONSERVATION

Environmental factors affecting flora and fauna

In order to understand variations in flora and fauna better, it is important to understand the geographic variation in a number of environmental factors, including global distribution, regional variations in Norway as a whole, and local variations within smaller areas. The framework for the formation of a natural environment is provided by an area's climate, topography, geology, and soils. These physical factors either wholly or partly govern plant life, animal life, and how humans exploit nature. In other words, variations in plant and animals (i.e. biodiversity) are not random; rather, they are a result of diverse natural conditions.

Part of the explanation for this great diversity can be found in the variations in bedrock, geological deposits, and local landforms. The terrain's orientation (degree of exposure) and gradient contribute to local variations in climatic conditions. Geology or bedrock only directly affect plant life in areas where there are no geological deposits or where the deposits are thin. Otherwise, the bedrock is important with regards to quantity and quality of the deposits. In addition, the ecological conditions are influenced by the content of nutrient-rich rock types and the proportions of fine and coarse particles.

3.18 Ulvik in Hardanger, Norway. Photo: Fjellanger Widerøe AS (VF 9858)

Thematic data

3.19 During the last 15-20 years there has been an intensefocus in Norway on the value of cultural monuments, biodiversity and traditional cultural landscapes. This is supported by various types of agricultural subsidies. Nonetheless, it hasproved difficult to preservesome older manmade landscape elements because, among other things, land use or theowner's financial situation and interests change over time. The photo shows a section of 'Paradisgrenda', a former crofter's hamlet below Rollag Church, photographed in 1992 (above) and 2004 (below). In 1994, the area was granted 'nationally valuable cultural landscape' status. Today, the cabin has gone and the meadow above the road has become partly overgrown. Photo: Oscar Pushmann, NIBIO.

The different methods used by humans to exploit natural resources throughout the ages have resulted in various types of cultural landscapes that have also contributed to the increasing diversity. Farming, felling, the planting of trees that are not indigenous to the local area, grazing and browsing in outlying areas, burning heather, ditch digging, and other forms of exploiting rural resources have had a great influence on local landscapes and habitats. The process of change can occur quickly and this should be taken into account when using older mapping materials. Understanding geographic distributions and changes in nature due to cultural factors – the diversity of habitats and species, the complexity of explanatory variables, and the dynamic forces of nature – is a demanding task. One has to be aware of this complexity when working on such issues. However, this also opens up for exciting interdisciplinary geographic analyses and results that reveal interesting geographic covariations between biodiversity and the physical environment.

In efforts to preserve biodiversity, it is important that those who manage an area are aware of its natural qualities. Such information can be collected from the

3.20 The Svalbard poppy is a plant species protected pursuant to the Norwegian Nature Conservation Act. Photo: Ingvill Richardsen.

3 – GEOGRAPHIC DATA

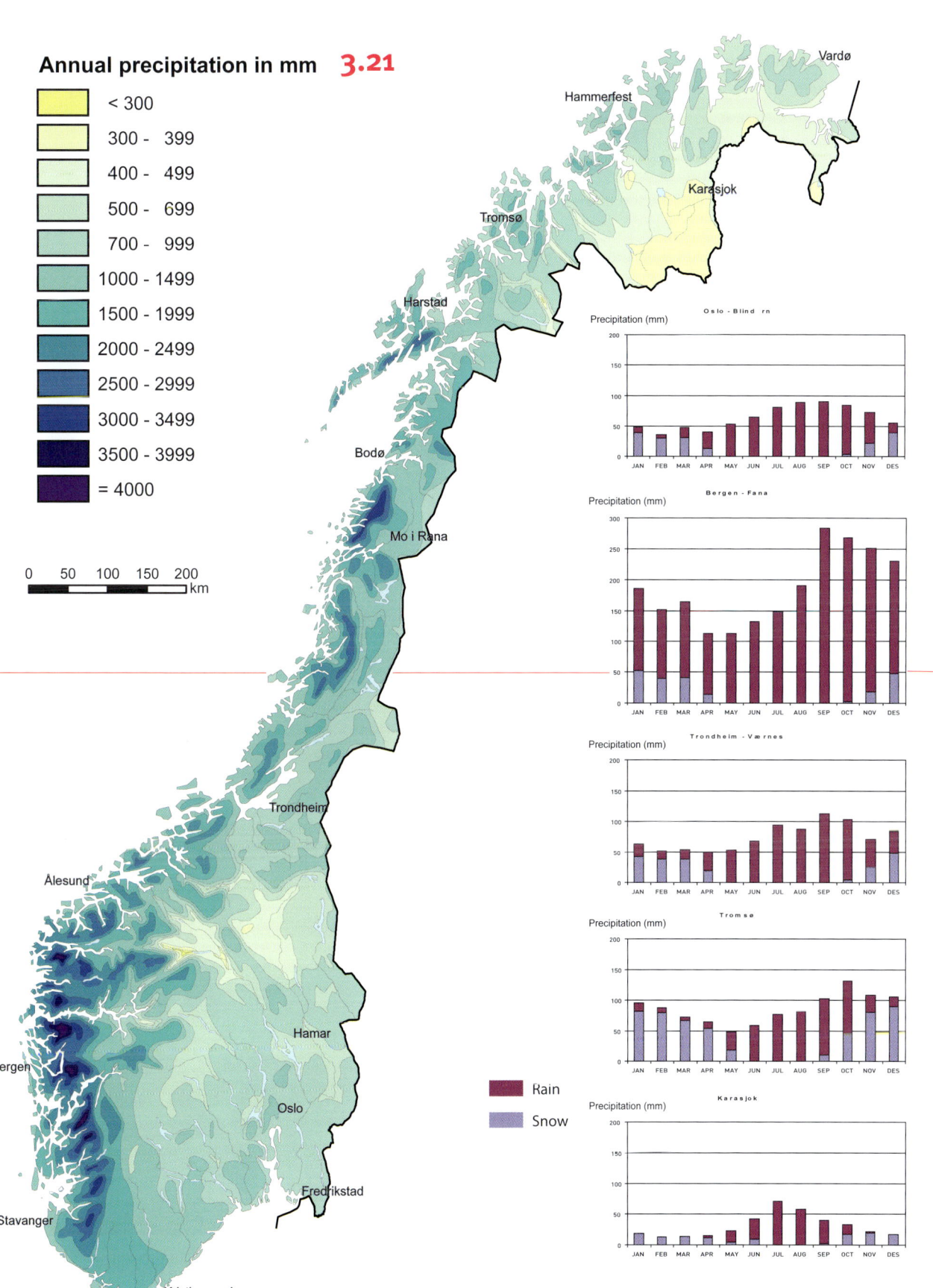

Annual precipitation in mm 3.21

- < 300
- 300 - 399
- 400 - 499
- 500 - 699
- 700 - 999
- 1000 - 1499
- 1500 - 1999
- 2000 - 2499
- 2500 - 2999
- 3000 - 3499
- 3500 - 3999
- = 4000

3.21 Map of annual precipitation. The zone of maximum precipitation in Western Norway is among the areas that experience the most precipitation in Europe. This is due to humid air masses arriving over the land from the west and being pushed upwards by the high coastal mountains and dropping precipitation. The inland regions of Eastern Norway and the Finnmark Plateau are sheltered from the prevailing winds that produce a lot of the precipitation along the coast from Western Norway to Lofoten. Annual precipitation varies in Southern Norway between 278 mm at Skjåk in Nord-Gudbrandsdalen to 3,575 mm at Brekke in outer Sogn. The figures refer to the average values for the period 1961-1990. Source: The Norwegian Meteorological Institute.

following sources and others: vegetation maps and data relating to game or other species, habitats, conservation areas, and wilderness areas.

The following subsections present more detailed descriptions of the various forms of biodiversity mapping, possible methods that can be used in the context of land use, and biological data recorded through various mapping projects and made available through ongoing data administration.

Nature in Norway
Nature Types in Norway (Natur i Norge, NiN) is a comprehensive classification and systematization system for nature in Norway, and was initiated by the Norwegian Biodiversity Information Centre (NBIC). NiN was relaunched in 2015, with a comprehensive areawide division of Norwegian nature. NiN consists of descriptions of natural variations on all scales, from landscapes to small detailed life forms. In addition to being a tool to describe variations in nature, NiN is also a database for mapping habitats. Further mapping of variations and habitats will therefore follow the established structure of NiN (Artsdatabanken, 2015).

Vegetation maps
Vegetation maps communicate information about plant life, and extensive ecological information can be deduced from such maps. In short, vegetation maps describe the natural vegetation in a particular location.

A number of municipalities have established vegetation maps as a general basis for planning, but such maps are also being established for places where more significant encroachments on the natural environment require environmental impact assessments, such as in the case of hydropower developments. These types of surveys are conducted by university and university college departments, consultancy firms and, not least, by NIBIO. Vegetation mapping is done in accordance with the habitat types classified in the NiN database.

Most vegetation maps are produced to meet the needs of as many user groups as possible. The information contained in such maps is quite extensive and will often contain far more information than a single user needs. A vegetation map is a database in visual form, in which all

3.22 Satellite images provide an overview of habitats and nature types. Resources satellites often contain various sensors and record differences in reflections from the ground. Image analysis allows you to produce details that may not be particularly visible to the human eye, including different types of vegetations, humidity conditions and tree damage. This image was taken over Rena in Østerdalen.

3.22

the information about vegetation and the environment is stored. Derived maps can be produced when one wants to simplify the content and focus on specific subtopics and users, such as grazing potential, forest productivity, soils, or the vegetation's durability.

Habitats

The Environment Agency's data set 'Viktige naturtyper' (Important habitats) constitutes an important basis for spatial planning and management with regard to biodiversity. The data set contains a set of priority habitats that are particularly important for biodiversity. The information can be used as a basis for assessing the value of biodiversity, along with data relating to localities for endangered species, information about game, and freshwater mapping.

Prior to 2015, habitats were surveyed in accordance with the following manuals: *DN-håndbok 13-2007, Kartlegging av naturtyper – Verdisetting av biologisk mangfold* (Mapping habitats – assessing the value of biodiversity) (Directorate for naturforvaltning, 2007) and *DN handbook 19-2007, Kartlegging av marint biologisk mangfold* (Mapping marine biodiversity) (Direktoratet for naturforvaltning, 2007). Together with the introduction of the NiN classification system in 2015, these manuals now constitute the standard for further mapping in Norway.

Assessing the value of biodiversity

To ensure a wide ecological perspective in environmental planning, it is necessary to consider individual locations in a wider context. Following a survey of various natural elements, one could end up with a large amount of material on natural conditions. It would therefore be important to examine whether an area with one set of values could be linked to other areas, or whether an isolated area is insufficient to preserve biodiversity.

Map data that relate to different natural conditions could be used in extensive assessments; for example, the value of each area could be assessed. Assessments of the value of areas are also used to simplify and clarify the message when background data are very extensive and complex.

Thematic data

3.23 Vegetation map produced by NIBIO.

Wildlife and species mapping

Wildlife and species data can be mapped by individuals, by different groups and associations, and by the public.

The registered data can be localizations points for individual observations of a species or area boundaries for areas used by or occupied by each species or species group.

Species data are coordinated by the Norwegian Biodiversity Information Centre, and made available via the Norwegian Environment Agency's web portal Naturbase. The data sets contain documented localities (habitats) for the different species and groups of species. Access solutions also show an endangered category, a scale indicating whether a species is endangered, vulnerable, rare, or requires monitoring. Functional areas are recorded for amphibians, reptiles, birds, and mammals, such as grazing areas, migration routes, hibernation areas, roadside feeding areas, leks, or breeding areas. Following an overall assessment, the results are presented in Naturbase data sets relating to species of national interest with regard to management.

The Norwegian Environment Agency has set limits and stated which species and functional areas should have limited public access, as there has to be a balance between transparency and secrecy in administrative and information strategies. On the one hand, the proliferation of sensitive species data, especially data produced in the form of maps, can lead to illegal or unwanted attention at sites with rare species or species that are vulnerable to interference. On the other hand, information on vulnerable sites can enable managers and local people to pay attention to the species' habitats, such as forests, and enable them to be aware of any unsolicited activity (Artsdatabanken, 2015).

3.24 A vegetation map is a resources map that besides providing information about the occurrence and distribution of specific vegetation types can also provide information about other components of nature.

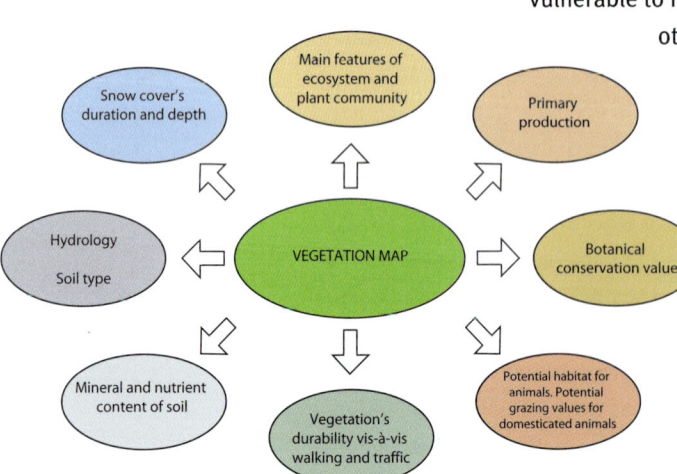

3.24

3.23

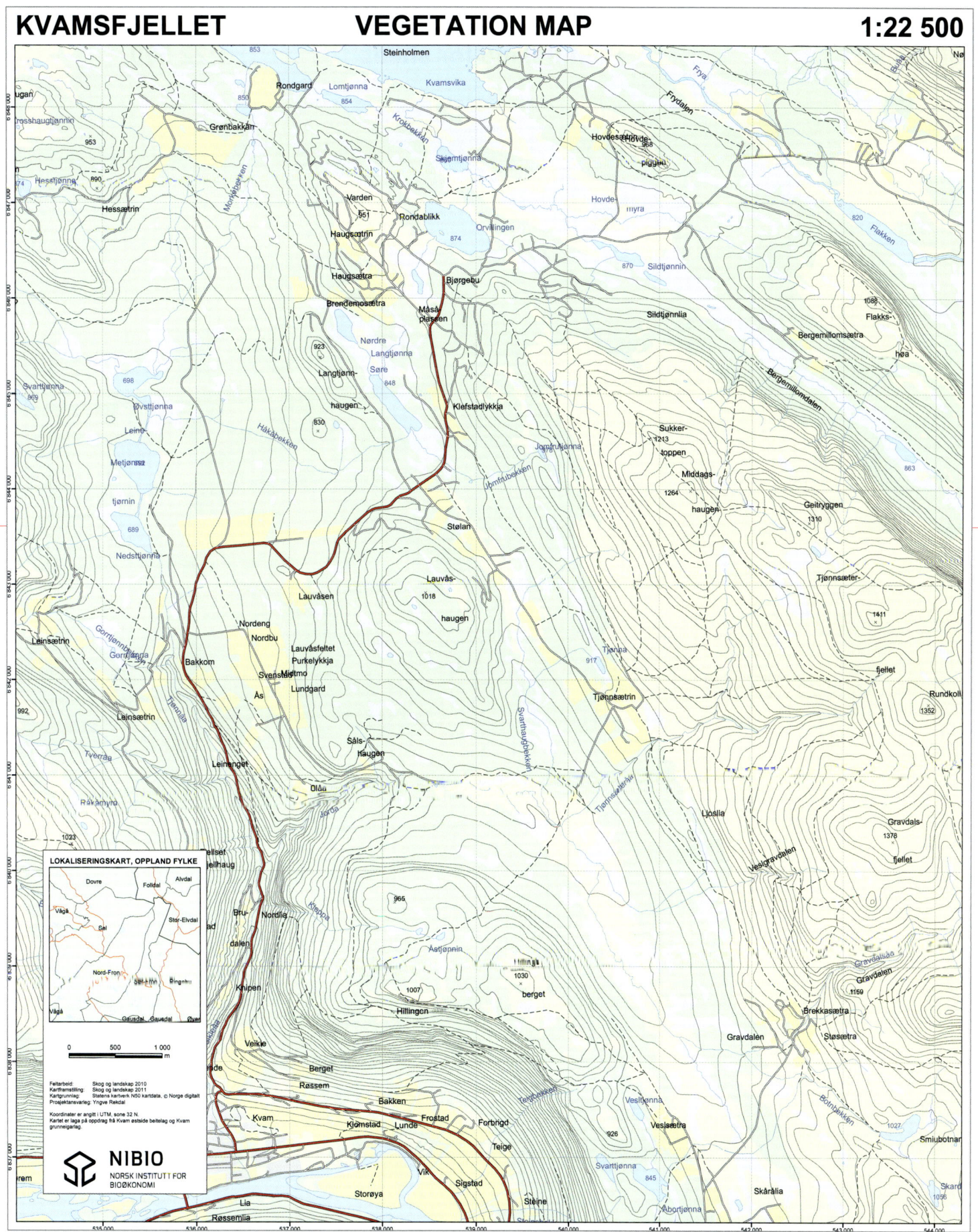

100 Thematic data

3.25

KVAMSFJELLET Pastureland for sheep 1:22 500

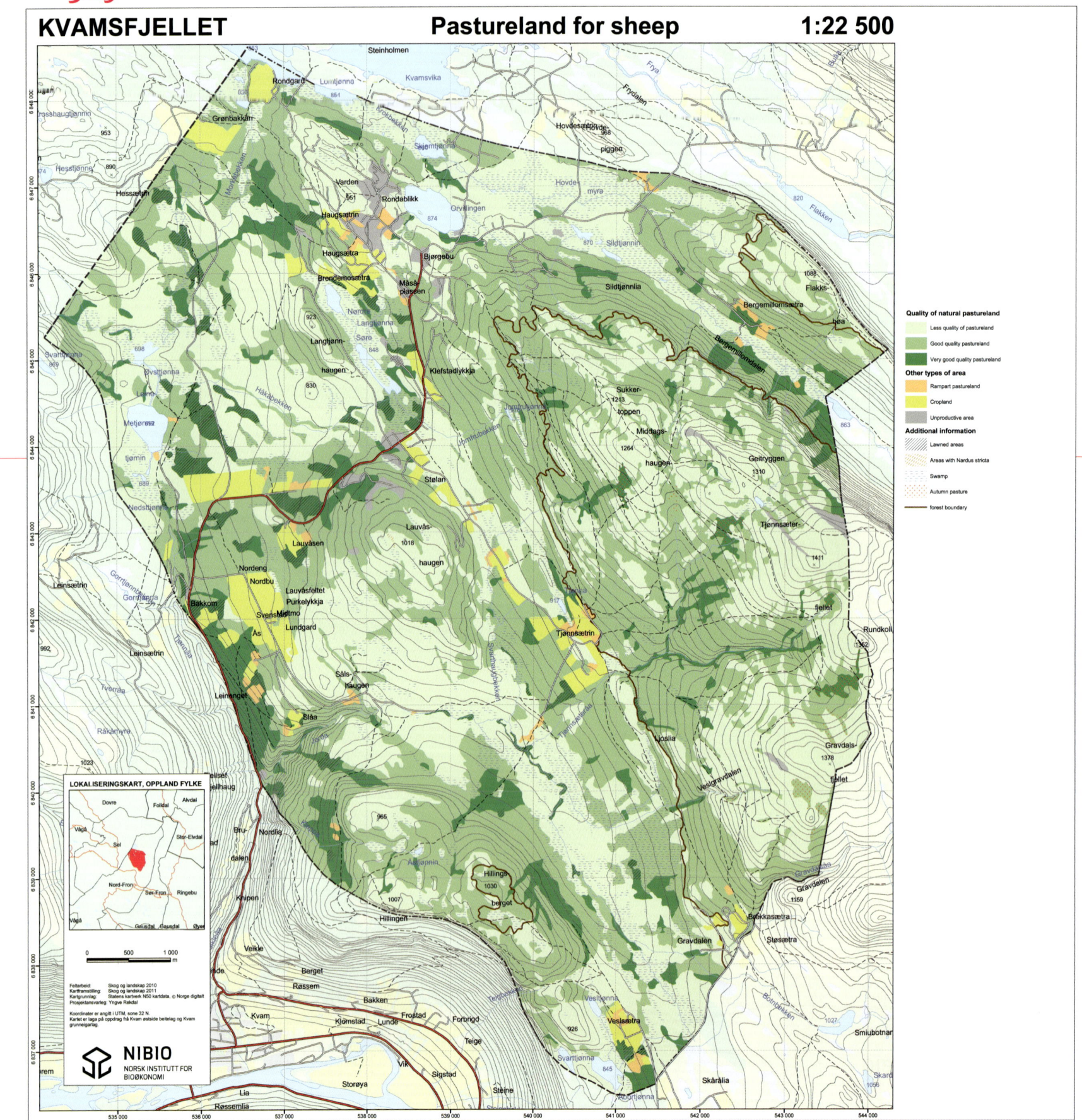

3 – GEOGRAPHIC DATA 101

3.25 Thematic map showing land cover grazing classes for grazing sheep. The map is derived from the vegetation map of the area. Source: NIBIO.

3.26 Hayfields: Voss, Seljord, Nore and Uvdal, Flesberg. The attention being paid to plant life dependent on farming has increased in the last 15 years. The beautiful, and increasingly rare, species-rich flower meadows have been particularly emphasised. As cultural environments they represent thousands of years of land use tradition, while some of the plant species also have long utilitarian traditions. However, preserving species-rich hayfields is a major challenge since they are particularly vulnerable to changes of use. The only thing that can preserve them is traditional haymaking. The hayfields in the photos were listed as regionally/nationally valuable between 1992 and 1994. Today, three of them have deteriorated markedly with respect to the range of species they contain and are of little biological value. Photo: Oscar Puschmann © NIBIO.

3.26

Thematic data

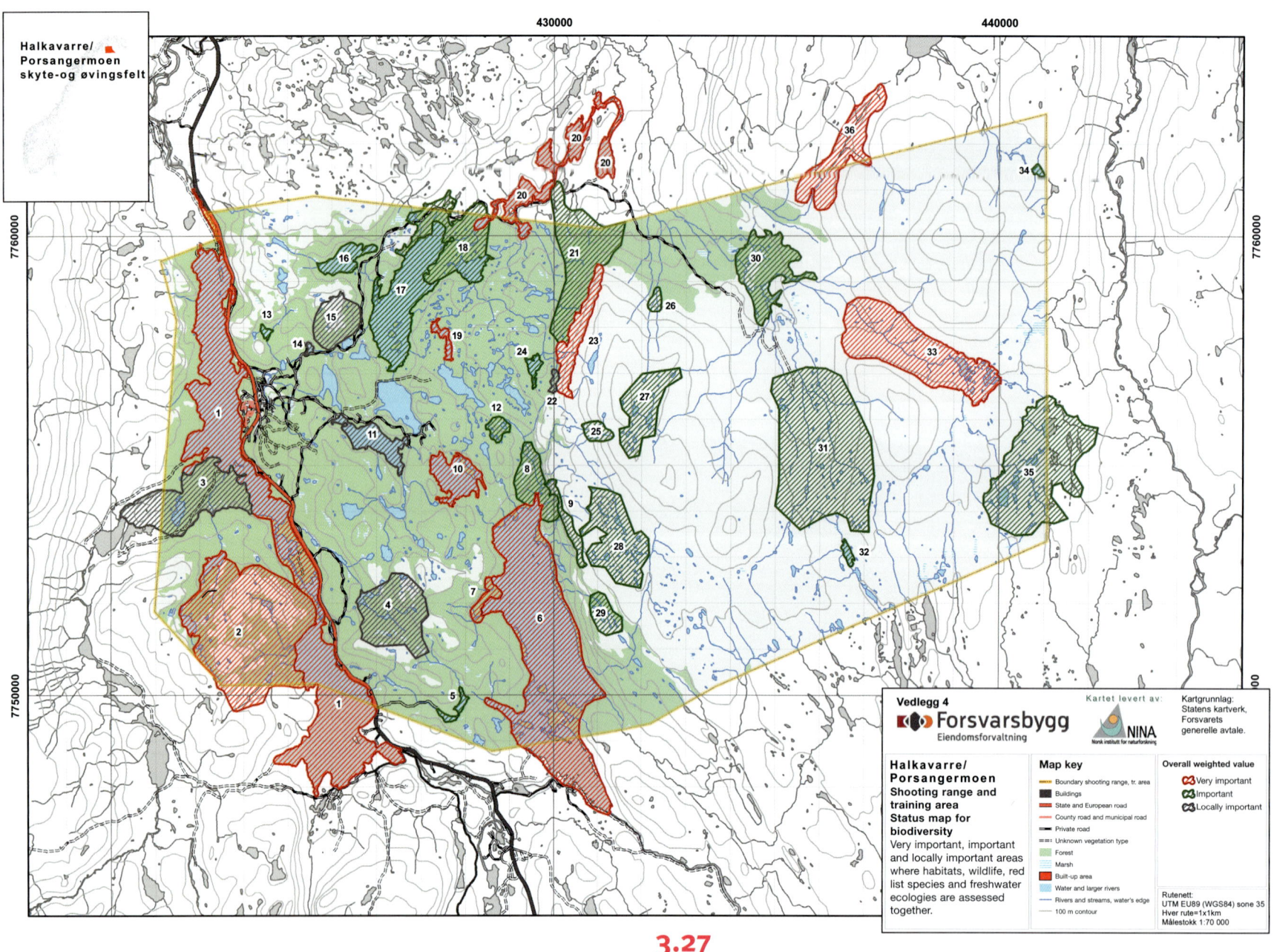

3.27

3.27 Biodiversity status map for Halkavarre in Finnmark.

Nature conservation areas
The term conservation area is usually applied to areas that are subject to formal protection. These are areas with very special environmental qualities, either on a national or international level, and one should be especially aware of them when planning or executing administrative procedures. Often, the boundaries of conservation areas will not correspond to natural boundaries between different habitats; rather, they will follow property boundaries or other forms of administrative boundaries. Thus, conservation areas can be differentiated from other nature-based divisions.

There are hundreds of conservation areas in Norway, both large and small, and c.14% of the terrestrial mainland area is protected. A number of laws form the basis of conservation legislation. Of these, the Nature Diversity Act of 2009 (Naturmangfoldloven) is the most important and provides various means of protection. In addition, the Svalbard Act of 1925 (Lov om Svalbard), the Wildlife Act, the 1992 act relating to salmonids and freshwater fish (Lov om laksefisk og innlandsfisk m.v.), the 1978 Cultural Heritage Act (Kulturminneloven), the 2008 Planning and Building Act, and the watercourse law are used in connection with the protection of nature and landscape.

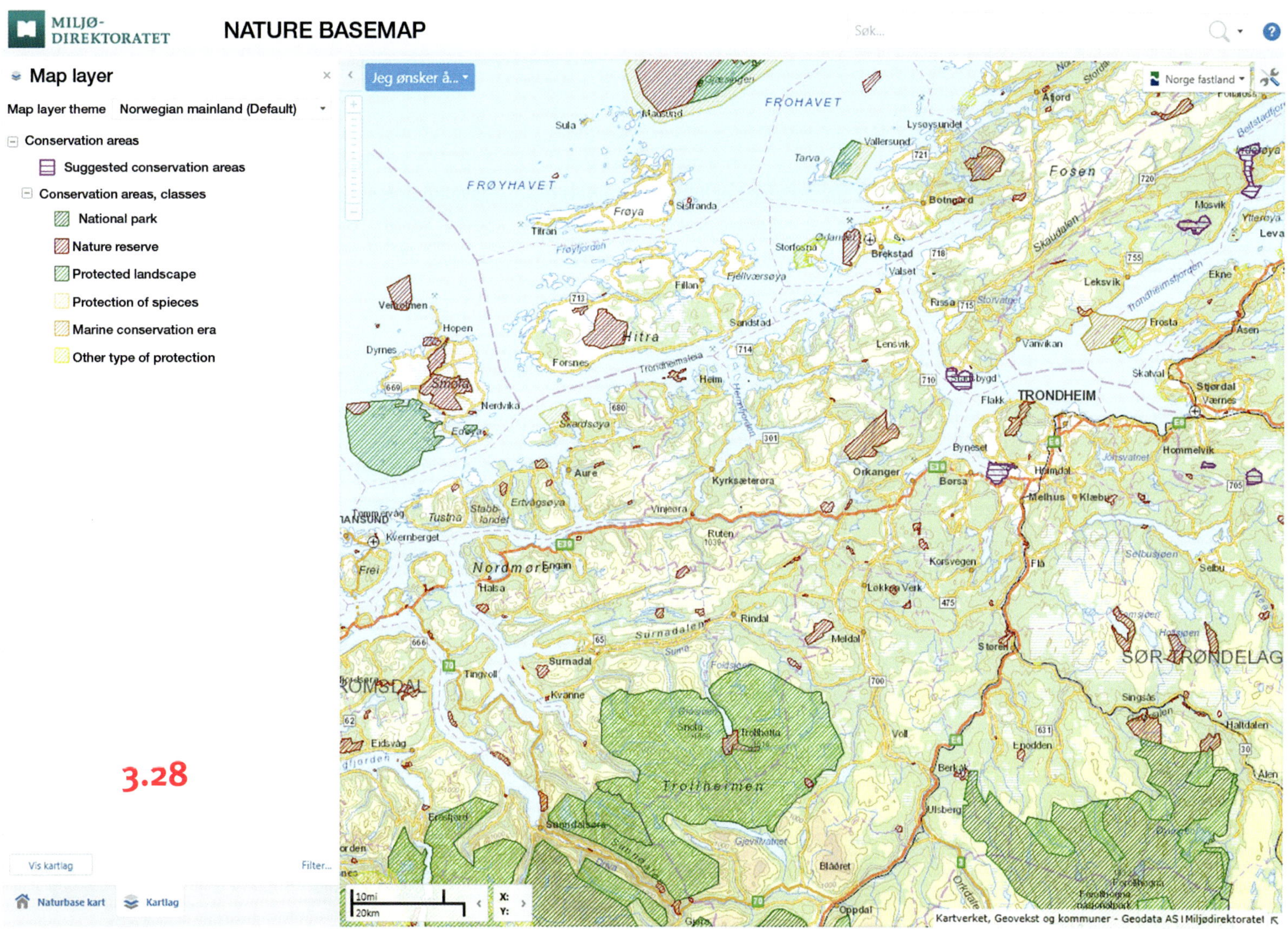

The Norwegian Environment Agency's geographic database of protected areas contains data for natural areas with legal protection status, as well as areas with administrative protection. All areas protected under the Nature Diversity Act are included too. Data for the following different types of protected areas are included too:

- *national parks*
- *nature reserves*
- *habitat areas*
- *protected landscapes*
- *protected marine areas*.

Certain areas may be of major local value but not of such great value that they are protected under the Nature Diversity Act. In order for a municipality to protect an area in some other way, a commonly used alternative is the Planning and Building Act of 2008, which demarcates areas as special nature conservation areas (*spesialområde naturvern*). Similarly, municipalities can protect local cultural monuments by designating them as special cultural conservation areas (*spesialområde kulturvern*). Individual municipalities should have data or maps covering such areas.

Whole watercourses or stretches of watercourses can be protected against further hydropower developments by a resolution adopted in the Norwegian parliament (Stortinget), based on the *Verneplan for vassdrag* (Conservation Plan for Waterways), Plans I–IV (produced in the period 1972–1991). The Norwegian Water Resources and Energy Directorate's data set shows protected catchment areas and provides a good overview of rivers and watercourses for which special rules apply with respect to their use (Norges vassdrags- og energidirektorat, 2015).

3.28 Naturbase is the Norwegian Environment Agency's data system for environmental data. Conservation areas constitute one of many environmental themes administered by the agency. Source: Miljødirektoratet, Naturbase.

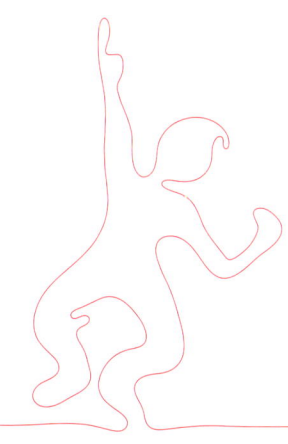

Major infrastructure development – untouched nature

The Norwegian Environment Agency administers the geographic data set Inngrepsfrie naturområder i Norge (INON) (undeveloped natural areas in Norway). The data set shows the extent to which an area is affected by technical encroachments. The map data cover a number of years and can be used as a basis for assessing the development in an area over time. The maps merely inform about technical encroachments 'eating' into nature from all sides. INON is an important indicator of the scale of large, contiguous natural areas, and the data are important for enabling follow-ups of national targets on the management of spaces. Ideally, INON data should provide information about social development in relation to the consumption of natural resources (Miljødirektoratet, 2015). The Norwegian Environment Agency delivers its own mapping service for INON.

Undeveloped natural areas are divided into zones based on their distance from the nearest major infrastructure development: ⟵ 1 km, 1.3 km, 5.3 km, and ⟶ 5 km. Forest located more than 5 km from such developments is characterized as 'wilderness' forest. In this context, major technical encroachments are defined as:

- public roads and railway lines, except tunnels
- forest roads
- tractor roads, agricultural roads, installation site and summer farm roads, and private roads more than 50 m
- old traffic routes for use by tractors and/or all-terrain vehicles
- approved trails over open terrain (Finnmark)
- power lines built to carry voltages of 33 kV or more
- larger ski lifts, ski jumps, and ski slopes
- larger stone tips, quarries, and open quarries
- water reservoirs, regulated rivers and streams, power plants, pipelines, canals, embankments, and dikes.

3.29 Wilderness located 5 km or in a direct line from heavy technical interventions. Source: Maps for 1900 and 1940: Bruun, Magne, NOU 1986: 13. Maps for 1988 and 2013: Miljødirektoratet/Miljøstatus.no

→ **Wilderness areas in Norway**

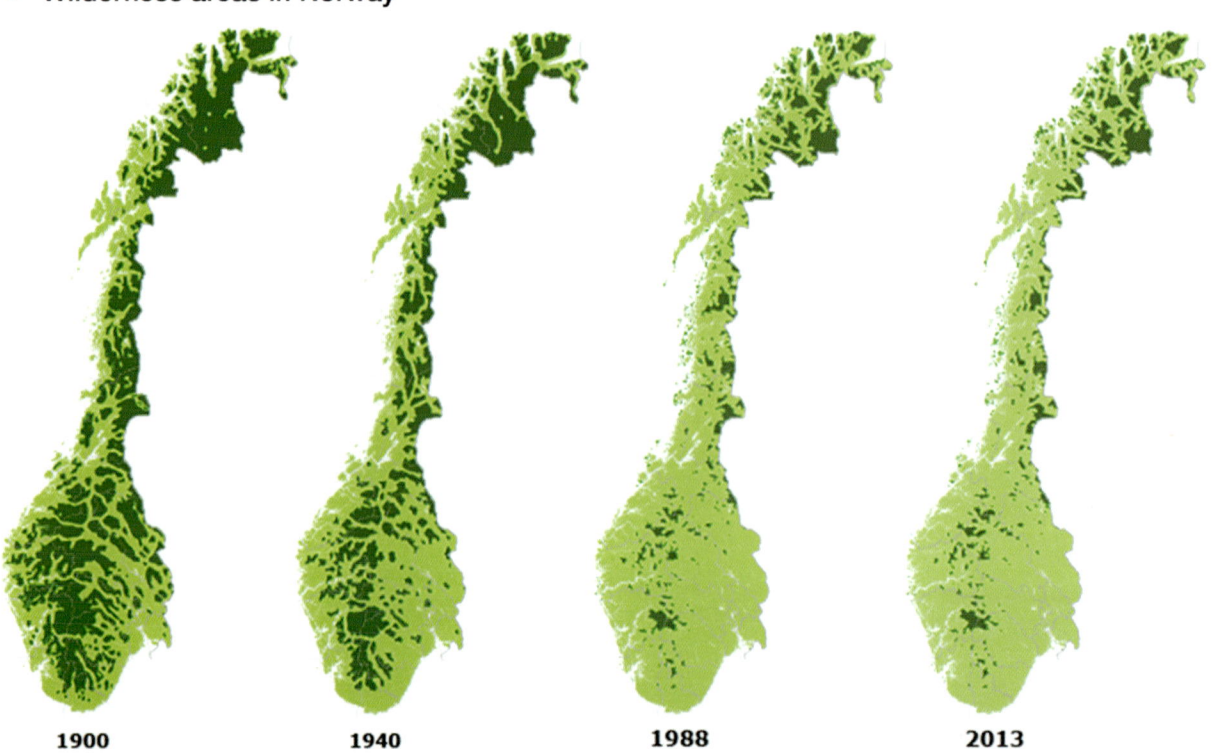

1900　1940　1988　2013

■ Wilderness: nature lying within 5 km or more in a direct line from major infrastructure development

3.29 Source: Norwegian Environment Agency, 2014/miljøstatus.no

3 – GEOGRAPHIC DATA 105

3.30

AGRICULTURE
Changes in the use of forestry and farming land
The land in Norway is made up of c.45% mountains and mountain plateaus, 39% forest, 7% freshwater and glaciers, and 6% marshes and wetlands, and only 3% is used for farming and 0.24% is urban land (Kartverket, 2015). Significant changes are taking place in the use of forest and farming land. Thousands of farms have closed down in recent years and their land currently lies fallow, while new cultivation is taking place in other areas. Many towns and urban areas are located in areas with valuable farmland, where there has been a substantial reduction in the amount of cultivated land. However, there are differences between the municipalities, and through policy measures some municipalities have succeeded in managing their land use to the extent that farmed areas near urban areas have been preserved. This has often had unfortunate consequences for urban areas where the residential areas and areas with other functions have been developed far from the urban centres, with the associated needs for transport.

Earlier, outlying land, grassland, shrubland, and grass-rich bogs and mountain areas were used actively. Today, the pressure on outlying land from grazing and haymaking has significantly reduced. The areas have become overgrown or have changed in character. These major changes in farming are due to new

3.30 Only 3% of Norway's total land area is used for agriculture. Photo: Marek Slusarczyk, ScanStockPhoto.

economic conditions, centralized settlement patterns, changed occupational patterns with farming as a secondary activity, technological innovations, and a changed view of land use. Today, agriculture is not just seen as a producer of food, useful plants, and forestry products, but is increasingly playing a role in producing cultural landscapes and managing biodiversity.

Land use information is important for both planning and administration. Various types of maps show how natural resources and land types are distributed, the suitability of different land with respect to agricultural production, and the development of land over time.

AR5 Land Resource Map

AR5 is a detailed land resource map produced by NIBIO. The data set divides the land area by land type, forest quality, wood types, and soil conditions. AR5 describes Norway's land resources at a scale of 1:5000 for forested areas, and is adapted for use at scales ranging from 1:500 to 1:20 000. The main division in AR5 are land types, whereby all land is classified as one of the following 11 area types: fully cultivated agricultural land, surface-cultivated farmland, pastureland, forests, marshes, open land, water, snow and glaciers, communication lines, undeveloped, and not mapped. Other characteristics, such as forest quality, wood types, and soil conditions, are classified to the extent that they are relevant to the land type. AR5 is considered part of the national map database data (FKB data) and is updated through Geovekst projects. The data set Dyrkbar jord (Arable land) is derived from AR5 and shows areas where cultivation can be practised in compliance with the regulations for cultivated soil and the requirements in terms of climate and soil quality for crop production.

The land resource maps AR50 and AR250 (at scales of 1:50 000 and 1:250 000 respectively) provide simplified information and are suitable for overviews (Vaaje-Kolstad, 2011).

NIBIO has made all of its data accessible via a separate map portal called The sourse.

Soil quality – soil data

Together with climate, soil quality is crucial as far as cultivation potential is concerned, and it affects the land's susceptibility to soil erosion and herbicides' effect on nutrient leaching. Soil quality data can therefore be used in the context of soil conservation, in agricultural planning, and in the work on reducing the negative environmental impacts of agriculture.

NIBIO delivers a separate data set for soil quality. Soil quality is categorized on the basis of soil properties that are important for agronomy and with respect to farmland gradient. The data set is suitable for use in planning and studies of development projects that affect farmland. The data set is at a scale 1:5000. There are data for c.50% of the agricultural area in Norway (geonorge.no, 2014).

3.31 Agricultural resources map from the Norwegian Forest and Landscape Institute (NIBIO) presents agricultural resources.

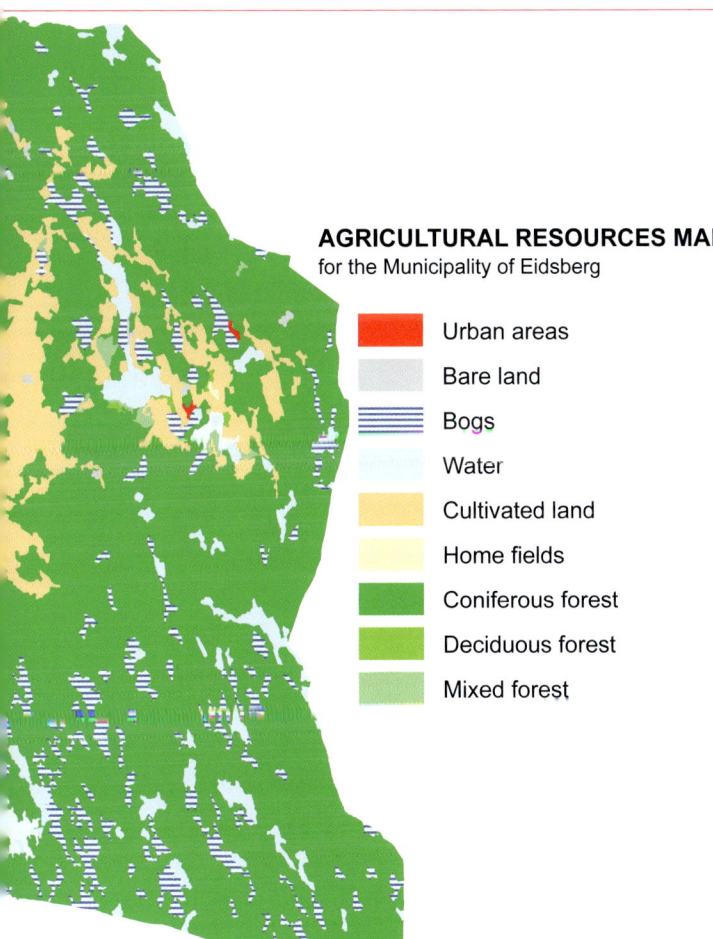

AGRICULTURAL RESOURCES MAP
for the Municipality of Eidsberg

- Urban areas
- Bare land
- Bogs
- Water
- Cultivated land
- Home fields
- Coniferous forest
- Deciduous forest
- Mixed forest

Other map data suitable for an overview of farmland and soil properties include real estate data from the Land Registry, orthophotos, land resource maps, grazing land maps, and data sets showing summer farms that are in operation.

Individual farmers can generate their own maps by using map data from NIBIO's Gårdskart (Farm map), which is accessible via the Internet. The service connects several geographic databases and information from agricultural registers, so that farmer can gain a good overview of their property, the soil quality, land size, and the various strips of land included in the property.

Rough pasturelands

Rough pastureland is an important part of the resource base for Norwegian agriculture. NIBIO statistics relating to organized grazing show that in 2014 almost 1.5 million sheep, 69,000 cattle, and 15,000 goats were grazed on rough pastures (NIBIO, 2014). This makes rough pastureland the second most important industry in outlying areas after forestry. In mountain areas in Southern Norway, the landscape is greatly characterized by summer farm and extensive grazing.

In the past, resources in outlying areas were under greater pressure than today. Before animal feed and

3.32 Most agricultural areas have changed gradually in the last few centuries. Wide-ranging and extensive land use has been replaced by concentrated efforts in a few, more productive meadows and fields. This means that marginal and demanding farming land has fallen out of production. This can be seen in the photos from Valle in Setesdal taken in 1992 and 2002. In the photo from 1992 we can see that old meadows and fields with cleared slopes, stonewalls and barns have been laid out for grazing. Small spruce trees show that the overgrowing in the area is in its start phase. The photo from 2002 shows how quickly this can take place and how the view can be reduced. Photo: Oscar Puschmann © NIBIO.

Thematic data

3.33 Coniferous forest consists of spruce and pine trees. Birch trees also form large forests, both as mountain birch forests and in areas between coniferous forests and moorland, and in low-lying districts in Western Norway and Northern Norway. Deciduous forests can be found in areas where the climate is particularly favourable. Forestry takes place in areas with productive forest, which covers approx. 22% of the land. The tree line in central Southern Norway lies at about 1,000 metres above sea level. In Western Norway the tree line is no higher than 300 metres above sea level, while in the most northerly regions of the country forests disappear because the tree line is lower than sea level. Source: NIBIO.

3.34 The map shows the proportion of felling classes 4 and 5 that exists in the Municipality of Kongsvinger. Felling class 5 is forest that is mature and ready for felling. Source: NIBIO.

fertilizers became common, farmers were dependent on rough pasturelands to maintain the production levels of their livestock and agricultural products. Changes in agricultural policies have led to a stronger focus on new types of commercial exploitation of resources from outlying area. There is also a growing interest in and demand for economic goods based on these areas.

NIBIO, in collaboration with various county agricultural departments, has developed an information system for grazing in outlying areas – beitebruk i utmark (IBU) – with maps and statistical data. This shows where the sheep graze, the number of sheep in each grazing class, the animal density, and percentage of loss per grazing class.

Forestry

In total, 26% of land in Norway is covered by productive forest. In addition, there are upland birch woodlands, wooded marshes, and low-productive forests on land with thin soils. More than one-third of Norway is covered by all types of forests. The amount of forest has changed little since 1920 to the present day. However, there has been a massive growth in forest resources in the form of timber: In the period 2011–2014, the tree volume increased from 300 million m³ to 929 million m³, and the annual production of timber increased from 10–11 million m³ to c.26 million m³ (SSB, 2015).

Forestry is a major industry in Norway, and various information products, registers, and geographic databases are used in this sector. In addition, NIBIO delivers a number of forest-related data sets. Its land use resource data (AR maps series) cover large areas, including forested areas. The data sets include information about the quality of land used for forest production. NIBIO also delivers its own data set giving an overview of forest resources.

General forestry plans are produced for municipalities or landowner associations. Forestry plans are also produced for individual properties, with valuations of stands in forests. These often contain very detailed

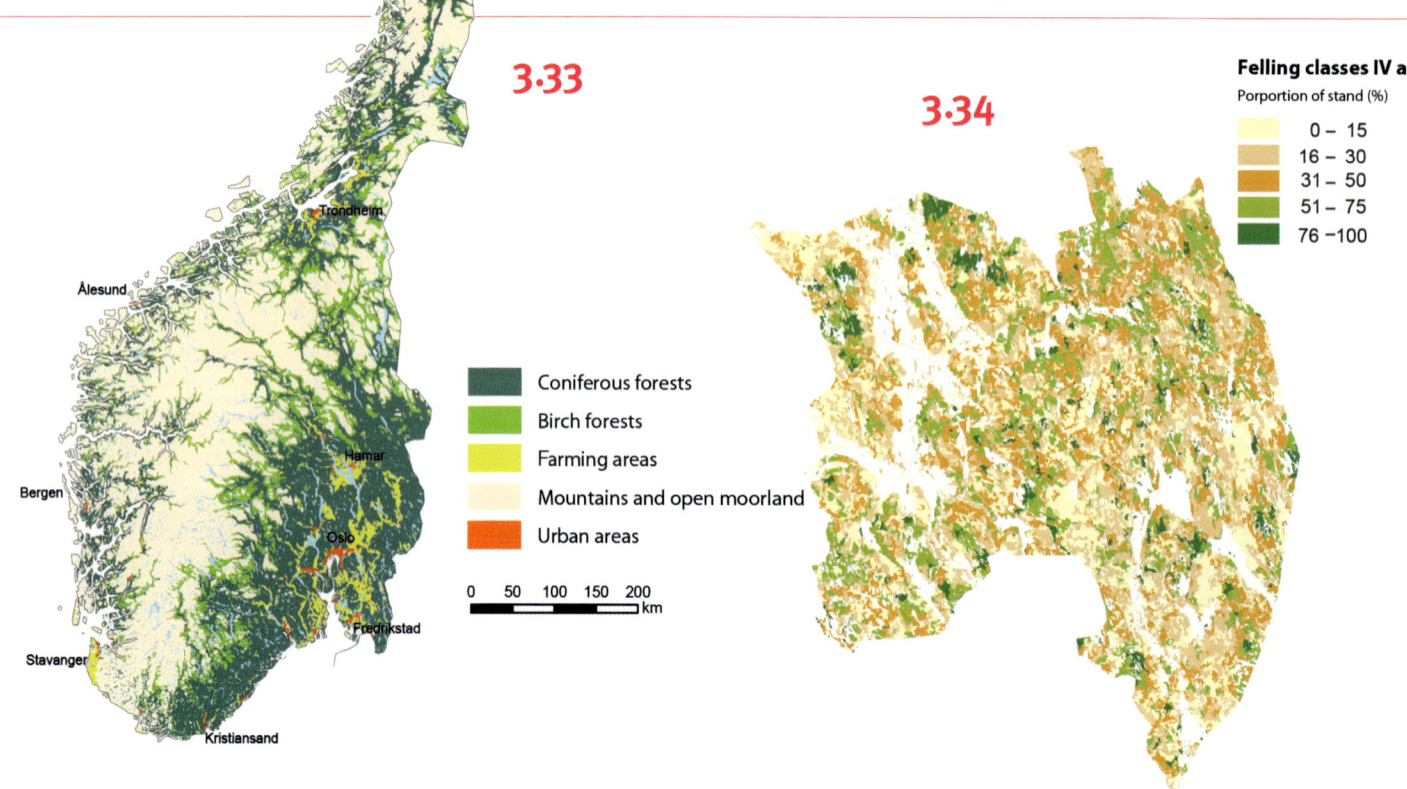

3.33

3.34

Coniferous forests
Birch forests
Farming areas
Mountains and open moorland
Urban areas

0 50 100 150 200 km

Felling classes IV and V
Porportion of stand (%)
- 0 – 15
- 16 – 30
- 31 – 50
- 51 – 75
- 76 –100

information about the felling class of the stands, from which one can deduce information about the value of forestry resources on the property, as well as information about the distribution of vulnerable species. Forestry plans also contain geographic overviews of forest roads (e.g. master plans of forest roads).

A protection forest is forest that functions as a windbreak for nearby farming and other forested areas as well as urban areas. Municipalities establish protection forest data in which restrictions are stipulated concerning which forests should be preserved as protection forests. In these areas people are advised against felling or advised to limit the amount of felling so that the climatic impacts on the surrounding areas are limited. The county's governor's agriculture department in each county is responsible for mapping protection forests (NIBIO, 2015).

The forestry industry has to take account of the role forest plays as a habitat. The industry has launched 'living forest' strategies in connection with the administration of biodiversity. This entails taking into account the role of both the natural environment and forest in the context of recreation. Forestry plans now also contain detailed data relating to biodiversity and habitats with special qualities. This is reflected in a separate standard, PEFC Norway (Norsk PEFC Skogstandard), the Norwegian version of the worldwide forest certification system to ensure sustainable forestry.

Reindeer husbandry
Reindeer husbandry is practised over large areas of land in outlying areas. Sámi reindeer husbandry takes place in the counties of Hedmark, Sør-Trøndelag, Nord-Trøndelag, Nordland, Troms, and Finnmark. The Sámi people have

3.35 Reindeer grazing data show grazing areas, migration routes and herding routes. From Porsanger in Finnmark.

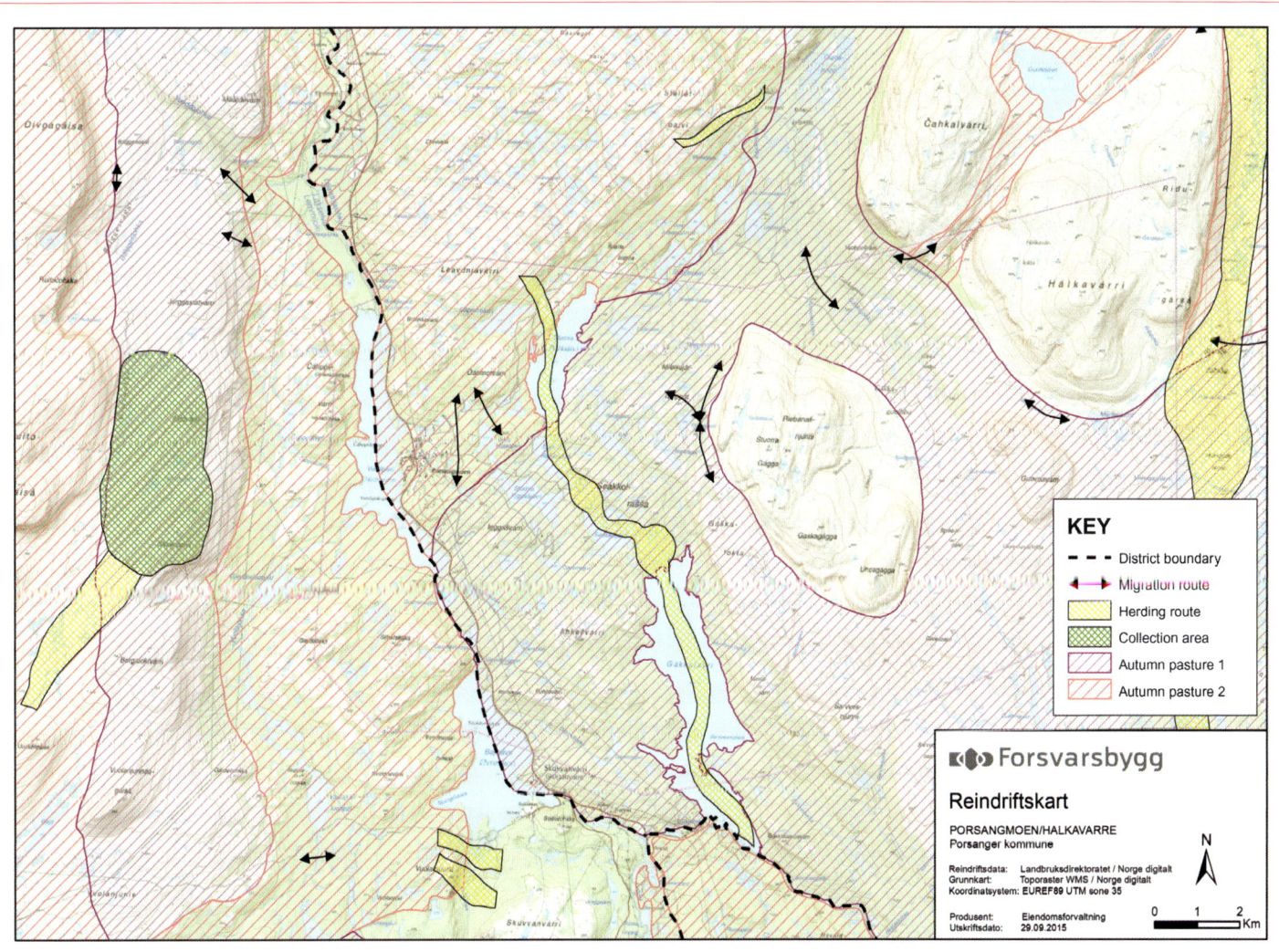

indigenous status in Norway and reindeer husbandry constitutes the core of their culture. Preserving the reindeer husbandry industry is therefore central to Norway's international obligations with respect to indigenous peoples. Non-Sámi, tame reindeer husbandry is practised in parts of Southern Norway, especially in the county of Oppland. Reindeer husbandry requires a lot of land because the grazing areas are extensive. Access to the industry's own data is therefore useful in land use planning. Reindeer husbandry data exist for all of the areas in which Sámi reindeer husbandry takes place. The maps have been drawn up by the reindeer herders themselves through the reindeer husbandry area boards under the guidance of the Norwegian Reindeer Husbandry Administration. Map materials and digital data with a similar content and structure have also been established for areas where non-Sámi reindeer husbandry takes place.

Reindeer husbandry geodata relating to grazing areas and migration routes are used by many actors at local and regional level. Information about reindeer husbandry combined with other information can highlight conflicts and possible solutions in planning and building application contexts, as well as for the administration of outlying areas. Reindeer husbandry and other industries with domesticated animals grazing on outlying land have to coexist in harmony. Reindeer husbandry data are also important when considering the location of fish farming facilities, building holiday cabins, developing recreational areas, and planning military exercises.

Reindeer husbandry utilizes wide expanses of land due to the reindeers' grazing areas and migration routes. This makes it difficult for land use planners and authorities in other sectors to determine which areas are most valuable for reindeer husbandry. The Norwegian Reindeer Husbandry Administration therefore wants to develop a classification system for the value of land, in order to make the prioritized areas more readily apparent.

NIBIO delivers several data sets showing land use for reindeer husbandry. These include a seasonal grazing map, showing which areas reindeer use in the different grazing seasons, as well as dedicated data showing migration routes, reindeer migration routes, and herding areas.

WATERCOURSES

The Norwegian Water Resources and Energy Directorate (NVE) is both an agency and responsible institution for energy and hydrological data, and provides large quantities of important data related to water resources. In addition, it administers the web portal NVE Atlas, where all NVE data are presented in a common solution.

In Norway, the lakes and rivers, including related wetlands and groundwater, are of great importance for plants, animals, and humans. Some of the challenges to the environment in these water bodies are watercourse regulations, acid rain, and run-off containing nutrients from agricultural fields, pollutants, and the impact of invasive species. The Norwegian Environment Agency coordinates the integrated ecosystem-based management of water, with the objective of ensuring that all water has a good ecological status and a good chemical status. There are three basic geographic databases for inland waters in Norway (NVE, 2014), and these are described below.

Elvedatabasen (river database) is a nationwide river network database established and managed by NVE. The database is built up as a logical network, where the direction of flow is shown for all lines, and is thus a watercourse database showing the network structure. Elvedatabasen is established on the basis of N50 map data from the Norwegian Mapping Authority. Mathematical centrelines are made through lakes and river areas in the map data. The generated centrelines are linked to the rivers, which are represented on the map by simple lines, such that the combined data represent a large river network.

REGINE (catchment database) is the national hydrographic division for watercourses in Norway and contains a comprehensive register of watersheds. It divides Norway into over 20,000 units, showing how precipitation ends up in the various waterways. All waterways are registered with a separate watercourse number, which allows the data to be combined with other geographic and statistical data. The database has a hierarchical structure, which makes it possible to see which waterways are interconnected.

3 – GEOGRAPHIC DATA   111

Part of the NVE Atlas

Key
- Main river
- Depth chart
- Lake database
- Aributaries catchment
- GeocacheGreytone

Scale 1:160 000
NVEAtlas 10/2/2015

3.36 The map portal NVE Atlas provides a good overview of the Norwegian Water Resources and Energy Directorate's map data.

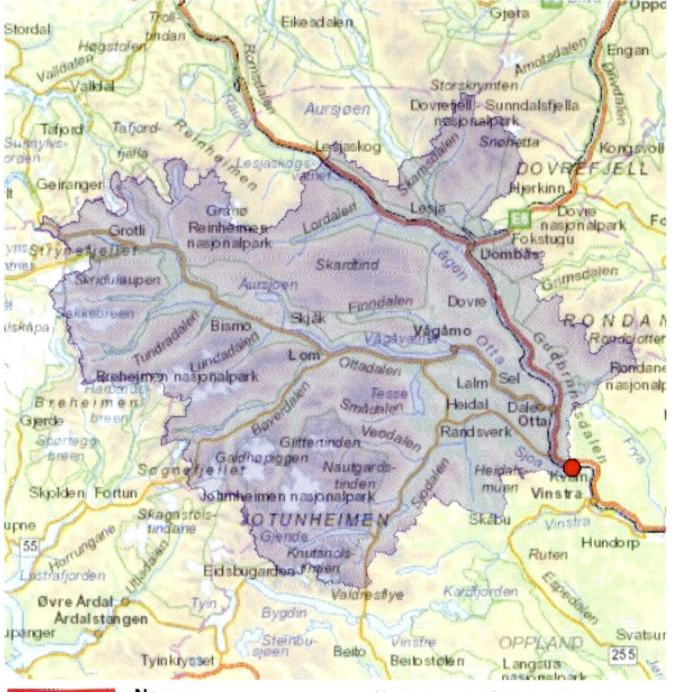

Lavvannskart

Vassdragsnr.:	002.DG4
Kommune:	Nord-Fron
Fylke:	Oppland
Vassdrag:	VORMA-LÅGEN

Vannføringsindeks, se merknader

Middelvannføring (61-90)	22.9 l/s/km²
Alminnelig lavvannføring	4.1 l/s/km²
5-persentil (hele året)	4.1 l/s/km²
5-persentil (1/5-30/9)	19.4 l/s/km²
5-persentil (1/10-30/4)	3.4 l/s/km²
Base flow	8.3 l/s/km²
BFI	0.4

Klima

Klimaregion		Ost
Årsnedbør	850	mm
Sommernedbør	374	mm
Vinternedbør	475	mm
Årstemperatur	-2.1	°C
Sommertemperatur	4.4	°C
Vintertemperatur	-6.8	°C
Temperatur Juli	6.4	°C
Temperatur August	7.5	°C

Feltparametere

Areal (A)	7781.0 km²
Effektiv sjø (S_{eff})	0.1 %
Elvelengde (E_L)	115.5 km
Elvegradient (E_G)	3.2 m/km
Elvegradient$_{1085}$ (G_{1085})	3.1 m/km
Feltlengde (F_L)	132.8 km
H_{min}	258 moh.
H_{10}	733 moh.
H_{20}	917 moh.
H_{30}	1045 moh.
H_{40}	1151 moh.
H_{50}	1253 moh.
H_{60}	1350 moh.
H_{70}	1445 moh.
H_{80}	1551 moh.
H_{90}	1702 moh.
H_{max}	2463 moh.
Bre	3.7 %
Dyrket mark	2.0 %
Myr	1.3 %
Sjø	3.1 %
Skog	20.1 %
Snaufjell	65.9 %
Urban	0.1 %

Norges vassdrags- og energidirektorat

Kartbakgrunn:	Statens Kartverk
Kartdatum:	EUREF89 WGS84
Projeksjon:	UTM 33N

Nedbørfeltgrenser, feltparametere og vannføringsindekser er automatisk generert og kan inneholde feil. Resultatene må kvalitetssikres.

De estimerte lavvannsindeksene i denne regionen er usikre. Spesielt ofte er 5-persentil (sommer) for liten. Indekser som ikke er beregnet skyldes manglende parameter(e). Lavvannskartet gir usannsynlig stort estimat av en eller flere

Det er generelt stor usikkerhet i beregninger av lavvannsindekser. Resultatene bør verifiseres mot egne observasjoner eller sammenlignbare målestasjoner.

I nedbørfelt med høy breprosent eller stor innsjøprosent vil tørrværsavrenning (baseflow) ha store bidrag fra disse lagringsmagasinene.

3.37 A flood map produced using the mapping service NEVINA, which is provided by the Norwegian Water Resources and Energy Directorate. Such information can be used as a basis to support license applications and various types of hydrological calculations.

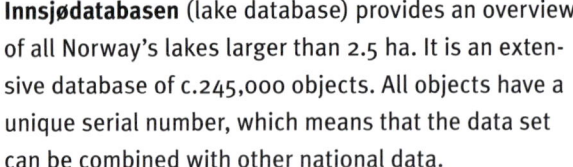

3.38 There are c.1500 large and small hydroelectric plants throughout Norway. Photo of Storfallet hydroelectric power plant on the river Søkkunda in Stor-Elvdal. Photo: Arne T. Hamarsland, NVE.

Innsjødatabasen (lake database) provides an overview of all Norway's lakes larger than 2.5 ha. It is an extensive database of c.245,000 objects. All objects have a unique serial number, which means that the data set can be combined with other national data.

ENERGY
Hydropower
In Norway, hydropower is the largest renewable energy source and by far the largest energy source. Norway is Europe's largest producer of hydroelectric power, and there are c.1500 small and large hydropower plants throughout the country. Although hydropower is considered an environmentally friendly energy source, it has a significant environmental impact and affects plant and animal life. To ensure uniform national management of water systems, a comprehensive plan for water resources was established as early as 1984. The initial goal of the plan was to control the order of the development of hydropower resources, but now it is being converted into a sorting scheme that is increasingly based on the value of rivers or sections of watercourses. The change is reflected in the regulations on the framework for water management (Vannforskriften 2006) and the introduction of the EU Water Framework Directive in Norway. Comprehensive plans of watercourses are delivered as separate geographic data sets through Geonorge (NVE, 2011, a).

Wind power
Give its elongated coastline and very harsh climate, Norway has good prospects for generating wind power. Wind power currently represents 1.5% of Norway's total power production. NVE delivers an overview map showing Norway's mainland and offshore wind resources. It also delivers its own data sets showing wind turbines and wind farm sites.

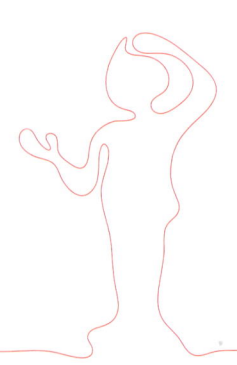

WEATHER AND CLIMATE

Weather maps are familiar to most of us. The Norwegian Meteorological Institute (MET Norway) has an extensive database with weather data from all of its weather stations, up to the present day. It is possible to find precipitation statistics and maps dating back to 1960. The data are made available through the web portal eKlima, where users can also create their own maps and reports from the data. In addition, the Norwegian Meteorological Institute also supplies the website yr.no, where information about weather is presented as real-time observations and weather forecasts in the form of cartographic time series.

The Norwegian Meteorological Institute, Norwegian Water Resources and Energy Directorate (NVE), and the Norwegian Mapping Authority collaborate on administering the map portal seNorge.no, which offers daily updated maps showing the weather, water, snow, and climate conditions. The seNorge.no portal is designed to strengthen the three national goals for society: flood warning and emergency preparedness against watercourse accidents, reliable supplies of electricity, and weather forecasting and dissemination of climate information.

POLLUTION

Pollution is stray waste. Many forms of human activity lead to the release of harmful substances or materials into soils, air, and water, and all sectors in society and all companies have a responsibility to minimize pollution. Environmental protection authorities monitor the environment and how industry and other businesses use purification measures, make adjustments to their processes, and adopt other mitigating measures. In general, the increased focus on the environment and the technological developments has led to considerably less pollution per produced item or activity, but volume growth in terms of consumption and traffic has resulted in significant levels of pollution and impacts on people and the environment.

Drawing attention to the problem of pollution does not just concern emissions and purification. Instead, the causes of the problems need to be remedied and therefore it is often necessary to examine the underlying driving forces and physical or structural conditions in society. The relationships in cause-and-effect chains provide complex explanatory models. Frameworks such as DPSIR (Driving forces, Pressure, State, Impact, Response) (European Environment Agency, 2003) identify various aspects of processes – the driving forces, the concrete impacts, the status of conditions in the environment, the impacts or effects, and the responses in the form of measures. All of these conditions can be geographically distributed and will change over time. Uncovering their distribution patterns may provide important insights on which to base further action. GIS can be used to examine changes over time, disseminate results, and provide a useful basis for the measures that should be implemented.

In order to gain a good overview of pollution and other environmental data in Norway, the website *Miljøstatus in Norge* (Environment.no) is recommended as one of the most important suppliers of environmental data.

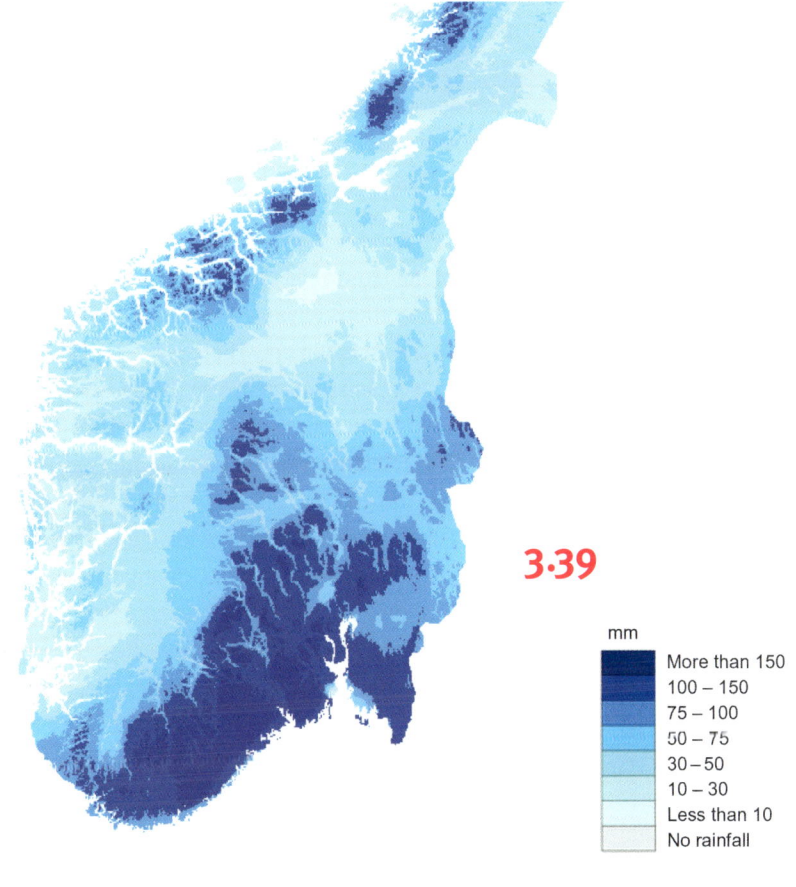

3.39

mm
More than 150
100 – 150
75 – 100
50 – 75
30 – 50
10 – 30
Less than 10
No rainfall

3.39 Precipitation levels in southern Norway during the extreme weather event Petra in September 2015. Source: SeNorge.no

There are a number of categories of pollution, including the following:

- *ground and water pollution*
- *air pollution*
- *noise*.

Ground and water pollution
Contaminated soils can be defined as having elevated concentrations of substances that are harmful to health or to the environment. There can be various sources and causes of soil contamination. Often, contamination is caused by emissions from old industries and other economic activities or by leakages from landfills. In some areas, the ground can be so heavily polluted that there is a risk of the pollutants spreading and entering the food chain. Living or staying in such areas for any length of time should be avoided. In places where groundwater is used as drinking water, it is particularly important to prevent pollutants from being released into the ground. Contaminated materials can affect to earth to the extent that important properties of the ecosystem can be lost. This in turn means that living conditions become more challenging or the ecology of an area can be altered.

When excavation or construction work starts in a contaminated area, there may be an increased risk of pollution spreading. In some cases, contact with contaminated soils can be harmful to health. Hence, anyone wanting to build on contaminated sites or to carry out clean-up operations in such areas must obtain permission from a relevant local authority. The Norwegian Environment Agency has prepared a list of state classes for ground pollution that set limits on acceptable levels of pollutants in soils according to different land uses (Statens forurensningstilsyn, 2009). The agency delivers its own data sets related to contaminated sites. Data on contaminated sites are also held in the Land Registry.

Air pollution
Many pollutants are released into Norwegian nature via air and precipitation. Heavy metals such as lead and mercury are emitted as particles into the air from facilities where metals are extracted or processed and the particles can remain airborne for a long time. Deposited heavy metals are subsequently taken up by the vegetation. By measuring the concentrations of heavy metal particles in moss it is possible to gain an insight into the level of pollution. Contamination by heavy metals has declined dramatically in recent decades, with the exception of mercury, which has the same concentration levels as before.

Many heavy metals pollutants in Norway originate from the continent. As the supply of heavy metals from Europe is reduced, local pollution levels are becoming relatively higher. Surveys of selected industrial sites in Norway have shown elevated concentrations of heavy metals in their surrounding areas.

The concentrations of heavy metals differ in their geographic distributions. In Southern Norway, primarily lead, cadmium, arsenic, zinc, and vanadium are found, whereas in Eastern Finnmark the deposits are mainly nickel and copper (Miljødirektoratet, 2011).

Noise
Noise pollution is among the environmental problems that affects the largest number of people in Norway. Environmental authorities are aiming to prevent or reduce noise problems to ensure that people's health and sense of well-being are maintained. The national target is for noise nuisance to be reduced by 10% by 2020 compared to 1999. This objective relates to general noise annoyance in Norway, but is far from being met. On the contrary, noise levels have increased significantly as more and more people are disturbed by road traffic noise.

3.40 Waste volumes increased by over 50% during the period 1995–2013. Photo: Arne Flaaten, The Norwegian Armed Forces Media Centre.

Road traffic is by far the greatest source of noise pollution in Norway and is responsible for almost 80% of all noise complaints. The number of residents exposed to noise above L_{den} = 55 dBA (A-weighted decibels) from road traffic increased by 22% (c.226,000) in the period 1999–2011. In total, c.1.7 million people in Norway are exposed to noise from various sources. In the European context too, there is a major focus on noise. In the EU, 125 million people are exposed to a noise level of L_{den} = 55 dBA outside the walls of their houses (European Environment Agency, 2014).

Noise has very different characteristics (e.g. frequency, rise time, and duration). Hence, sound from a shooting range is very different from noise from a railway or continuous fan noise from an industry. Generally, noise is greatest near the source.

The main guideline for noise control is *Retningslinje for behandling av støy i arealplanlegging* (T-1442/2012) (Guideline for the treatment of noise in spatial planning; T-1442/2012) (Klima- og miljødepartementet, 2012). This requires owners of noisy sites to prepare their own noise zone map, showing a red zone (building prohibited zone) and a yellow zone (building restriction zone). These zones can be illustrated very well with the use of GIS tools.

Noise maps can be used for many purposes:

- to map and find measures for housing in accordance with Chapter 5 of the regulations on the limitation of pollution (Forskrift om begrensning av forurensning (forurensningsforskriften)) produced in 2004:
- to identify the number of people affected by noise
- for land use planning for new residential areas, kindergartens, and recreation areas
- for the preservation of 'quiet' areas.

Most major owners of noise-emitting sources have systems for calculating noise zones and continually distributing the data to affected municipalities, so that the noise zones can be used in their land use planning. Specialist noise evaluations should underpin the construction of noise-sensitive environments, such as homes, hospitals, institutions, second homes, schools, and kindergartens.

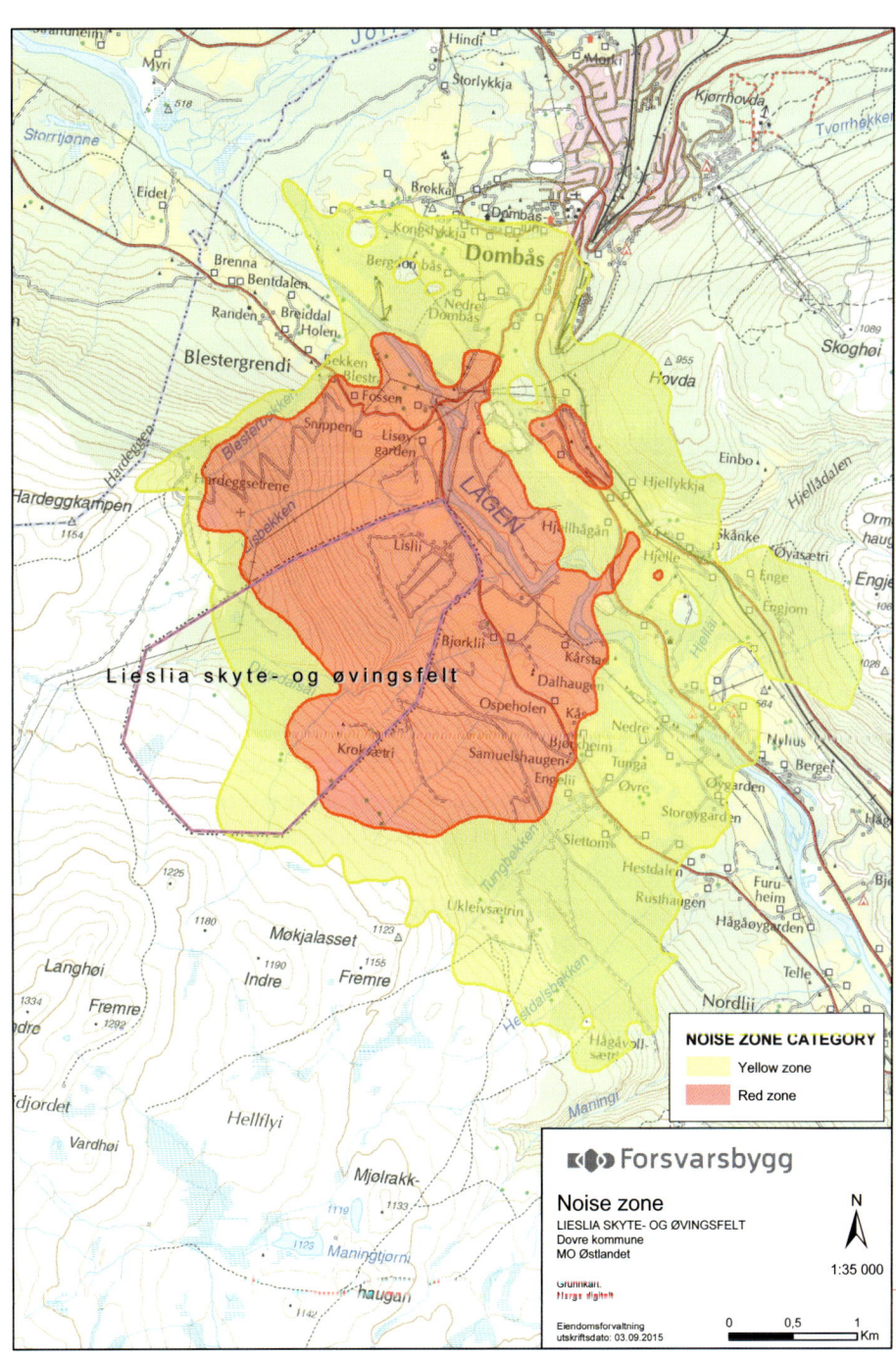

3.41

3.41 Noise zones at Lieslia firing range and training area in Dovre, Norway. Noise zones are drawn up using the computational tool Milstøy.

Thematic data

3.42 In 1980, the mining town Røros and Røros Church were listed on the World Heritage List as a unique cultural environment. Photo: Per Gunnar Ulveseth.

3.43 Conservation plan for Karljohansvern (former base of the Royal Norwegian Navy). Prepared by the Norwegian Defence Estates Agency.

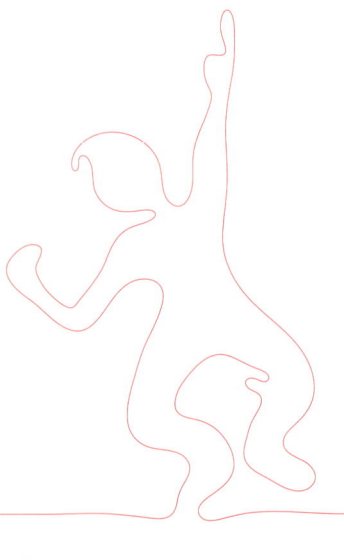

In studies of noise zones, the following sources of noise are taken into account: roads, railways, airports, industries, ports and terminals, shooting ranges, military training areas, motor racing tracks, and wind turbines.

CULTURAL HERITAGE SITES

What are cultural monuments and cultural heritage sites?
The Cultural Heritage Act of 1978 (Kulturminneloven, 1978) defines cultural monuments in Norway as all traces of human activity in the physical environment, including places associated with historical events, beliefs, and traditions. The definition of cultural monuments does not take account the monuments' age, protection status, extent, or condition. The cultural monuments' geospatial locations, in the form of sites, linear features or areas, are called cultural heritage sites.

Cultural monuments or heritage sites can range from a Stone Age dwelling site, a stave church dating from the Middle Ages, archaeological remains of quarrying in the 1700s, to a modern office building or block of flats. Alternatively, they could be the ruins of a dwelling, grave mounds, rural forts, hunting facilities, old roads, other types of ruins, houses, factories, bridges, canals, technical installations, Sámi cultural monuments, or underwater cultural monuments. A location mentioned in various sagas may be designated a cultural monument, as might natural deposits with cultural history associations.

Protection forms
With the exception of automatically protected cultural remains, all cultural sites and monuments are more or less unprotected in Norway. Assessments of whether cultural remains are worth protecting made by professional heritage specialists. In accordance with either the Cultural Heritage Act or the Planning and Building Act of 2008, or by entering into agreements with owners, cultural heritage be protected on the grounds of having use values, experience values, or serving as sources of information or scientific source materials for future generations. Cultural monument protection status stipulates the protection type (*vernetype*) is entered into a GIS data set: 'Listed' (*Fredet*) indicates protection under the Cultural Heritage Act (1978), and 'Protected' (*Vernet*) indicates protection under the Planning and Building Act (2008).

All sites dating from before 1537 (i.e. from before the Reformation) are *automatically protected* under the Cultural Heritage Act (§ 4). For standing buildings, the threshold for automatic protection is the year 1650. Sami monuments older than 100 years and cultural heritage in Svalbard dating from before 1946 are also automatically protected under Norwegian law.

The Directorate for Cultural Heritage (Riksantikvaren) may decide to list structures and sites that are not covered by the laws for automatic protection. The listing may include both the exteriors and interiors of buildings and monuments. Cultural environments may also be protected according to the decision to designate heritage as listed. State-owned buildings may be protected in accordance with specific regulations.

The historical value of state properties is evaluated through the work of sectoral-related *land conservation plans*. The cultural heritage sites and environments that are included in such plans can be protected in different ways, and are protected under the Cultural Heritage Act or the Planning and Building Act.

Municipalities have the opportunity to protect cultural remains and areas of cultural heritage through the

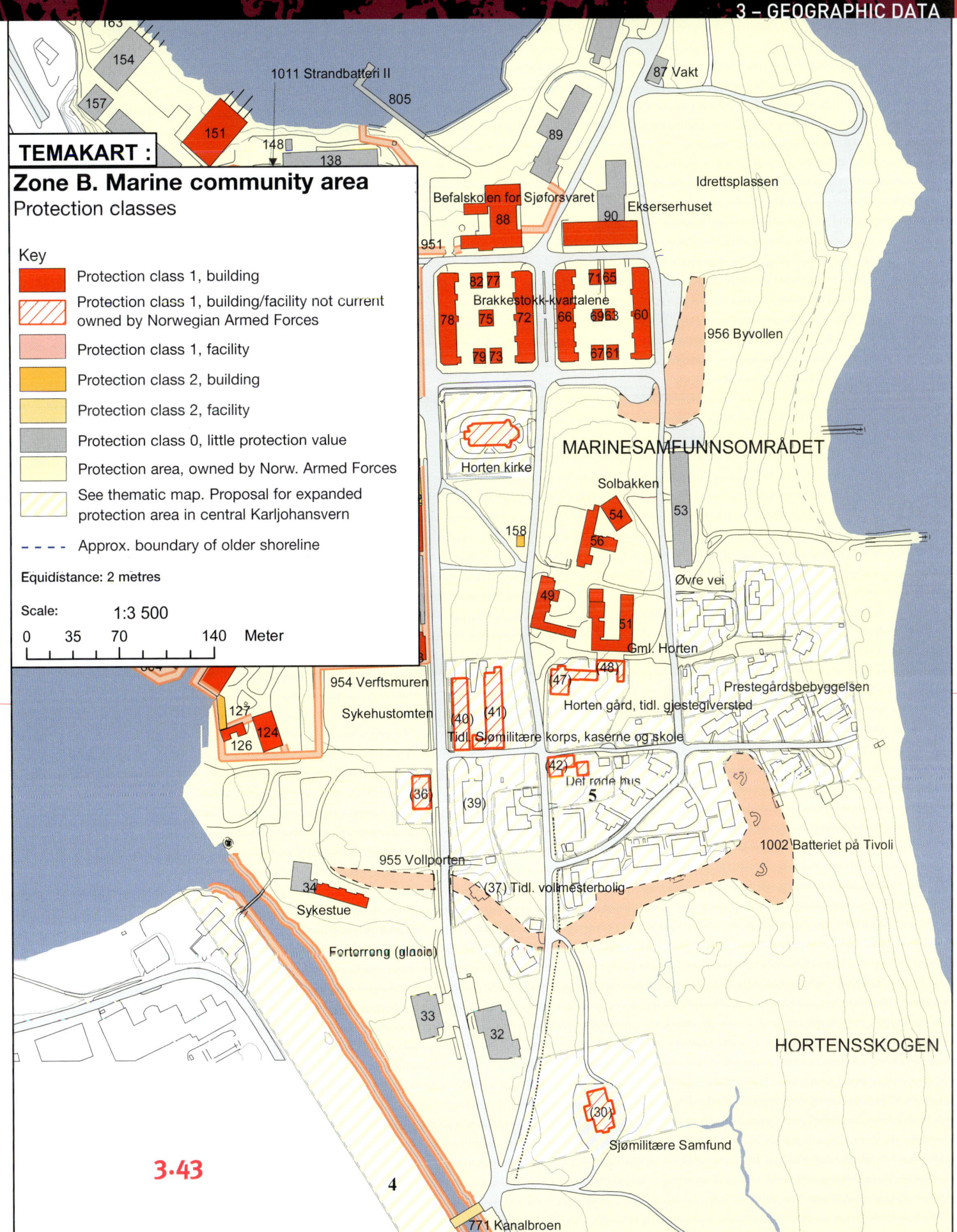

3.44 Oscarsborg Fortress is one of the fortresses most frequently visited by tourists in Norway. The fortress was protected in 2014.

planning provisions of the Planning and Building Act, by introducing municipal land use plans and zoning.

Cultural heritage sites and monuments designated for protection include only external spaces and the exteriors of buildings. Some municipalities are in the process of preparing municipal master plans for the protection of cultural heritage and areas with cultural monuments.

Cultural heritage data

All categories of cultural heritage data are collected in the national cultural heritage database *Askeladden*. The database is administered by the Directorate for Cultural Heritage and contains data on cultural monuments and environments protected under the Cultural Heritage Act or the Planning and Building Act, or that have been evaluated as worthy of protection. The web portal *Kulturminnesøk* provides public access to much of the data in the Askeladden database.

The data set *Fredete bygninger* (protected buildings) contains information about buildings and churches that are protected by law as well as churches with listed building status.

Archaeological sites are physical traces of earlier societies and cultures, and include, for example, burial mounds, animal burial pits, and rock art.

The data set Arkeologiske kulturminner (Archaeological heritage) contains three types of heritage geometries: archaeological sites that are automatically protected are registered with a *protection zone* of at least 5 m around individual monuments. A collection of *individual monuments* constitutes a *locality*.

The data set for cultural environments includes monuments that are part of a larger integrated environment with a collective cultural history value. *Protected cultural environments* are nationally important cultural environments, which the Directorate for

Cultural Heritage has protected under the Cultural Heritage Act (§ 20). *World Heritage Sites* are cultural environments on UNESCO's list of cultural environments worthy of protection.

In addition, the Directorate for Cultural Heritage has registered data of cultural environments that are not formally protected. The data set *Nasjonale interesser i by* (National interests in cities) includes cultural heritage environments of national importance in cities, towns, and other densely populated areas. The data set *Tette trehusmiljøer* (Dense wooden building environments) shows areas with adjacent or adjoining old wooden building in cities and other densely populated areas, where there is a need for special fire-protection measures.

Also, the *SEFRAK register* includes all buildings constructed in the period 1537–1900 (1537–1945 Finnmark).

COASTAL AND MARINE DATA

Many activities take place in coastal zones where the sea and land meet, including fishing, fish farming, transport, trade, and so forth. These activities result in pressure on the available land and resources, and may conflict with nature conservation and other environmental interests. A primary industry, such as fishing, depends on sustainable management and balancing the extraction of resources against the available natural resources in order to ensure long-term returns. The Norwegian coast consists of different types of landscape that are often open and vulnerable to constructions and encroachments. Increasing attention is being paid to coastal landscapes as a valuable resource for the tourism industry and local residents.

Conflicts over the use of common land use and resources include building or road construction in coastal zones, privatization and fencing-off areas that have previously been open to the public, the establishment of wind power facilities in open countryside, fish farming in areas where other types of traditional fishing are practised, or fish farming both in areas used for outdoor recreation and in nature conservation areas. The issue of construction and coastal zones under intense pressure is particularly prevalent in the Oslo fjord region, on the southern coast of Norway, and around the major cities. From the county of Rogaland and northwards, interested parties typically include those involved in traditional fishing and fish farming, and more recently also transport and offshore oil activities. For example, when someone is trying to establish a new fish farming facility, conflicts may arise between fish farming interests and environmental protection considerations. Fish farming facilities often require a lot of space, given the no-traffic zones and no-fishing zones that apply to them.

The wide range of business activities, other activities, interested parties, and the interaction between nature and culture makes the use of geographic data of particular interest. GIS makes it possible to visualize the use of and need for space with regard to the fishing industry, the fish farming industry, and maritime transport. GIS methods and relevant data can be used in coastal zone planning to display various sectors' interests and reveal any conflicts, and to produce a basis for improved cooperation between various administrative levels.

The Norwegian Coastal Administration, the Directorate of Fisheries, the Institute of Marine Research, the Norwegian Mapping Authority and its Mapping and Cadastre Division, as well as other agencies all provide data relating to coastal and marine areas. *MAREANO* is a major national programme that focuses on the establishment of a geographic data set for marine areas. Examples of geographic data that exist nationally and locally include:

- *Nautical charts for navigation purposes, including lighthouses and beacons*
- *Depth data (bathymetric data)*
- *Waterways, ports, harbour areas, anchoring areas, 'emergency ports' (which are used, for example, for vessels in distress)*
- *Pipes and cables, on land or on the seabed*
- *Fisheries data*
- *Ice conditions*
- *Aquaculture*
- *Physical and chemical conditions in the sea, currents, temperatures, nutritional content, and the like*
- *Seabed conditions and sediments.*

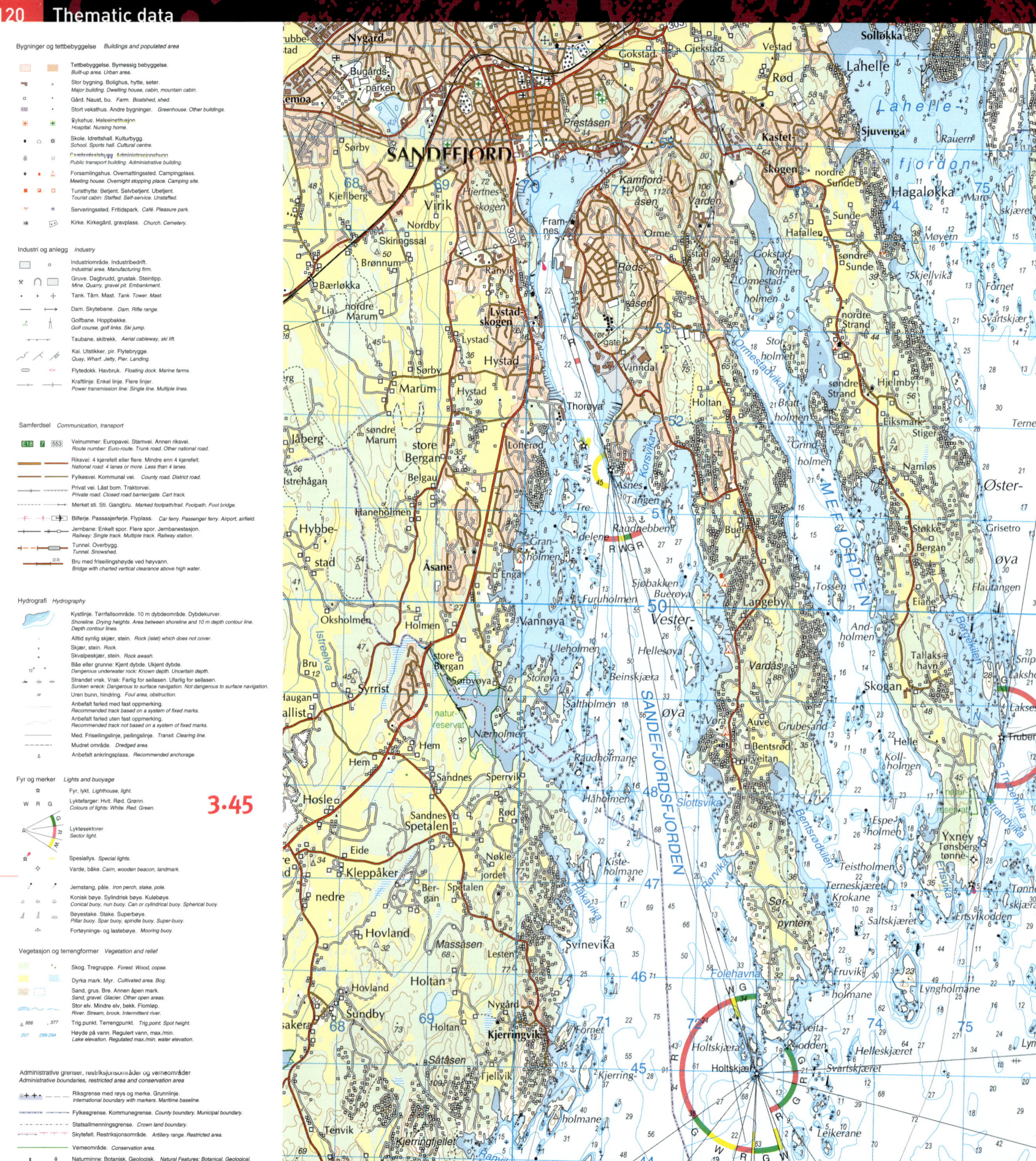

These data may reflect the various specialist interests of coastal industries. In order to perform an impact analysis in a coastal zone, one also needs a base map and other thematic data that show of other sectors' activities and resources.

OUTDOOR RECREATION

In recent decades there has been steadily increasing pressure on areas used by the general public for recreational activities. This pressure is partly due to large-scale commercial activities. However, the construction of second homes and cabins, and tourism are imposing limits on the public use of recreational areas. Therefore, to ensure people's access to recreational areas in the long-term, it is important to present information about the most important areas of interest with respect to outdoor activities in the individual municipalities. Major conflicts can arise between private ownership rights and use by the general public. In addition, conflicts over public access and commercial activities, such as fishing, forestry, and recreational activities, are common. There is a risk that nature and biodiversity may be disturbed in areas of high natural value. However, there are also examples of good facilitation measures that lead to compromises being reached closer to urban centres. One example is the reversal of the old port facilities into attractive beaches such as at Kristiansand or Hamar.

Some recreational areas are secured through support from the public. The Norwegian Environment Agency provides support for the acquisition of property or other agreements on the use of properties so that the public can be guaranteed access to various types of recreation areas, such as swimming areas, beaches and other types of shorelines, inland water s, and nearby areas for walking and hiking. The Norwegian Environment Agency's data for government-secured recreational areas is helpful in planning. The agency has also released guides for the valuation and mapping of recreational areas (Miljødirektoratet, 2013) and green infrastructure (Miljødirektoratet, 2014).

The municipal plans and development plans contain information about which areas are set aside for green spaces. The LNF (Landbruk-, Natur- og Friluftsområder) are larger green areas set aside for public recreation

and leisure. Some municipal plans have more detailed data relating to the paths, trails, and recreational areas. Planning data are therefore a source of information in the context of recreation.

Some municipalities may have different specialist data relating to outdoor activities and recreation. Some have delivered their own data on recreation areas, with additional information about services and access facilities (e.g. toilets, jetties, and diving boards, as well as ramps for the disabled).

Touring maps

Although many people in Norway currently use map services on mobile phones or tablets, paper maps are still widely used. Many customized maps designed for recreational purposes have been produced and there are several providers of such products.

Topographic maps of Norway at a scale of 1:50 000 are relatively detailed and show terrain with contour lines, lakes, rivers and streams, roads, ski trails, and other types of trails. The map series is suitable for walks in woods, on mountain plateaus, or in the mountains.

3.45 Combined land and sea map at a scale of 1:50 000. Published by the Norwegian Military Geographic Service. Series M711LW.

3.46 Hiking in the mountains is a traditional form of recreation in Norway. Photo: Inger Storm-Furru.

3.46

122 Thematic data

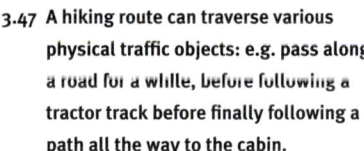

3.47 A hiking route can traverse various physical traffic objects: e.g. pass along a road for a while, before following a tractor track before finally following a path all the way to the cabin.

3.48 The ortophoto shows areas where children play and children's own records of which paths and roads they use. Source: Norsk Form.

For trips in more urban areas and densely populated areas, it will generally be necessary to have maps at a larger scale, with more details, and generally on an extended theme.

Touring maps are especially designed for walkers and hikers and contain more detailed information about footpaths, ski trails, cabins, inns, overnight accommodation, and so forth. They often show the distances between cabins, and highlight sites and cultural monuments. Touring maps often differ from the normal map-sheet system followed in Norway in that they encompass the natural boundaries of normal hiking areas. Common map scales for touring maps are 1:50 000 or 1:25 000.

Hunting and fishing maps show hunting areas, fishing areas, and good fishing spots for sports anglers. Some landowners and landowner associations produce their own maps for their own use.

Nautical charts show water depths, the contours of the seabed, lighthouses, and sectors for safe navigation, as well as harbours and other places for mooring. Maps for leisure sailing are nautical charts that contain extra information in more convenient packaging.

Hiking routes, paths, and ski trails
Some municipalities publish separate maps showing local routes or trails for hiking, walking, or skiing.

In Norway, the term *sti* (path) refers to the physical imprint that a path leaves in the landscape. In other words, a *sti* is an object in the same way as a tractor track is an object. The term *rute* (route) is used to describe routes in the abstract sense, not the physical sense. A hiking route can run from A to B, such as from a car park to a cabin, and along the way it will traverse various physical traffic objects (e.g. a road) for while, and then perhaps following a tractor track before finally following a path all the way to the cabin. The term

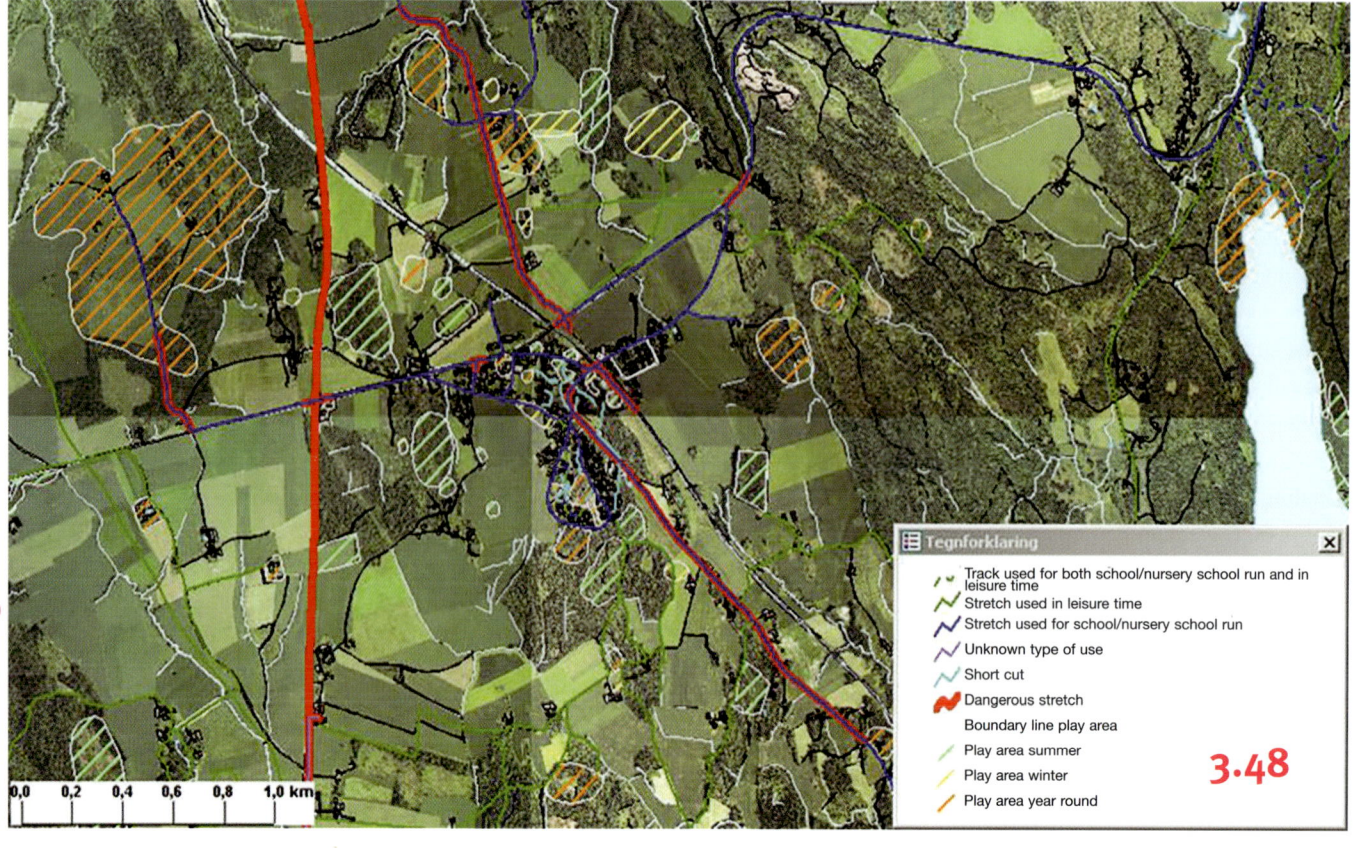

rute is also used to refer to a ski trail, a cycling route, or a bridle path. Special paths that have been modified, such as cultural trails, nature trails, and training tracks, can have special information linked to them in the digital data relating to hiking routes.

Ski trail data often differentiate between wider and narrower cross-country ski trails, and have additional data on, for example, lighting and the number of ski 'lanes' per trail. Similarly, cycle routes are categorized into national, regional, and local routes. Motorized forms of transport are permitted on snowmobile trails across undeveloped areas in the natural environment. These trails are used both for recreational purposes and commercial activities, such as in areas where reindeer husbandry takes place.

Registration of children's play areas
In urban planning, an important task to provide facilities for enjoyment, fun, and safe communities. It is therefore important to know which areas children and young people use and whether they are used as play areas, trails, or shortcuts. These are collected under a separate term called *barnetråkk* (children's paths), and a special a guide has been developed for mapping this type of land use (Norsk form, 2010). Young people can register as users and access the digital tool Barnetråkk and use the maps actively to highlight the areas they use. The resulting data sets are relevant for municipal planning. There are many examples of play areas that adults might regards as forest without any value or as areas that be undervalued from a classical planning perspective. For more information about registered data on Barnetråkk, see barnetråkk.no or kartiskolen.no. The registrations are used by the Norwegian Public Roads Administration (NPRA) in connection with environmental impacts when planning new roads (Vegdirektoratet, 2014) and can also be used as a basic for establishing green posters for densely populated areas or municipalities (Direktoratet for naturforvaltning, 2003).

Land use planning

PLANNING PROCESS
Through their responsibility for preparing municipal land use plans, municipalities have great authority to control the disposition of land resources. Land use in a municipality is clarified by balancing various interests against local democratic processes and interaction with affected government authorities. The municipal council will have to decide whether development should take priority over the preservation of farmland and nature, where and how towns and residential areas are to be further developed, and the infrastructure to be established. They will have take into account national interests, but in cases when the municipality considerations conflict with government interests, the conflicts will be resolved through a process for handling objections at county governor level and ultimate-

3.49 An example of how different provisions for succession, planning requirements, and criteria for the localization of developments can be linked to different parts of a construction site. Accessed from Veileder T-1491/2012: Kommuneplanens arealdel (Municipal Master Plan for Land Management).

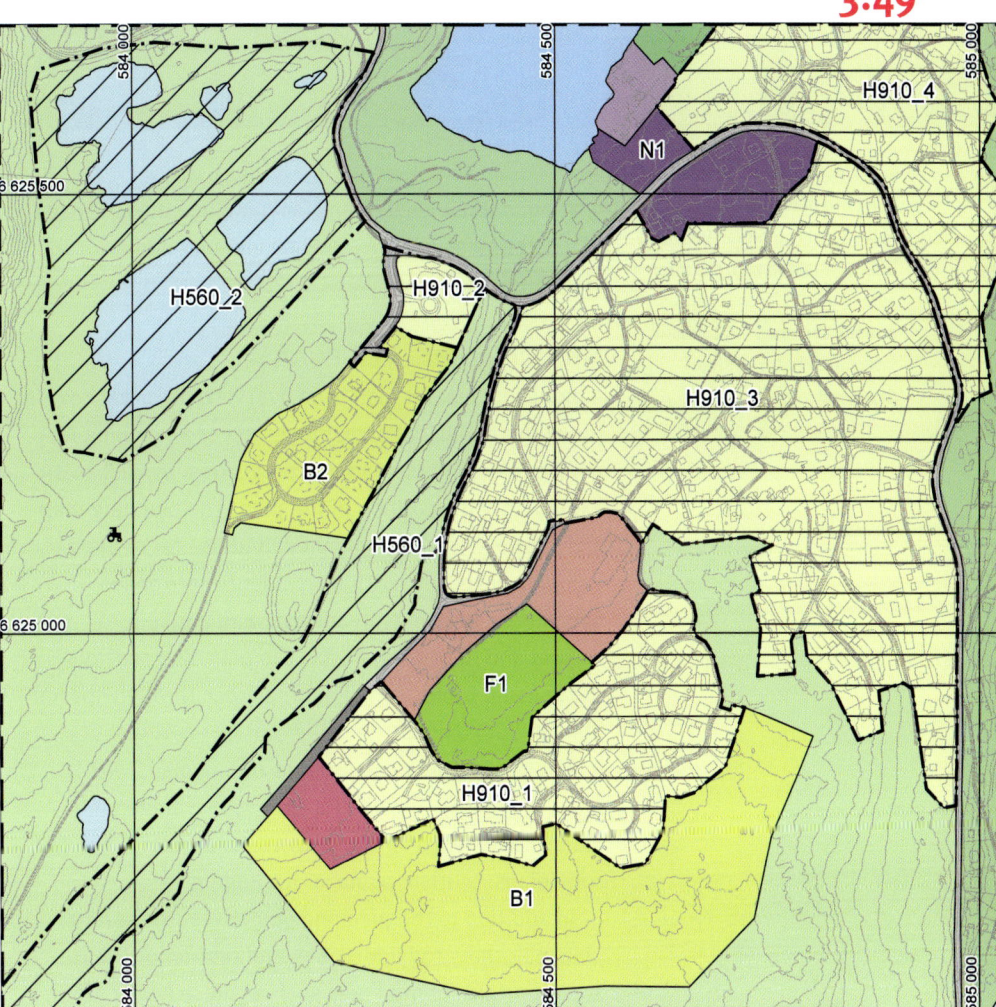

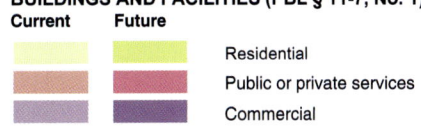

Land use planning

3.50 Land use planning is a process in which development interests are coordinated with the interests of other affected parties and the community's common interests. Photo: Øystein Sanderud.

ly at ministerial level. Municipal land use plans consist of master plans (with a land use element and a social element) and zoning plans.

MUNICIPAL MASTER PLANS

A municipal master plan is an overview plan that covers the entire municipal area. It consists of a planning map with legally binding planning provisions, as well as an explanatory description of the plan. The planning map will show existing and planned land use for defined land use purposes, such as residential purposes, commercial buildings, roadways, and agricultural, and natural and recreation areas. The plan also controls the use and protection of marine areas through special planning provisions.

The municipality can prepare zoning plans with a focus on either a theme or a geographic area in the municipality, such as city or town centre plans, coastal zone plans, tourism plans, or heritage plans.

In addition to the planning objectives, the zoning plan may show considerations and restrictions that are important for land use, such as noise zones, flood zones, or cultural considerations. Data on restriction zones are often generated by relevant professional bodies and made available and downloadable via Geonorge.no.

DEVELOPMENT PLANS

A development plan is a more detailed type of plan that is prepared for areas where the municipal master plan has set requirements for this. Similarly, municipal development plans stipulate objectives that determine future land use. Development plans also provide for expropriation or other measures. A development plan can be prepared as an area development plan or as a detailed development plan.

An area development plans is drawn up by the municipality or by others in close cooperation with the municipality. Detailed development plans are prepared for the implementation of development projects. Detailed development plans are mainly prepared by private or public authorities, but must be administered by the municipality.

Planning data

Planning data are the main thematic data in land use planning, and their value increases when they are combined with base maps, environmental data, and resource data. Active use of various thematic data in a planning process allows for a thorough inspection of land values and sectoral interests, which in turn may uncover any land use conflicts. Municipalities have to maintain a planning register with a master plan that makes it possible to gain an overview of permitted and planned uses of land in Norway under the Planning and Building Act (2008). In accordance with mapping and planning regulations, the municipalities ensure that data from digital planning records are available for download via Geonorge. It is important that the municipal planning database has a uniform digital form, so that planning data can be used across sectors, either as downloadable data sets or as map services.

The Norwegian Mapping Authority's web portal *sePlan* provides easy access to the municipal land use planning registers, with links to the municipal planning databases and individual plans. Planning data are often relatively complicated and it is necessary to gain an insight into what are legal content and values in order to prepare and use planning data. The Planning and Building Act defines the categories of land use objectives that can be used in municipal master plans and development plans. The web portal for planning – planlegging.no – provides guidance on how land use plans have to be prepared with respect to the legal provisions of the Planning and Building Act, legal cartography, and technical structure of data defined through the SOSI features catalogue, and product specifications for land use plans and digital planning registers.

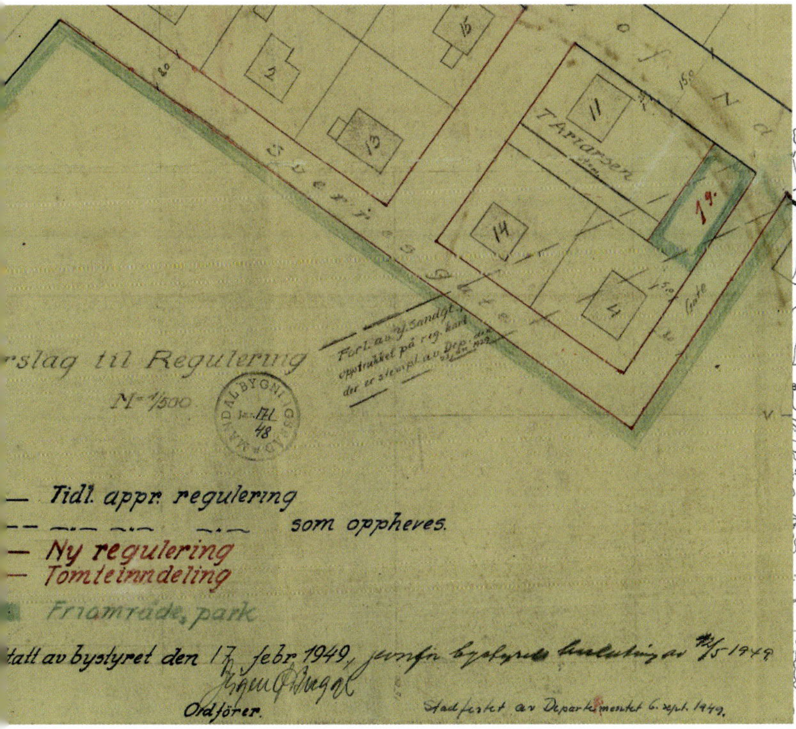

GEOMETRY FROM ORIGINAL DATA SOURCES

Municipal master plans and development plans are legal documents, and inaccurate boundary data plans can lead to uncertainty and conflicts. When the master plans and development plans are established, geometry from other sources is sometimes used (i.e. points, lines, and areas from other databases are copied and replicated in the digital plans). It is highly desirable that this happens even when private development proposals are prepared, because planning data will be used in conjunction with the municipality basemap. Often, map symbols in municipal and land use planning are coarser or based on inputs other than municipal geodata, which will create disparity in boundaries and cause problems later on. Different boundary data may appear similar in drawings on a smaller scale, but if there is a discrepancy in the database in the detailed view, this will lead to uncertainty when the plans are interpreted. When it is important that the depicted boundaries of areas in a plan correspond to a real property boundary, the person preparing the plan should ensure that the boundaries have the same accuracy and geometry as recorded in the Land Registry. The same applies also to other data.

In the digital municipal plans and land use planning, the boundaries of different areas with restricted access should be specified. Such boundaries exist in several other thematic data sets and can often be used directly in the planning process. Nature conservation areas and protected cultural environments are examples of such data sets that can be embedded in the digital data sets for the municipal master plan (i.e. the land use element of the plan). It is important that the protection boundaries in specialist databases are replicated in the current plan databases.

ENVIRONMENTAL IMPACT ASSESSMENTS

An environmental impact assessment (EIA) is a description and assessment of a development project's potential consequences for the environment, natural resources, and social conditions in general. The Planning and Building Act (2008) contains provisions for when and how impact assessments should be prepared. The

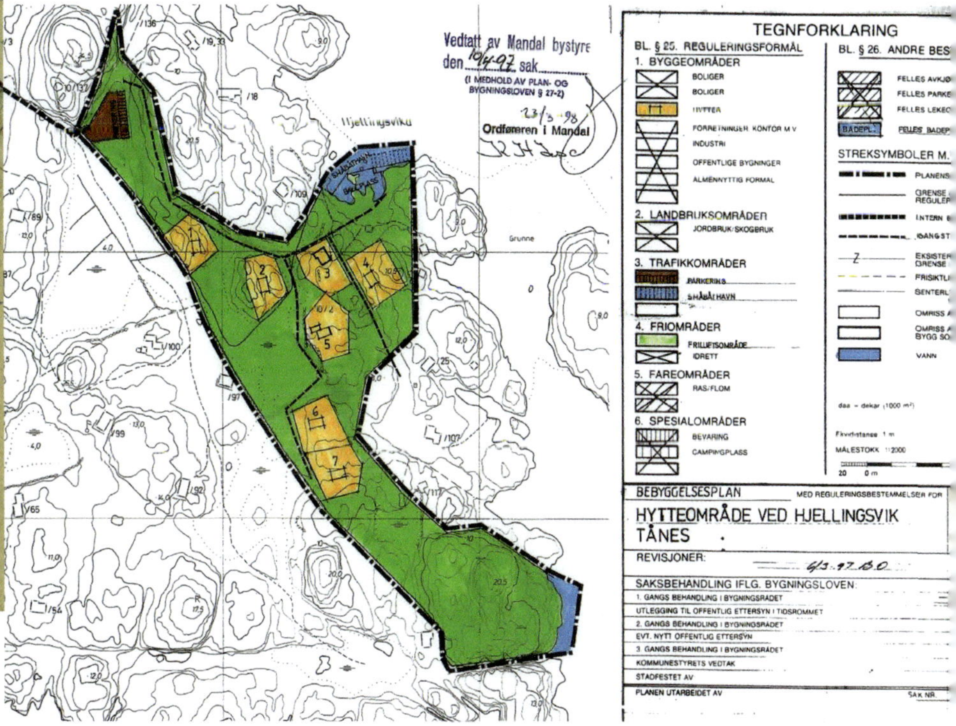

3.51 Analogue land use plans have to be digitized and vectorized in order to meet Norge digitalt's goal of a comprehensive data set of digital plans. Source: Mandal kommune.

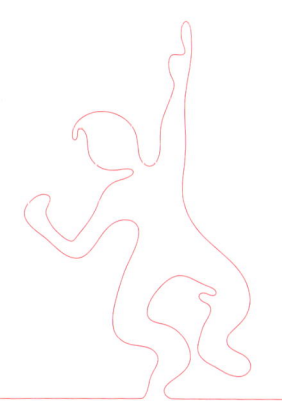

Land use planning

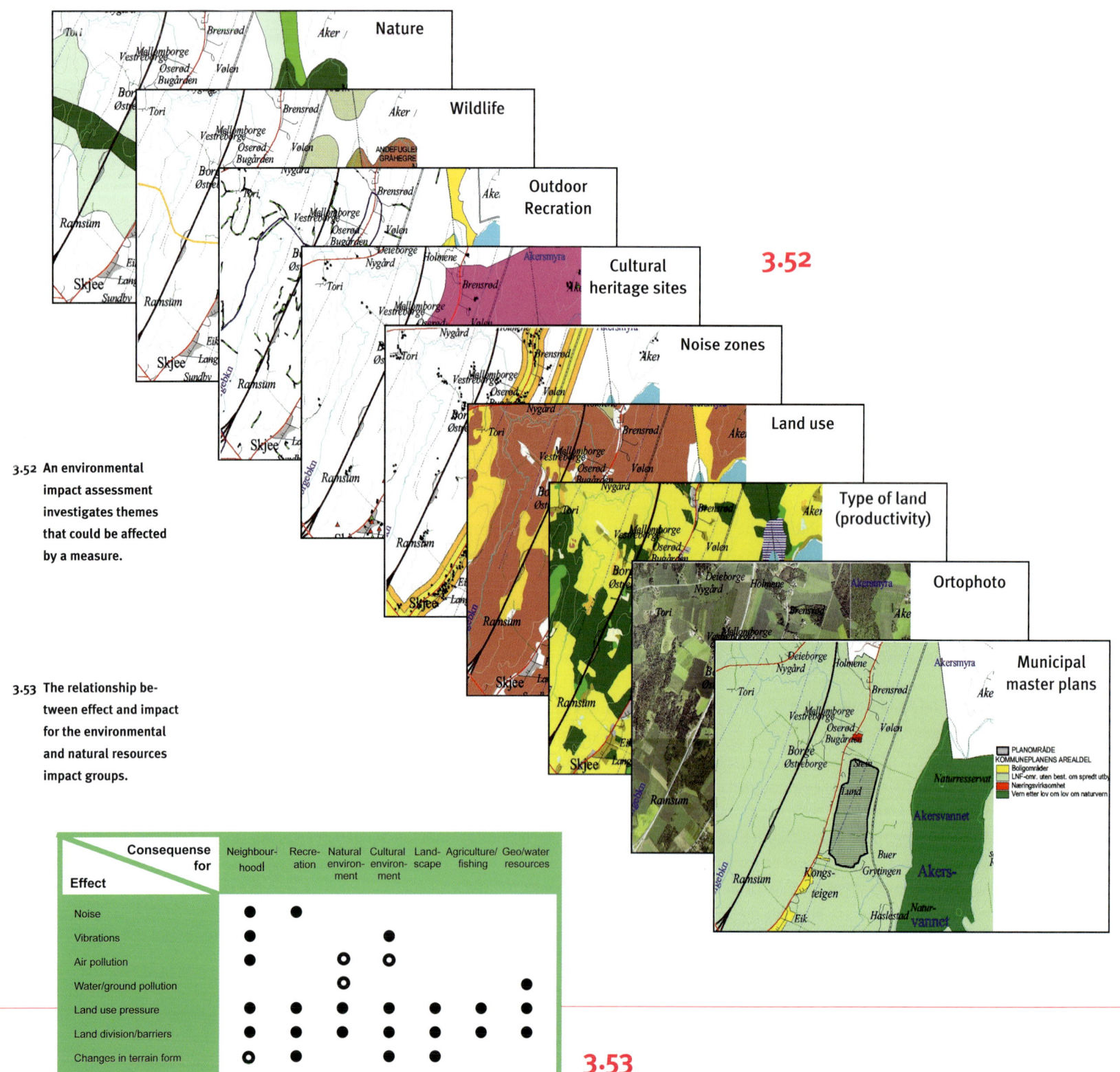

3.52 An environmental impact assessment investigates themes that could be affected by a measure.

3.53 The relationship between effect and impact for the environmental and natural resources impact groups.

purpose of an environmental impact assessment is to study and acquire knowledge about the environment, natural resources, and society, so that relevant considerations are included as premises in the planning and decision-making on the same level as technical and financial considerations.

In terms of their methodology, environmental impact assessments entail an assessment of cause-and-effect relationships, since the impact of a measure can be examined by understanding these relationships. The measures are always the cause in environmental impact assessments and therefore the relationships between measures and their impact are studied. Such relationships may be simple, unambiguous, and immediate, but they may also be complex and long term. Ecological and social phenomena often have multiple dimensions, and situations can therefore be complex.

There are guidelines and handbooks for how environmental impact assessments should be conducted in accordance with the Planning and Building Act. The Norwegian Public Roads Administration's *Handbook V712, Konsekvensanalyse* (environmental impact assessments for road projects (Vegdirektoratet, 2014) has become recognized as a 'standard' and is also used in environmental impact assessment contexts relating to developments other than roads. The handbook describes the method for assessing both value and scope as a basis for classifying environmental impacts.

Work is currently being done to specify more concretely defined procedures and methods for both land use planning and environmental impact assessments. This requires the provision of reliable thematic data. The need for thematic geodata in environmental impact assessments differs from other important types of use. There is a need for a wide range of themes on a par with that needed in a municipal master plan process. However, at the same time there is a need for richness of detail similar to that needed at the level of a land use plan. In the efforts to identify these conditions, it is effective to use thematic geodata.

Geodata available via Norge digitalt may cover topics that environmental impact assessments should focus on. There are data relating to values for various types of land use and resources, and the information relating to rural areas is particularly well organized. Examples are type assignment and evaluation of agricultural land, forestry areas, reindeer husbandry areas, areas with ores, soils, water and energy resources, and fisheries. The public map database is established as a set of thematic geodata organized for planning and building work.

Civil protection and emergency preparedness

The Planning and Building Act (2008) requires local authorities to carry out *risk* and *vulnerability analyses* (RVA analyses) prior to any development. In addition, the Civil Protection Law (Sivilbeskyttelsesloven) of 2010 requires the municipalities to prepare comprehensive risk and vulnerability analyses in order to identify what adverse events might occur in their respective communities, as well as the likelihood of such events occurring and their likely impact.

On the basis of these RVA analyses, municipalities have to prepare contingency plans so that they are best equipped to deal with any incidents that may occur. The Norwegian Directorate for Civil Protection (DSB) has prepared a guide for civil protection in land use planning, which describes how this should be done and how GIS can be used to great advantage in this respect (Direktoratet for samfunnssikkerhet og beredskap, 2011).

Various thematic map data are relevant for RVA analyses for land use planning, and they can be divided into several subgroups. It is common to distinguish between map data related to natural hazards and map data on hazards to people and businesses. Some examples of thematic map data related to natural hazards that are relevant in a RVA analysis are presented below.

Radon
Radon is a naturally occurring gas that migrates through bedrock, and can occur in high concentrations inside buildings. Persons exposed to high radon levels over time have an increased risk of developing lung cancer. The Norwegian Government has introduced a special strategy to reduce exposure to radon in Nor-

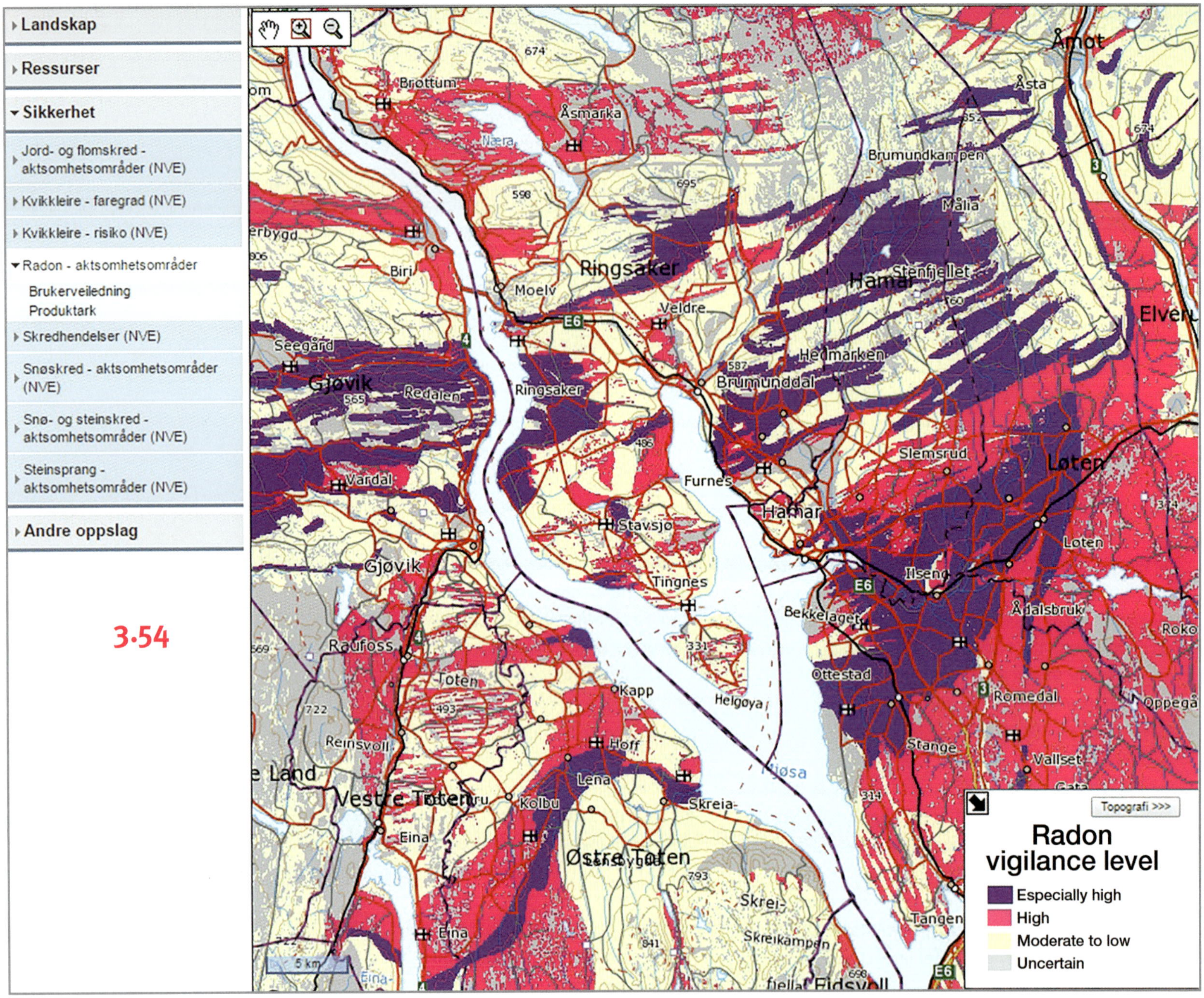

128　Civil protection and emergency preparedness

3.54

Legend (map panel, left):
- Landskap
- Ressurser
- ▼ Sikkerhet
  - Jord- og flomskred - aktsomhetsområder (NVE)
  - Kvikkleire - faregrad (NVE)
  - Kvikkleire - risiko (NVE)
  - ▼ Radon - aktsomhetsområder
    - Brukerveiledning
    - Produktark
  - Skredhendelser (NVE)
  - Snøskred - aktsomhetsområder (NVE)
  - Snø- og steinskred - aktsomhetsområder (NVE)
  - Steinsprang - aktsomhetsområder (NVE)
- Andre oppslag

Radon vigilance level:
- Especially high
- High
- Moderate to low
- Uncertain

3.54 Radon risk levels. Map: Norges geologiske undersøkelse and Norge digitalt.

way, and this implies the importance of identifying risk areas. The Norwegian Public Roads Administration and the Geological Survey of Norway have prepared a national vigilance map for radon, showing which areas are more at risk of exposure to radon than others. The results stem from indoor air measurements and knowledge of geological conditions. Especially areas with alum shale are categorized as areas where particularly high care should be taken. The data set (in the vigilance map) is generalized to a scale of 1:50 000 and should not be used to evaluate radon levels for individual building plots or small areas of housing. Nevertheless, it can still help municipalities to make an initial evaluation of radon risks (NGU, 2015).

Flood zone data
Floods, erosion, and landslides are natural processes both within and along Norwegian watercourses, and these process help to shape the landscape. Floods can cause damage to structures, infrastructure, and farming. The risk of damage from watercourses can be averted by land use planning that takes flood risks into account.

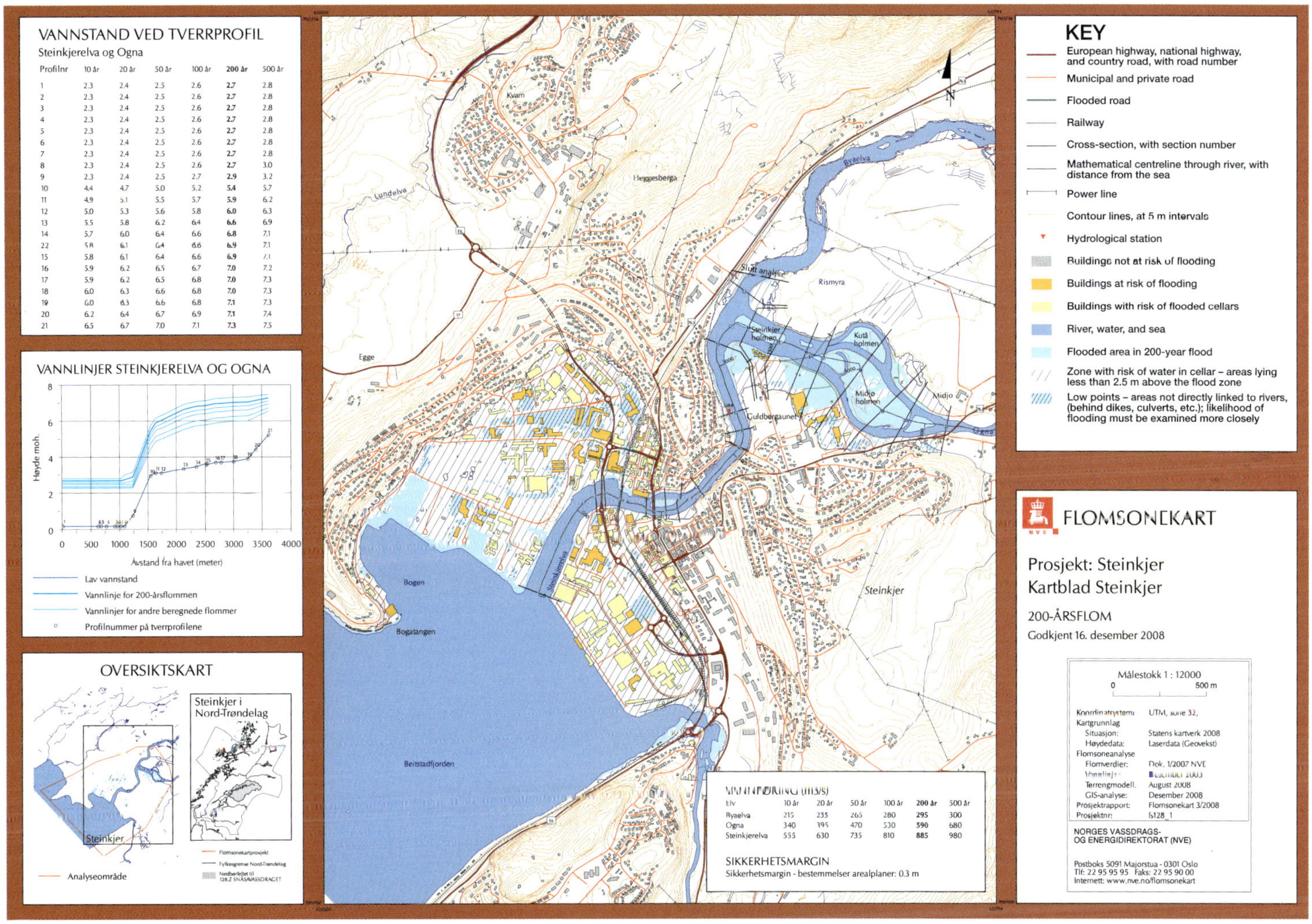

3.55 Flood zone map of Steinkjer, showing which areas would be flooded during a 200-year flood. Source: Norwegian Water Resources and Energy Directorate.

3.56 Flood hit farms in Storsjøen in Odalen. This flood in Eastern Norway in 1995 was the largest since Stor-Ofsen in 1789. Photo: Torbjørn Kjosvold/ The Norwegian Armed Forces Media Centre.

Physical safety measures, as well as good emergency preparedness in the case of major floods and dangerous situations along watercourses, can all help to prevent damage to structures. This requires knowledge of which areas may be vulnerable to such damage.

The Norwegian Water Resources and Energy Directorate identifies areas that may be prone to flooding and makes relevant data available. When planning and processing building applications, the risk of flooding has to be taken into consideration according to the probability of a given flood height. If there is reason to believe that an area is prone to flooding, that area should not be used any ways that might incur unacceptable risks to human life or damage to property (NVE, 2011, b). Acceptable risk levels are differentiated according to the values that could be endangered. Residential and commercial buildings have to be protected against a 100-year flood,

130 Civil protection and emergency Preparedness

while larger public buildings and critical infrastructure have to be protected against a 200-year flood. The municipal provisions require that building heights in flood-prone areas are in accordance with Norwegian Water Resources and Energy Directorate's guidelines.

Quick clay landslide risk zones
A risk of quick clay landslides can exist in areas with marine deposits (i.e. clay originally deposited in seawater). During the last glacial period (the Ice Age), the land was depressed by the weight of the ice, and when the ice melted it took some time before the land adjusted. As a result, the sea reached far inland in Eastern Norway and Trøndelag. Meltwater rivers transported mud that was later deposited as clay containing salts, in the fjords and coastal areas. The land has been rising ever since the major inland ice fields melted, and this has resulted in areas that previously were seabed becoming dry land. Quick clay can form in such areas.

Once the salt in clay has leached out, the clay becomes unstable and can eventually turn into quick clay. The clay can then suddenly transform from solid clay into liquefied clay when it is subjected to great pressure and dilution. In Norway, quick clay can be found in low-lying areas in Eastern Norway and Trøndelag, most often below 200 m a.s.l. Large quick clay landslides can result in several hundred thousand cubic metres sliding downwards, with houses, farms, and roads sinking and being carried away with it. Famous examples of this type of landslide in Norway include the Verdal landslide (1893) and the Rissa landslide (1978). The former resulted in 112 deaths. The risk of landslides of this magnitude has to be analysed and the information should provide a basis for planning and land disposition.

In connection with planning applications for developments, developers are obliged to ensure that the area they plan to develop is not exposed to any risk of landslides or other natural hazards. The Norwegian Geotechnical Institute (NGI) is the largest academic environment involved in research on landslides and has developed methods for mapping and classification of risk zones.

In 2009, the Norwegian Water Resources and Energy Directorate was assigned the role of national landslide agency, with nationwide responsibility for minimizing the risk of all types of slides. The directorate delivers multiple data sets and services related to landslides, including data showing the prevalence of quick clay. It is also responsible for the mapping portal Skrednett (skrednett.no).

Landslides and avalanches
Many areas in Norway are prone to landslides and avalanches. The country's topography is characterized by steep terrain, and its climate is characterized by rain, snow, wind, and freeze–thaw processes lead to problems with rocks falling, landslides, and avalanches.

The NGI's vigilance maps of avalanches and rock falls indicate areas that are potentially exposed to natural hazards. In order to identify areas where snow and rock avalanches can be triggered, elevation models are used to calculate the terrain gradient. Discharge

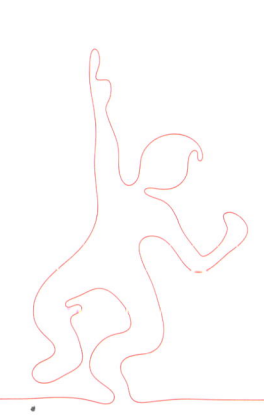

3.57 Information about snow avalanches
Most snow avalanches rupture during bad weather, i.e. periods of heavy snowfall accompanied by strong winds. A pronounced rise in temperature may also lead to the occurrence of a snow avalanche.

- Snow avalanche hazard increases with:
  – Heavy snowfall (20 cm or more per day).
  – Winds resulting in drifting snow.
  – A rapid increase in snow temperature, caused by sunlight radiation, rain or warm air.
- The danger is usually greatest when snow-fall is accompanied by wind.
- Although the surface of the snow appears to be hard and firm, a loose layer may occur farther down in the snowcover, which can lead to avalanche ruptures.
- Such loose snow layers are often formed as a result of extensive cold spells with little snow.
- Sudden booming noises in the snow are an indication of avalanche hazard.
- Exercise extreme caution if you observe other avalanches in the terrain.
- Mountain sides and slopes lying on a lee side of the wind accumulate larger amounts of snow. Potential avalanche hazard is greatest in these areas. Avoid areas which appear to have more snow than other areas on a mountain side.
- Choose route in the vicinity of protruding rock, boulders, etc. Dense forest vegetation may also impede avalanche release.

Snow Avalanche Map
This map indicates areas where snow avalanche may occur.
The maps are produced for the Norwegian Military Geographic Service and should not be used for other purposes without reservation.
This map is divided into the following zones:

| Zone 1 | 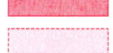 | **Avalanche starting zones.** Terrain steep enough for avalanche release. |
| Zone 2 | | **Avalache run-out zones** Terrain lying below starting zones, which may exposed to avalanche. |

Limitations of the map:
The map only indicates rupture zones which may be indentified on the basis of contour lines. Therefore avalanches may occur on smaller slopes which do not appear on the map.
Zone 1: Some steep areas may have been excluded because dense forest vegetation minimizes avalanche hazard.
The map does not give any information about the frequency of avalanche occurrence. In some areas avalanche occur annually, while in other areas several years may elapse between avalanches.

Classification of avalanche hazard on exercises where avalanche team is established:
Avalanche hazard 0: NO AVALANCHE HAZARD.
Avalanche hazard 1: LOW AVALANCHE HAZARD
  Snow cover is mainly stable. Low probability of avalanche release.
Avalanche hazard 2: MEDIUM AVALANCHE HAZARD
  Risk of avalanches being triggered by personnel or vehicles on steep slopes (>30°), i.e. in zone 1 on avalanche maps.
Avalanche hazard 3: HIGH AVALANCHE HAZARD.
  Naturally occuring avalanches may run into zone 2 on avalanche maps. Avalanche hazard is greatest on slopes in the lee of the current wind direction or exposed to sunlight radiation or warm air.
Avalanche hazard 4: VERY HIGH AVALANCHE HAZARD
  High probability of avalanche release. Avalanche hazard in zones 1 and 2 on avalanche maps.

Prepared by the Norwegian Geotechnical Institute. Published by Norwegian Military Geographic Service.

3.57 Avalanche map for use by the Norwegian Armed Forces. Prepared by the Norwegian Military Geographic Service.

zones are calculated using an empirically based model developed by the NGI. Site visits are made to potential areas and the risks are assessed by specialists before the discharge zones are drawn onto the maps. The landslide and avalanche hazard zones indicate areas that require vigilance regarding the possible need for relocation or exploitation.

The Norwegian Public Roads Administration (NPRA) has to deal with landslides and avalanches regularly because many roads pass through areas that are prone to them. Tunnels and other protection measures are used to reduce the risk to traffic along the roads. The NPRA registers all landslides and avalanches to gain an overview of the most hazardous stretches of road. The information is made available through the National Road DataBase (Nasjonal vegdatabank, NVDB), and is used by land use planners and for evaluating other land uses.

The Norwegian Armed Forces establishes hazard maps showing landslides and avalanche zones in areas used for military exercises. These cover areas other than residential areas and transport routes, and include maps of valleys and mountain areas.

In the context of land use, it can also be relevant to consider other forms of natural hazards, such as areas prone to flooding during storm surges (high tides) or to high winds.

Hazards caused by human activities and businesses
A number of geographic data sets are suitable for use in RVA analyses. Many environmental hazards are caused by humans or are linked to particular businesses or industries. A business risk describes human activity that may represent a risk to others. Geographic data that shed light on such conditions may include data relating to transportation routes for hazardous materials, such as airports, ports, bridges, roads, tunnels, and railways. This can apply to chemicals that can result in serious explosions or poisoning and other forms of contamination, but may also apply to electricity distribution with power plants and networks of electricity pylons.

Municipalities map special fire objects and the NPRA provides overviews of where serious accidents occur (i.e. traffic hazards). Mapping the production and storage areas for hazardous waste, petroleum products, explosives, chemicals, gases, and radioactive materials is done by local and regional authorities in connection with RVA analyses.

All sectors of society may be affected by natural disasters, industrial accidents, or other emergencies. For geographic analyses and assessments of who may be or has been affected by such a crisis, it is thus relevant to have a wide variety of thematic data available, such as the following:

- *population data with details of where people live, work, and travel; data can relate to individual addresses, road sections, blocks, or grids*
- *particularly vulnerable objects such as kindergartens, schools, hospitals, and retirement homes*

- *transport themes, with data for roads and railways*
- *agriculture, with farming and forestry areas*
- *data relating to different businesses and industries*
- *power supplies with dams, power stations, transmission grids, and transformer substations*
- *water sources, catchments, and technical installations*
- *cultural heritage and areas with special natural qualities.*

Depending upon the type of hazard, all of the above data types could be vulnerable objects in emergency situations. It is therefore important that geographic data are readily available not only when accidents occur, but also for long-term planning and prevention.

3.58

Statistical data

One of the advantages of modern GIS is that they can be combined with map data from different sources, and thus provide an overview and insights that otherwise would not be possible if the data were viewed separately. Moreover, it can take a long time to find answers from detailed studies of traditional tables and figures. Compilation of data in a GIS is an effective way to gain an overview, to identify any pattern and trends in time and space, and possibly identify the reasons behind values listed in tables. Data sets, analyses, and results can be quickly presented in various map products, which in many cases are more intuitive and easy to understand than charts and tables.

Themes that are important to portray geographically are often linked to social data, such as demographic, social, and economic conditions, or patterns of development, city and town functions, and social infrastructure. Community data can be established in separate records, but often it is possible to use various existing registers as sources. These register data can then be linked to already established geographic objects using a common identifier, such as a municipal number. The geographic and administrative level on which work is done will determine the choice of resolution and level of detail of the basic data to be used. Common geographic divisions for social data are municipal and districts, and data are also available on address, property, and building levels. In addition, statistical data can be provided as grids, for example 1 x 1 km or 250 x 250 m grids.

SOURCES OF STATISTICAL DATA

Thematic social data are supplied by several different agencies. Possibly the largest provider of social data is Statistics Norway (Statistisk Sentralbyrå, SSB). Through Statistikkbanken from SSB (ssb.no/statistikkbanken), users can access tabular data in various statistical areas. By connecting these data to geographic divisions, the data can be presented on maps. Other agencies that provide statistical data include The Brønnøysund Register Centre (Brønnøysundregistrene), municipalities, and various ministries.

3.58 Hammerfest, Finnmark, is a modern fishing port and industrial city. In recent years it has become a centre for oil and gas extraction from the Barents Sea. Hammerfest was burnt to the ground during the German withdrawal at the end of the WWII. Today its buildings are still characterised by the reconstruction architecture prevalent in the 1950s. Photo: Oscar Pushmann, NIBIO.

3.59 The economic regions in Southern Norway. Statistics Norway has drawn up a new standard regional division between counties and local authorities. The country is divided into 90 economic regions and the criteria used to define the regions are linked to the region's economic factors such as the job market and commerce. The new division is primarily intended to represent an appropriate publication level for statistics. The economic regions are made up of entire local authorities, and the regions cannot cross county boundaries.

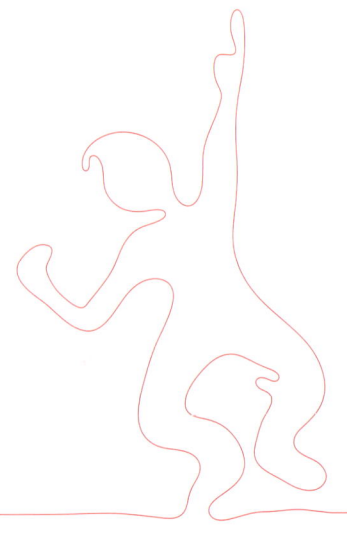

Municipalities will often know the breadth of available data and the application procedures for accessing detailed information. User requirements indicate the need for detailed social data with a high geographic resolution, but the more detailed information that is established, the more sensitive the data becomes. Hence, there are special procedures relating to how access to data is granted and the opportunities to use the most detailed data. The following are a few examples of such data:

- *Population, employment, and commerce at basic statistical wunit, municipality, or county level*
- *Social data in grid systems*
- *Population data linked to addresses or properties*
- *Public service provision and/or social infrastructure at point level*
- *Administrative boundaries*
- *Classification of land types and function areas.*

ADMINISTRATIVE BOUNDARIES

The Norwegian Mapping Authority is responsible for the administration of various administrative boundaries. These include national borders and county, municipality, and other administrative boundaries. Geographic data on these administrative levels are used in a great number of contexts, such as to show the territorial scope of a municipality, to show thematic information, to limit other data, and, first and foremost, to show the areas for which public or private organizations administer have responsibility as well as which boundaries apply. Various types of administrative divisions exist that could constitute important data for social planning, both within regional planning and when working with cities and urban areas:

- **Municipality and county division.**
  The Norwegian Mapping Authority's database of administrative boundaries (ABAS) covers county and municipality divisions. A municipality number has four digits: the first two show which county the municipality belongs to and the second two are a serial number that identifies the individual municipality. From time to time, changes take place in the structure of municipalities and municipality boundaries. Hence, there is a need to be aware of such changes when using data from different years in the same GIS analysis.

- **Basic statistical unit division**
  Basic statistical units are stable statistical units that were established at the end of the 1970s as a result of collaboration between municipalities and Statistics Norway. Population data and statistics from various sectors can be linked to the basic units and provide a basis for maps that show the regional distribution of such phenomena. The basic statistical units are recorded in the ABAS database. A number of basic statistical units constitute a subarea. The basic unit identifier or code has four digits: the first two digits show the aggregated area to which the unit belongs, and the second two show the actual basic unit number.

- **Election ward**
  Election wards are intended to be an appropriate division of municipalities during elections, so that the voters are not too far away from polling stations. The division is intended to serve administrative purposes, but has also been used to some degree to produce statistics for planning purposes. The wards are established by the municipalities themselves. Election wards vary greatly in size. In some municipalities the school catchment areas, basic units, and election wards coincide. Election wards or groups of election wards are used as planning areas in some municipalities. Election wards are changed relatively often.

- **Parishes**
  The administrative division of the Church is a hierarchical regional division, from the large regions such as dioceses to more finely classified divisions such as deaconships, and farther down to individual parishes. A parish is the designation given to an administratively delineated district in which the population can attend the same church (parish church). No parishes cross municipal boundaries. The parish division is used for the administration of the Church.

- **School catchment areas**
  Most municipalities are divided into school catchment areas, which are also called school districts or school enrolment areas. A school catchment area is an area in which the population can attend an individual school at a primary or lower secondary school level. School catchment areas are established by the municipalities, and changes are made relatively often. A number of muni-

cipalities have also shifted to flexible school catchment areas, where their boundaries are not fixed but their delineation is agreed upon at the start of every school year. Each year, Statistics Norway establishes municipal tables for the population in each school district according to gender and age.

- **The sector-specific administrative areas**
Sector-specific management areas are regions and subregions used by governmental authorities in their operations and management. Commonly, the delineation of their boundaries follows municipality or county boundaries, and in many cases a sector-specific region is based on the aggregation of several municipalities. No overall national GIS database exists for such sector-specific administrative regions. Examples of regional administrative divisions include the Directorate of Norwegian Customs' collection districts, the Norwegian Coastal Administration's district boundaries, the Norwegian Water Resources and Energy Directorate's districts, the judicial system's courts, the police, the Norwegian Civil Defence's regional divisions, the Norwegian Labour and Welfare Organization's job centre districts, the Norwegian land consolidation agency's land consolidation courts and land consolidation districts, the Norwegian veterinary districts, and the Norwegian Home Guard's home guard areas.

POPULATION DATA AT BASIC STATISTICAL UNIT LEVEL

Statistics Norway has information available at different aggregations and spatial breakdowns. Many types of statistics are available at municipal level, and detailed data are needed for many purposes. Statistics Norway therefore delivers a limited set of demographic data at a basic statistical unit level.

In some contexts it may be appropriate to produce population data for each age group. Population data can, for example, be delivered with various demography-based attributes, such as total population figures by gender, by age group, or as changes in the last four years. The classification system used should be suited to its purpose, such as fixed classes or age groups that are

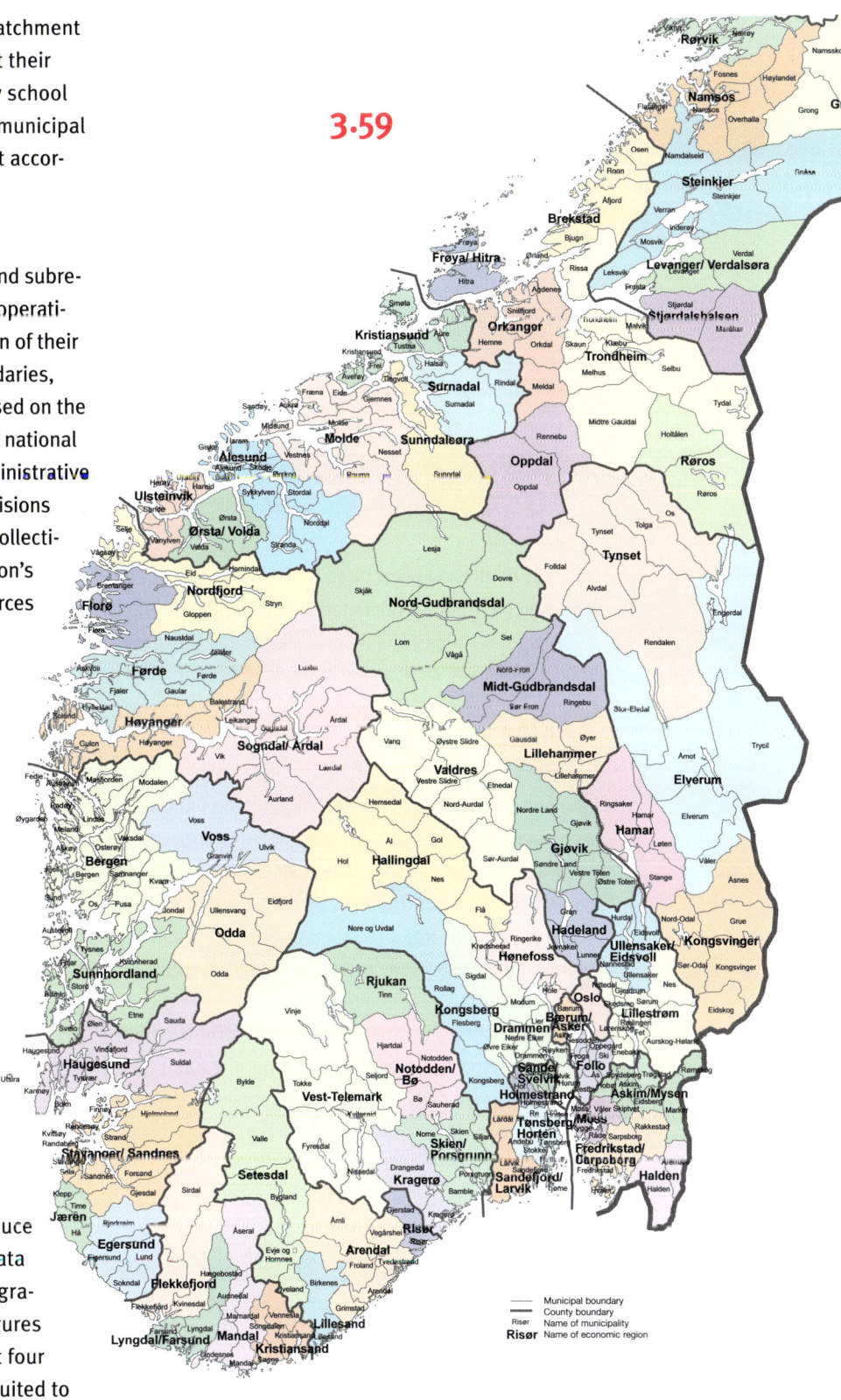

3.59

136 Statistical data

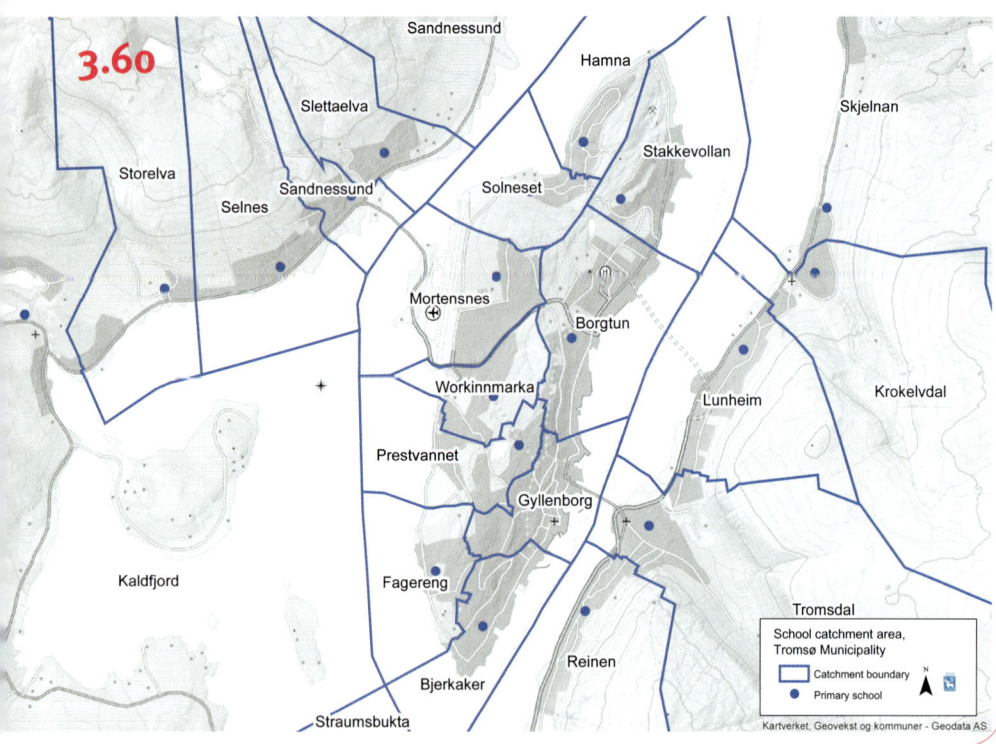

3.60 School catchment areas in Tromsø. Source: Urban Development, Tromsø Municipality.

relevant or already prescribed for use in a school context or in environmental impact assessments.

Some other types of data are available at basic statistical unit level, such as attribute data for a number of companies, services, and workplaces. The link between population data and workplace data is commonly analysed in order to gain an insight into the balance between the number of inhabitants and the number of employees in the various basic units. Such data can be used to highlight transport needs or to assess city centre and commercial structures within a municipality. The Norwegian Public Roads Administration is a frequent user of these data in environmental impact assessments. Examples of commercial and employment data delivered at a basic statistical unit level include company type, number of companies, number of employees, or totals with respect to main groups of businesses.

The Norwegian Public Roads Administration uses company data in its work on environmental impact assessments. It uses a standardized method to calculate the traffic impact of possible road construction alternatives or transport-related measures in cities and urban areas. The assessments are based on available data about the population and the population's activities, and the data are used at a basic statistical unit level. A data set is established as a link between population data and workplace data.

The link between population data and workplace data is made in order to obtain an overview of the number of inhabitants and number of employees in the various basic statistical units. Standard figures are then used for how many journeys to a residence, a workplace, or other place, are undertaken as car journeys (e.g. four journeys per day for a 'standard' residence).

If relevant data are available, it is also possible to calculate traffic loads on the basis of *daytime populations* (when people are at work or at school) and *nighttime populations* (when people are at home). Night-time populations are often used as a basis for planning in land use planning, even though day-time populations may be more relevant.

POPULATION DATA LINKED TO ADDRESSES

Many different agencies need data at address level, even though the data are subject to restricted public disclosure. When the emergency services respond to an incident and in the event of crisis management, detailed and fully up-to-date geographic information about the population as well as commerce and other societal activities can be used to achieve proper, efficient management of the situation.

Another good example of the use of such data can be found within the school sector. Many different types of school analyses are carried out in Norway. Municipalities and county councils analyse current school capacity, future school capacity, transport to schools, and the status of and need for outdoor areas. Typically, such analyses have to be repeated frequently, often annually. In other words, these are repetitive tasks with fixed structures. Despite being efficient approaches, standardized data flows and methodologies are generally not offered by national authorities. The population data used as the basis for the analyses often come from population registers and are geographically linked to addresses. A detailed geographic resolution such as this is necessary when carrying out school analyses. Data at a basic unit or a grid system level are insufficient. Rather, data relating to all school-age groups are required, and should also cover children attending kindergartens and primary schools (i.e. data for the age group 1–16 years). Such data with address-level resolution should not be available to the general public, but can be made available to planners in individual municipalities.

Geographic analyses are increasingly being used in school analyses. Two main groups of data are needed for this, in addition to information about a school's capacity. During such analyses, maps of school catchment areas are combined with population data.

On the basis of the catchment areas, schools within the areas, and the location of school children based on addresses, it is possible to provide results in the form of maps and statistics. The analyses that can be carried out using such data depend on the resolution of the data and how detailed they are. Relevant relationships that can be analysed include:

- *overall assessments of the needed capacity at primary schools*
- *assessing how many school places will be needed in the next school year and in the future*
- *highlighting changes in population or enrolment numbers due to immigration or emigration to or from a school catchment area*
- *changes in catchment areas, adjustment to boundaries to achieve optimal use of the capacity and resources*
- *indications regarding which catchment areas need to be dealt with individually*
- *restructuring of schools or remodelling of existing facilities*
- *temporary shifts in grades*
- *establishment of new schools or school closures.*

School analyses may also be relevant in other contexts. If there are relevant road data and data about the school's geographic location within its catchment area, as well as the children's addresses, it will be possible to perform spatial analysis to identify which children are entitled to public school transport.

3.61 Example of population data at a basic unit level. Source: Statistics Norway.

3.62 Example of employment and commerce at a basic unit level. Source: Statistics Norway.

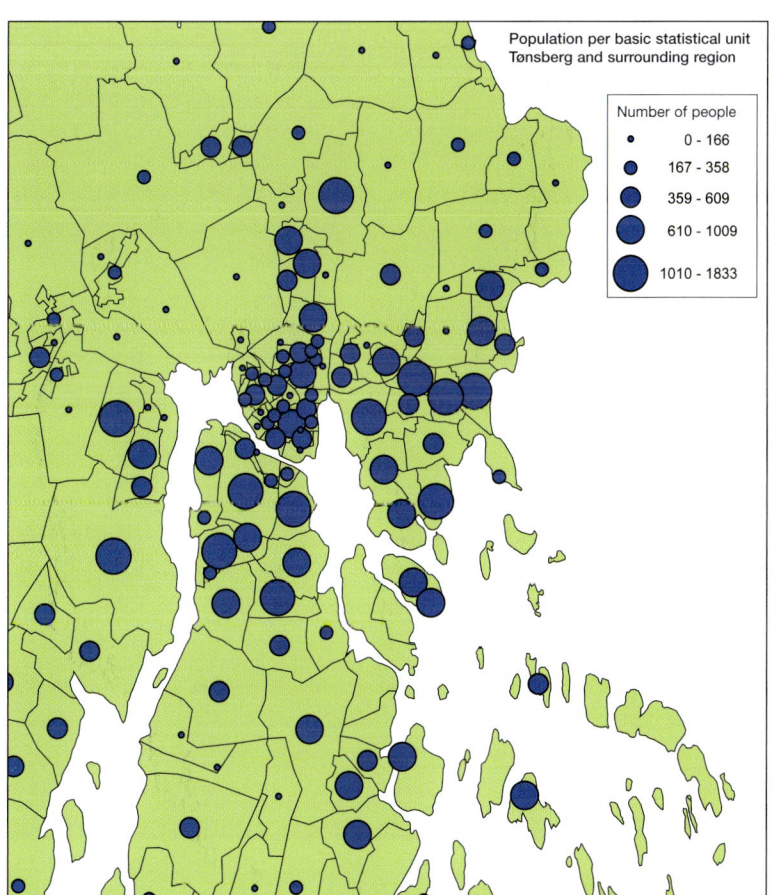
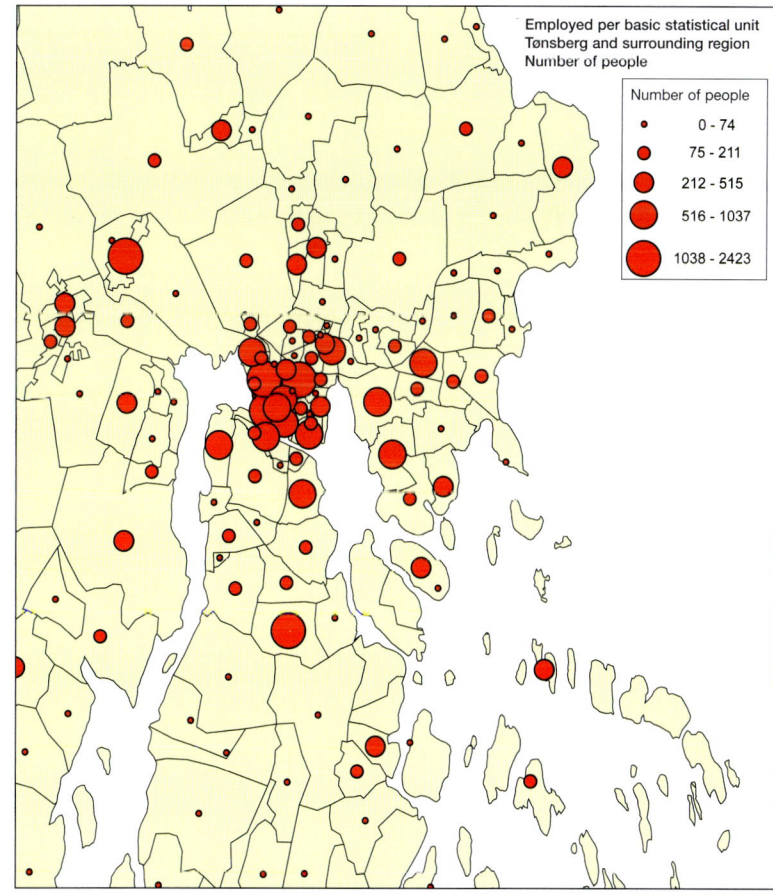

Øyvind Mauseth holds a bachelor's degree in forestry from Hedmark University College (now Hedmark University of Applied Sciences), and a cand.mag. degree in GIS from the former Gjøvik University College (now part of the Norwegian University of Science and Technology). He works as a senior engineer in GIS and map administration at the Norwegian Defence Estates Agency.

Øystein Sanderud holds a master's degree in geodesy from the Norwegian University of Science and Technology. He works as a senior engineer in GIS and map administration at the Norwegian Defence Estates Agency.

Ingvill Richardsen holds a degree in structural engineering from Narvik University College (now The Arctic University of Norway) and has since specialized in the use of GIS in public administration. She is currently a senior GIS engineer at the Norwegian Defence Estates Agency (NDEA).

The chapter is a revised version of the chapter written by Øystein Sanderud and Øyvind Mauseth in the 1st edition.

from the point. In buffer analyses this radius is called the buffer distance. Circular buffers can be created simultaneously around all features in a theme. One can choose whether all features should have the same buffer distance or whether the buffer distances should vary according to the attributes of one of the features. Moreover, it is normally possible to choose whether to transfer the attributes of the buffered features to the newly created buffer polygons. When many features are buffered simultaneously, the resulting polygons will often overlap. In such cases, one can choose either to amalgamate the overlapping polygons into one polygon or to keep the individual polygons. A new polygon formed by amalgamation cannot inherit the attributes of the amalgamated polygons since the attributes of the different features within those polygons might have different values. Buffers can be formed around points, lines, and polygons.

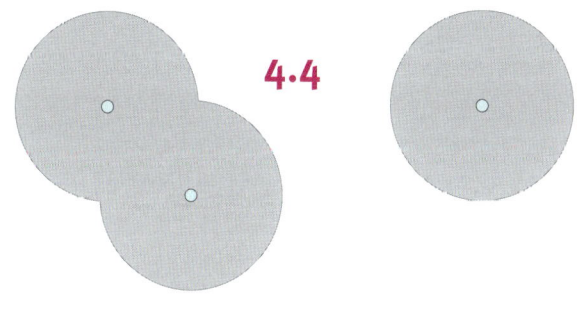

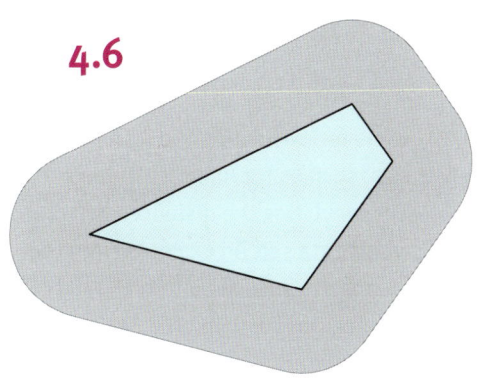

4.4 The formation of polygons around points based on buffering.

4.5 The formation of polygons around lines based on buffering.

4.6 The formation of polygons around existing polygons based on buffering.

Buffering points

A buffer can be created around a point as a circular polygon with a specific radius (i.e. the buffer distance) around each point. One example of when point buffering can be used is the determination of accident risks, such as when it would be dangerous to be within 1000 m of a burning gas tank. Establishing a 1000 m buffer around the tank's position would highlight the danger area. When combined with information relating to population and roads, the buffer area could be used to find out who would have to be evacuated and which roads would have to be closed. The same buffer analysis technique can be used to produce simple area coverage maps for signals transmitted from antennas (assuming one does not take into account the effects of topography on the signals).

Buffering lines

When buffering lines, one can choose whether to create a buffer on just one side of the line or on both sides. Examples of the use of line buffering include the identification of areas where trees need to be felled along a power line route or how much farming land needs to be acquired in order to establish footpaths or cycle paths alongside a road. In the first example, a buffer would be calculated for both sides of the power line and in the second example it would be calculated on only one side of the road.

When buffering lines, one also has to choose how to deal with the ends of the line. One can either create a semicircle with a radius corresponding to the buffer distance or the buffer polygon can be cut off at the end of the lines.

Buffering polygons

Buffering polygons involves calculating a zone with a given buffer distance from all sides of the polygon. One example could be the creation of a conservation area around a lake when both the lake and a 100 m wide belt around the lake are to be protected.

Multiple rings
In the above-mentioned examples, the GIS operation is performed with only one buffer distance per feature. However, it is also possible to have multiple buffer distances per feature in the same GIS operation, and most GIS tools have this capability. This can be useful when, for example, building is to be constructed near a navigation facility yet no building at all is permitted within a certain radius of that facility: within a slightly greater radius dispensations could be granted for low buildings that would not affect the facility. This could be resolved by performing two buffer analyses, but it is quicker to use two different buffer distances in one and the same operation.

Proximity
There are GIS tools for calculating the distances between features, such as for finding the nearest line to a point or calculating the distances between all points in a point data set. An example might be when a rescue team has received information about the geographical position of an injured person in the countryside and want to find a drivable road that will take them closest to the casualty.

OVERLAY ANALYSIS

Overlay analysis involves combining two or more themes to create a new theme based on the geography and attributes of the original themes. Depending upon how the attributes are to be weighted, the new theme can include attributes from the initial themes. The new theme can also include attributes derived from the original themes' attributes.

In order for an overlay analysis to be relevant, the themes should have completely or partially overlapping extents. Clearly, it will not be of any interest to merge landed properties in Norway with land use in England. In principle, in an overlay analysis two or more theme layers with overlapping extents are laid on top of each other. These theme layers can then be combined to form a new theme layer.

Overlay analysis based on vector data
Possibly the most common type of overlay analysis based on vector data is one that combines multiple polygon layers. Two different theme layers with polygons can be combined to form a new layer with new polygons that inherit the attributes of the original themes. In turn, new attributes can be calculated or classified (grouped) from different combinations of the inherited attributes.

The new theme layer will contain all of the polygon boundaries from the original layers and will therefore contain many more polygons than the original themes. Different vector themes can often differ in their geometric accuracy. When inaccurate data are included in an overlay analysis this can result in some unwanted effects. If bounderies between polygons that should correspond in the participating themes differs slightly, a number of small polygons will be created between these bounderies. An example could be municipal boundary and property boundary that should coincide. Most modern GIS therefore have methods for removing such unwanted polygons. The GIS operator will usually have the ability to control the parameters that describe how far apart the two boundary lines must be before they are considered as two different lines.

When two polygon themes are combined, one can choose how the two themes should influence the results. There are a number of possibilities, some of which are:

- Union – two or more theme layers are combined into one theme layer with attributes from both themes, and the resulting geographic area is the union of the geographic area of the two original themes.
- Intersect – two theme layers are combined into one theme layer with attributes from both themes, and the resulting geographic area is defined as the section where the two original themes overlap.
- Identity – two theme layers are combined into one theme layer with attributes from both themes, and the resulting geographic area is the same as the geographic area of the first original theme.

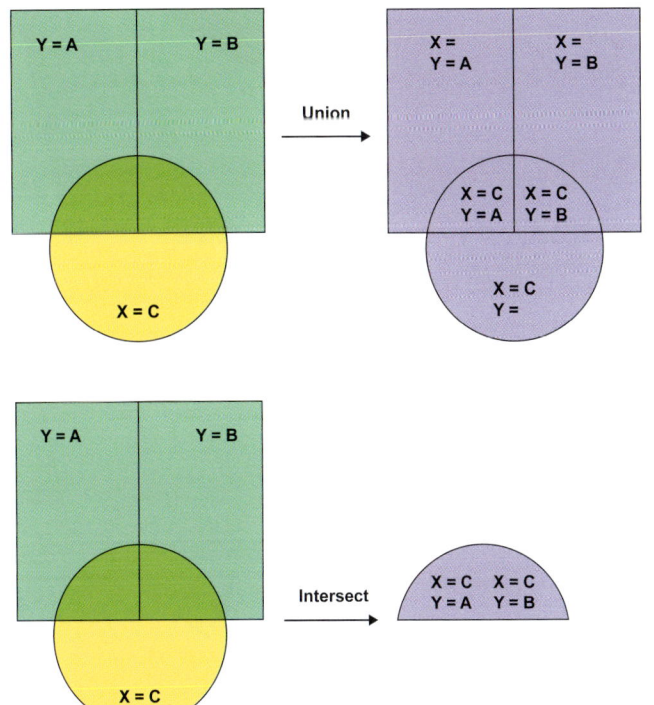
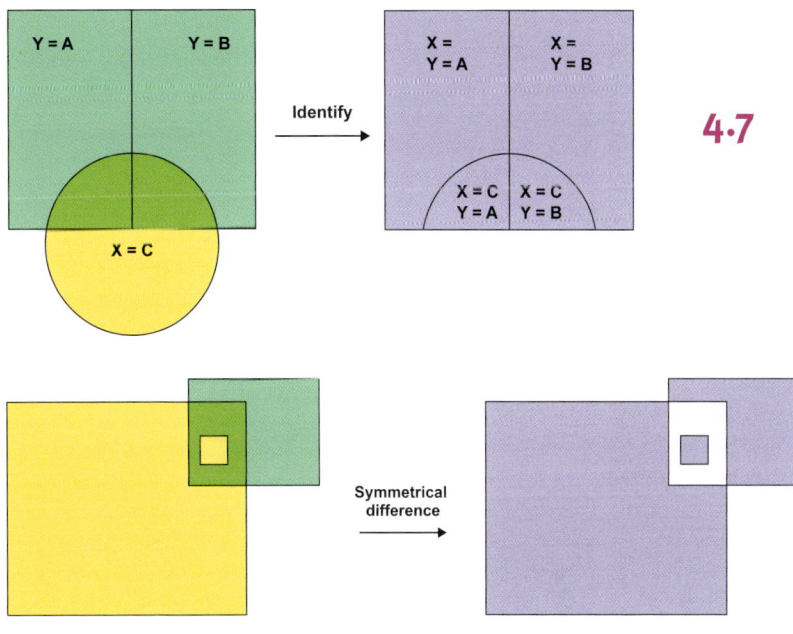

- Symmetrical difference – one theme layer is laid over another theme layer and area of the new theme layer will be where the original theme layers do not overlap.

Transferring polygon attributes to points and lines
In overlay analyses with points and polygons, the attributes from the polygons are transferred to points. It may be worth doing this if, for example, one has noise zones as polygons and different building types as points, and one wants to find out how many buildings of each type exist in each noise zone. Instead of analysing by building type and noise zone, one transfers the noise attributes from the polygons to the points and then derives statistics from the attributes of those points. In this way, one can quickly form a picture of the noise distribution between building types and the distribution of the noise in the area as a whole.

Lines – polygons
As in the case of points, polygon attributes can be transferred to lines.

Overlay analysis based on raster data
Overlay analyses can be performed with raster data. In this case, the attribute data are linked to each pixel in the image. In an overlay analysis with raster data, many layers can be placed on top of each other and the different layers' contribution to the result can be weighted differently. For example, such analyses could be used to find suitable areas for housing development. The criteria for a suitable *area might be:

- Preferably south-facing, though west-facing is also acceptable
- The area must not be too steep
- The ground conditions should be stable
- The area should not be exposed to a lot of noise.

The characteristics of the data sets are defined by setting the values as 0 = not acceptable, 1 = acceptable, and 2 = best, where the gradient and ground conditions are weighted as double the other criteria.
By combining the attributes and multiplying the weightings of all of the themes, a number in the range 0–12 will be obtained for each pixel describing

4.7 Overlay analysis.

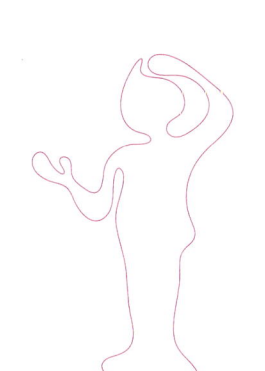

Analysis methodology

4.8 Schematic diagram of overlapping with polygons in which areas 1, 2 and 3 are assigned the codes 1A, 2A and 3A where they overlap with area A.

4.9 Schematic diagram of overlay analysis with points and polygons.

4.10 Schematic diagram of overlay analysis with lines and polygons.

the suitability of the area. If all criteria are fulfilled as best as the possible in an area, a value of 2 + 2 x 2 + 2 x 2 + 2 = 12 will be obtained.

CONVERSION BETWEEN VECTOR DATA AND RASTER DATA

The method for overlay analyses with raster data is often more effective than overlay analyses performed with vector data. For gradient analyses, the results are most often raster data and if these data are to be used in analyses together with vector data, the vector data will generally have to be converted to raster data to ensure data uniformity. Most GIS tools have functions for converting vector data to raster data. The raster data can later be converted back to vector data, if desired. Overlay analyses of raster data are based on the assumption that the image elements – the pixels – are the same size in all data sets.

Some examples of when overlay analyses can be useful are:

- finding the proportion of mature forest for each property in an area by combining property maps with felling classification maps
- identifying potential landslide areas by combining data sets for gradients and data sets for soil conditions
- calculating the number of noise-affected homes in an area by combining maps of building positions and noise zones.

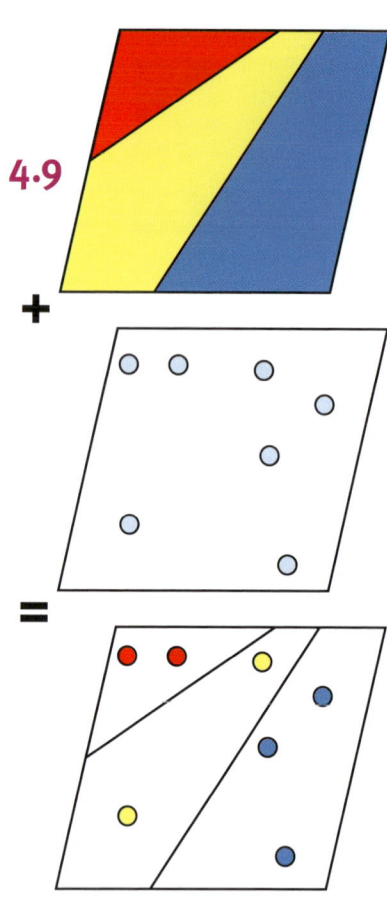

4.9

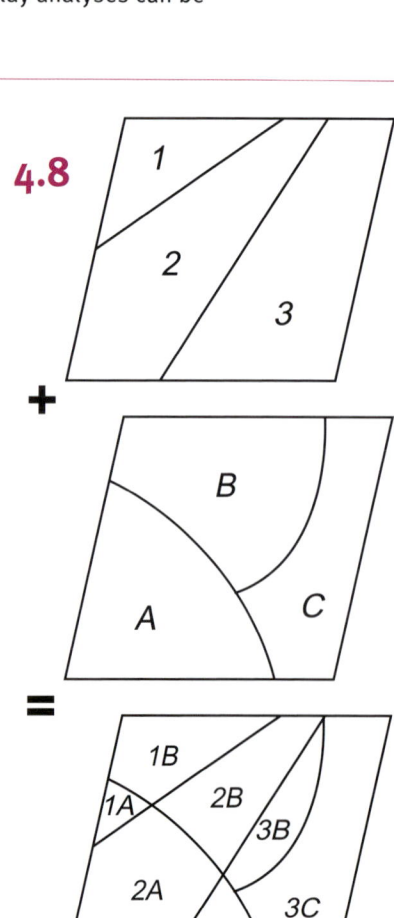

4.8

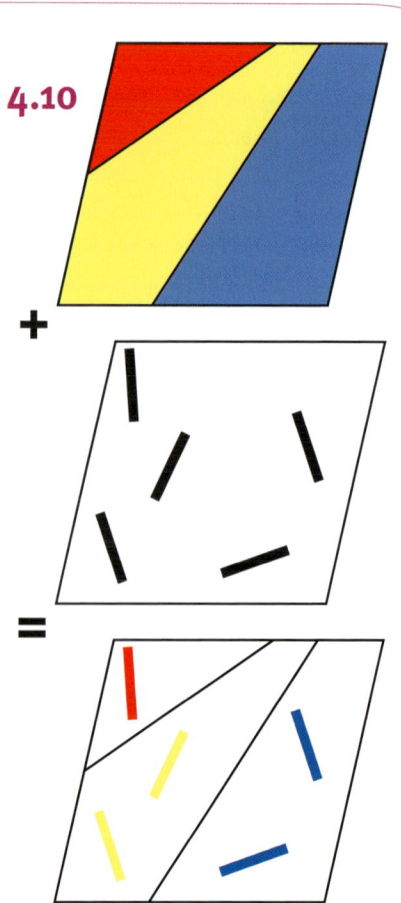

4.10

OTHER SPATIAL DATA MANIPULATIONS

- Select – Selection of a theme layer is done either on the basis of its attributes or through a spatial sampling based on the geography. A spatial sample can, for example, include all objects in a layer that are within the range of objects in another theme layer. The objects that make up the sample will retain their original attribute values.
- Clipping – The method can be compared to a spatial sample where the selection criteria are within a given area, but the result is clipped so that the sample only falls within the area being used as a working area.
- Splitting – Two theme layers comprising an initial theme and a split theme, are combined. The resulting theme is the same as the initial theme but is divided into areas according to the division in the split theme.
- Erase – Similar to clipping, but in this case just the area outside the clipping area is kept; in other words, some interior data are removed.
- Merge – Two or more features in a theme layer are merged into one feature, keeping the attributes from one of the features. Merging can be used for polygons, lines and points.
- Dissolve – Features with the same attribute values in a theme layer are joined to form one object. Boundaries between adjacent polygon features with the same attribute values are removed.
- Joining – One theme layer is combined with data from another table or theme layer, based on attributes with the same value in the theme layer and the table.

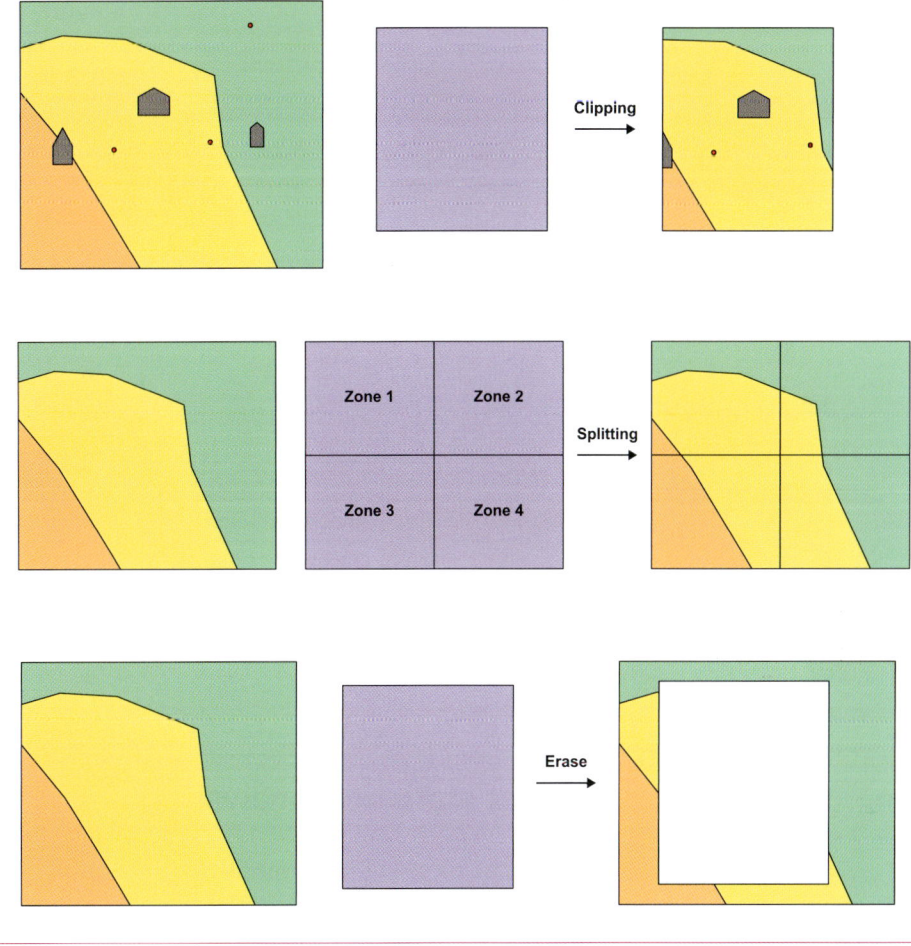

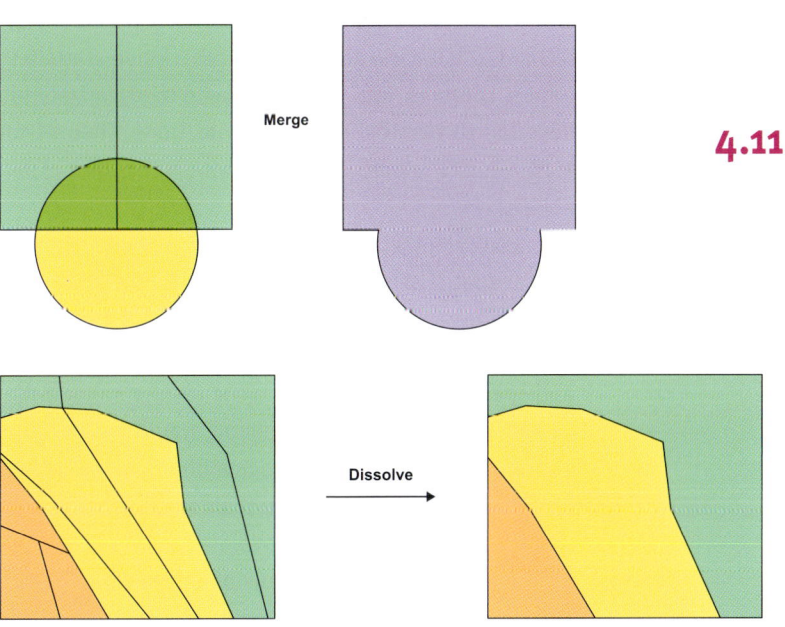

4.11 Clipping, splitting, erase, merge, and dissolve.

4.11

Municipalities

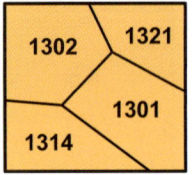

Municipality no.	Name
1301	Andeby
1302	Gåseby
1314	Fjærland
1321	Vingestad

Mayors

Municipality no.	Mayor's name	Gender	Political party affiliation
1301	Jon Pedersen	M	Ap
1302	Åse Konradsen	F	H
1313	Lars Hansen	M	Ap
1314	Vidar Larsen	M	Frp
1321	Amund Amundsen	M	Sp
1326	Liv Lund	F	Ap

4.12

Join ↓

Municipality by mayoral party

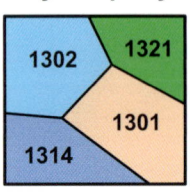

Municipality no.	Name	Mayor's name	Gender	Political party affiliation
1301	Andeby	Jon Pedersen	M	Ap
1302	Gåseby	Åse Konradsen	F	H
1314	Fjærland	Vidar Larsen	M	Frp
1321	Vingestad	Amund Amundsen	M	Sp

4.12 Joining.

Spatial statistical analysis

Spatial statistical analysis involves the use of statistical methods on quantitative and/or observed spatial data, which can also relate to time.

Traditionally, statistics consist of *descriptive statistics* and *theoretical statistics*. In descriptive statistics, tables, columns, and charts are used to organize and describe sizes and different compositions. Theoretical statistics are mainly based on mathematics and probability.

In traditional statistics it is assumed that data are free of spatial structure. This means, for example, they are not localized, clustered, or distributed, or that there are any spatial patterns in the data. If time and space are not parameters, the results might only be partially accurate. One example would be where a fire station records only the number of times there has been potential fire within their area of responsibility in one year. Such an overview would provide residents with information only about the number of outbreaks of fire per year. If additional information is recorded about where and when the fires occurred, it would be possible, for example, to perform an analysis of whether some places have more or fewer outbreaks than others.

Thus, spatial statistical analysis can be used to gain a better overview and to visualize phenomena. In such analyses, models and analytical methods are used to process the statistical and spatial variation in phenomena within an area.

If there are preconceived ideas about a phenomenon in an area, a spatial statistical analysis can be carried out to confirm or deny them. The analysis would identify any discrepancies and detect unusual patterns. Moreover, new and unexpected information might emerge for various phenomena such as diseases, high incidences of crime, and outbreaks of fire.

In short, spatial statistical analyses are about finding out whether certain data have any foundation or whether they form any patterns that could be used to draw conclusions.

The following section describes the methods used in gravity analysis and pattern analysis, commonly known as 'hot spot' analyses. Most GIS tools have more or less advanced functions to calculate these types of statistical analyses.

GRAVITY ANALYSIS

Gravity analyses are used to identify concentrations of particular phenomena. An example of a phenomenon could be outbreaks of disease in a defined geographic area. Classic examples of gravity analysis are described in Chapter 1 (Application), which elaborates on the outbreak of legionnaires' disease in Fredrikstad in 2005 and Dr John Snow's investigations into the spread of cholera in London in the mid-1800s. Figure 4.13 shows the results of Dr Snow's gravity analyses. The results of a similar analysis performed using the same database but with current GIS tools are shown in Figure 4.14.

In gravity analyses, the centre of gravity is calculated for a widespread phenomenon. The following are three different methods:

- *Average centre* (mean) – In this method, an average value for all x-coordinates and an average value for all y-coordinates are calculated. The average x- and y-values are the coordinates of the average centre.
- *Median centre* – the point with the shortest distance to the rest of the points in the database.
- *Central point* – the original point that are closest to the centre.

4.13 The map Dr John Snow produced in 1854 shows the deaths he recorded, depicted by black bar symbols. The triangle depicts the mean value, the square the median value and the circle the mode value (mode: most frequently occurring value in a dataset). The deaths were clustered around a particular water pump in Broad Street. Source: John Snow (1854).

4.14 The same data Dr John Snow collected has in modern times been recorded and presented on a map after calculating the mean and median values. Here you can clearly see that one arrives at the same answer using modern tools. Source: Cliff & Haggett (1988).

150 Spatial statistical analysis

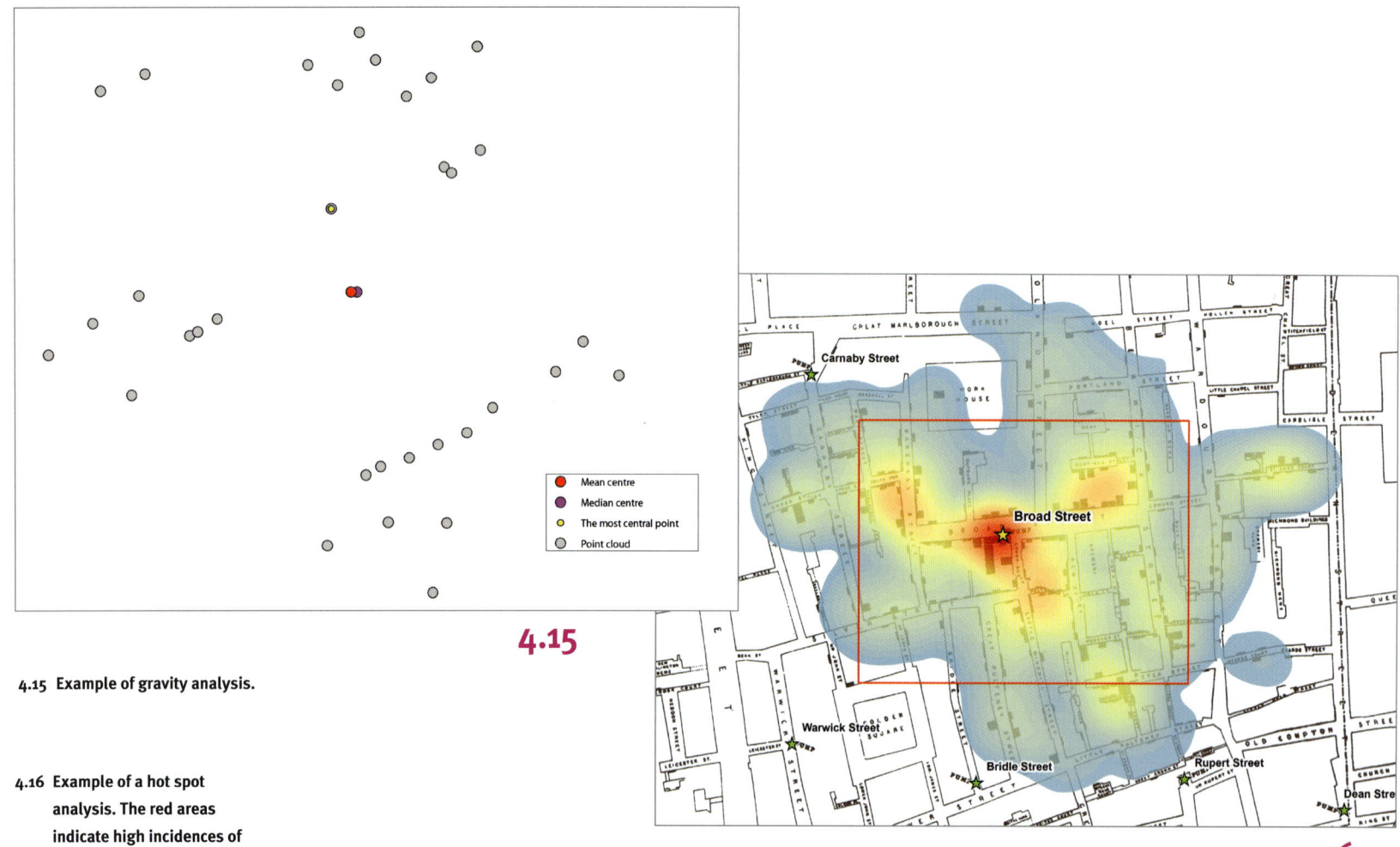

4.15

4.15 Example of gravity analysis.

4.16 Example of a hot spot analysis. The red areas indicate high incidences of illness, and the blue areas low incidences of illness. The example shows that an old database can be successfully used in conjunction with modern GIS tools.

4.16

PATTERN ANALYSIS

Pattern analyses can be used to detect and visualize different spatial phenomena.

An example of pattern analysis could be when a police district has an unusually high number of reports of burglary. The burglaries have occurred over almost the entire police district and they have been ongoing for seven months. First, the coordinates are recorded for all places where burglaries have been committed. An examination of the geographic distribution of the burglaries reveals that they are largely concentrated near a main road between two towns. On the map, the points representing the individual burglaries are supplemented with information about the day and time of the burglary, resulting in a new distribution pattern for the robberies. All burglaries have occurred on Mondays and Fridays. From the information on the times and a simple network analysis of driving times, the thieves' pattern of break-ins is revealed. The break-in pattern starts with high probability in village A on Monday morning and shifts towards town B, where further burglaries occur. On Friday morning the thieves move in the opposite direction, and commit further burglaries. Hence, there is a clear pattern of where and how the criminals act. The burglary sites do not have any distinct centres of gravity; rather, they are spread out over a relatively large geographic area. Nevertheless, a pattern has been revealed in the data. Without resolving the case, the investigators have gained access to useful information that they can use in their further investigations. For example, they could check the number plates of all cars driving along the relevant routes every Monday and Friday.

Pattern analysis is also termed *trend analysis*.

Figure 4.16 shows a density analysis of the part of London in which Dr John Snow carried out his surveys of cholera outbreaks. The Figure shows where the individuals who died had lived and the site of the water pump that was later revealed as the source of infection (marked with a yellow star).

Network analyses

Network analyses are becoming increasingly popular. Transport companies need to use the most cost-effective routes. Energy network companies want the best possible overview of their grid systems and their power stations in order to maintain and ensure the quality of their supplies. Increasingly, individual members of the public are using route planners either online or in their cars. Many types of geographic information can be described as networks, including the following:

- Transport networks, such as roads
- Pipe and cable networks, such as water and sewage pipelines
- Trail networks, such as ski trails
- River networks, such as rivers and streams.

The networks can be represented as network models in a GIS and the models can then be used for different types of network analyses.

STRUCTURE OF NETWORKS

Networks consist of *links* and *nodes*. A link is a connection between two nodes. A node can be connected to one or more links. In a road network, the links will typically be the roads and the nodes will be the junctions. In a sewage network, the nodes will be the manholes and links will bee the sewage pipes or drains between the manholes.

In addition to the geographic representation of a network, it will also be necessary to assign *resistance* to the links and nodes in the network. Examples of such resistance include speed limits on roads or expected waiting times at traffic lights on roads. If the speed limits and lengths of roads in a road network are known, it will be possible to calculate how long it would take to drive along any given route in that network.

A network can also include *barriers* that block a node or prevent a link from being used. Typical examples are when a road is closed due to roadworks or when houses in a particular neighbourhood suffer power cuts after an electricity cable has been broken.

In addition, a network can have directional restrictions such that a link can only be followed in one direction, as in the case of a one-way street. Networks also consist of *sources* and *end-users*. For example, in an electricity network the power stations will be the sources and all customers will be the end-users. Calculations of the electricity flow rates in such a network will be based on which sources exist and which end-users are using the network at any given time.

NETWORKS

The nodes and links in a network can be georeferenced and form a geometric network. In addition, a network will incorporate a logical network that contains information about how the network is configured. The logical network will describe which nodes are attached to which links. All elements in the logical network will be associated with a feature in the geometric network.

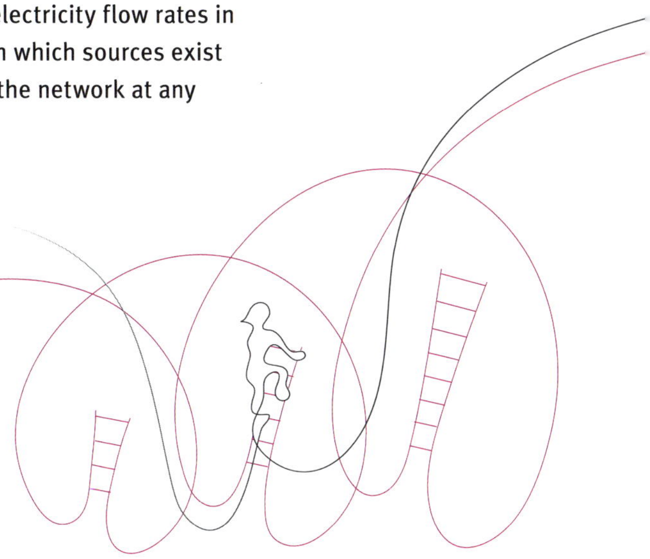

In a water supply network, the manholes are nodes and the water pipes are links. The manholes and pipes will be georeferenced and are stored as features with attributes in a geometric network. The attributes of manholes can include their height, size, and information about valves. The attributes of water pipes can be described in terms of their

152 Network analyses

4.17 The map shows the travel time from a point (Norwegian Home Guard centre) to various urban centres using a car, calculated on the basis of speed limits, ferry routes, etc, obtained from public databases.

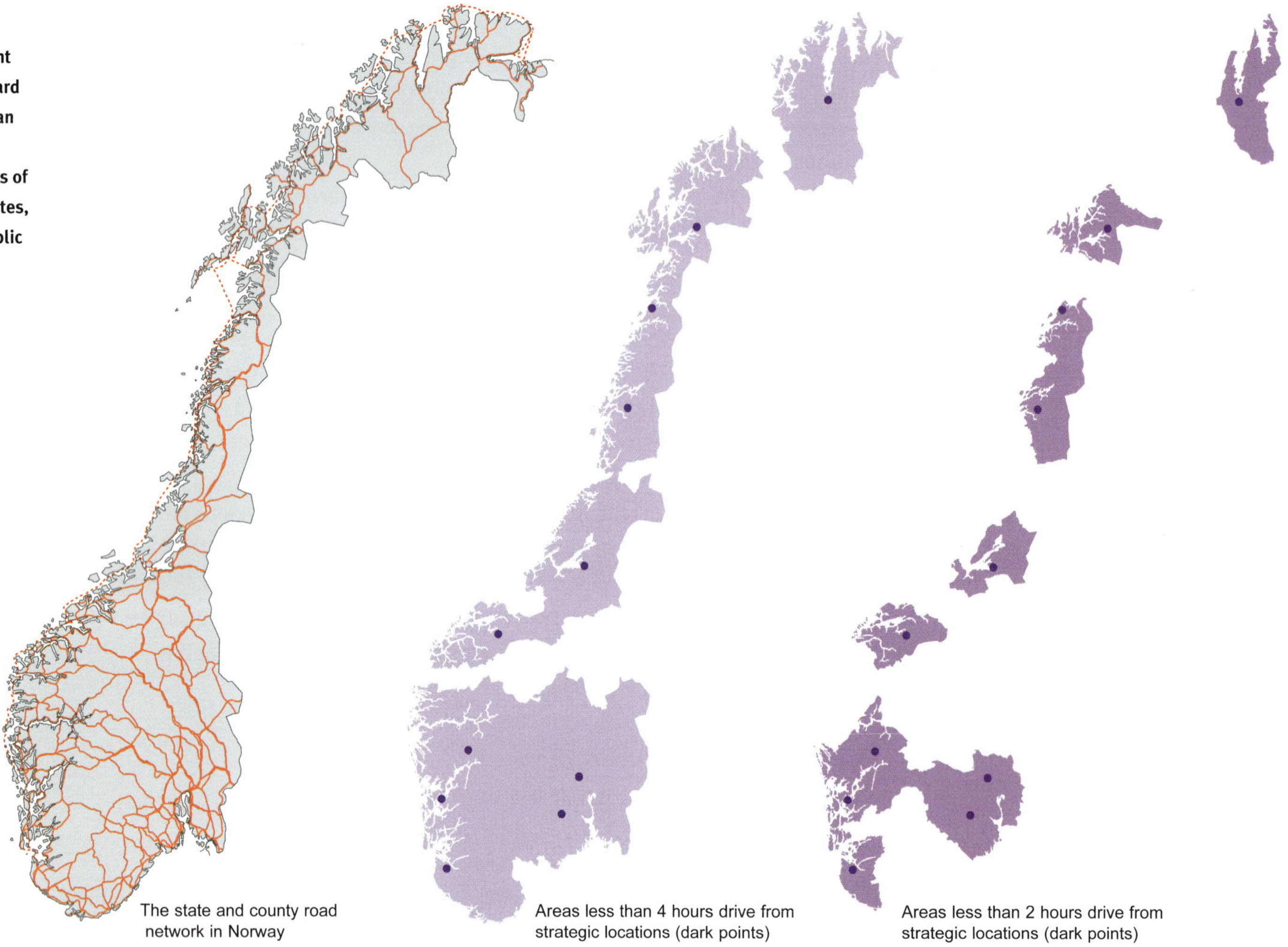

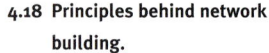

The state and county road network in Norway

Areas less than 4 hours drive from strategic locations (dark points)

Areas less than 2 hours drive from strategic locations (dark points)

4.18 Principles behind network building.

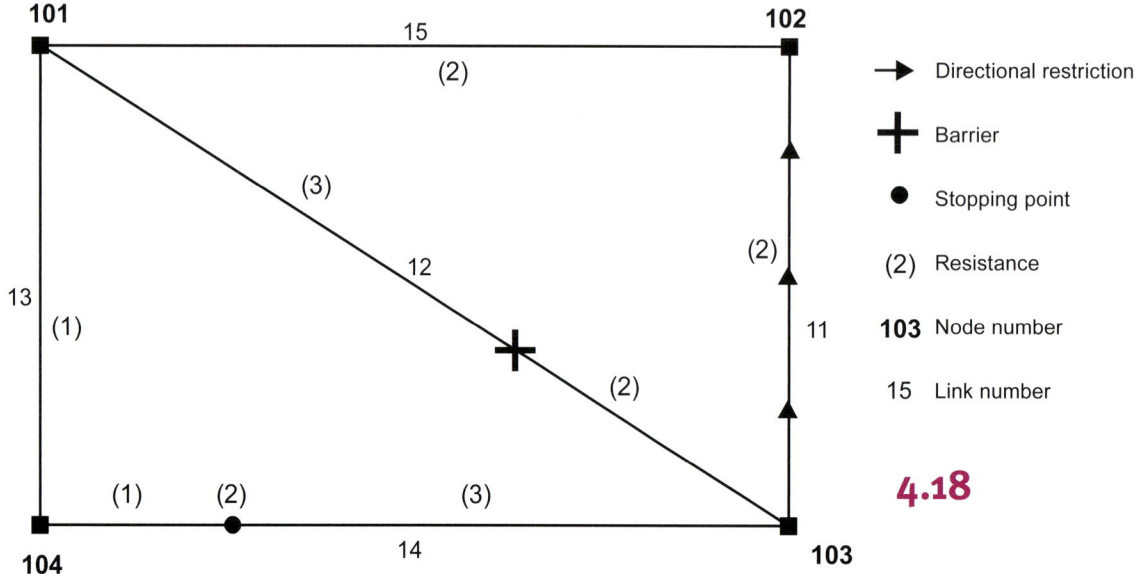

→ Directional restriction

✚ Barrier

● Stopping point

(2) Resistance

103 Node number

15 Link number

4.18

construction material, dimensions, and construction year. In addition, the nodes and links will have identity values that make it possible to incorporate them into a topological network. This means that it is possible to describe the manhole to which a water pipe is connected. In the event of a leak or a break in a water pipe, it will therefore be possible to analyse which parts of the water network will be cut off from the water supply. This is shown in Figure 4.19, which depicts a break in the older cast iron pipe between manhole 3 and manhole 4. This has resulted in a barrier in the water network. House 1 still has supplies of tap water, but house 2 has no water. The broken pipe' attributes indicate the manhole to which the pipe is connected, and this information can be used to determine which manhole to look for valve to close in order to shut off the water flow. In the depicted example, the water supply coming via manhole 3 has to be shut off.

NETWORK ANALYSIS

Many network analyses, including those for calculating the shortest routes, summarize the resistance in a network according to which route the network follows. An example is shown in Figure 4.18. In order to calculate the shortest route from node 102 to node 103 in Figure 4.18, we have to calculate the sum of the resistance along all of the possible routes between the two nodes. We therefore start at node 102 and first check along link 11, but this can only be followed from 103 to 102 because the route is designated for one-way traffic. We therefore check link 15 to node 101, and accumulate the resistance (2). From node 101, the route is followed along link 12 and the resistance (3) is accumulated. The total resistance is now 5. However, along link 12 there is a barrier that prevents the link from being followed all the way to node 103. Instead, we have to check whether it is possible to follow link 13 and accumulate the resistance (1). The total resistance up to node 104 is now 4. We continue in this way until all of the possible routes to node 103 have been checked. In the example, the route from 102 via 101 and 104 to 103 offers the least resistance. In the opposite direction, from node 103 to 102, link 11 (one-way traffic) could be used, and this route would therefore offer the least resistance. In this simple example, it is easy to see which route has the least resistance. However, in a complex network with millions of nodes and links, and with barriers such as one-way routes (e.g. the road network in Europe), it would be necessary to use a GIS tool to perform a similar analysis.

Optimized routes

An optimized route analysis is a type of network analysis used to find an optimal route based on the resistance in the network, as shown in the example in Figure 4.20. In a road network, the resistance can be represented by the length of a stretch of road or how long it takes to travel along a link. In some cases, other types of resistance may need to be optimized, such as costs. Also, road tolls, ferry charges, fuel, and other factors can count as forms of resistance in the network. Optimized route analyses can also include several points that are to be visited along the route. Such an analysis, involving multiple points, thus consists of a number of separate analyses, one for each pair of points. For example, an optimized trip from A to B via C, means finding an optimal route from A to C and then from C to B.

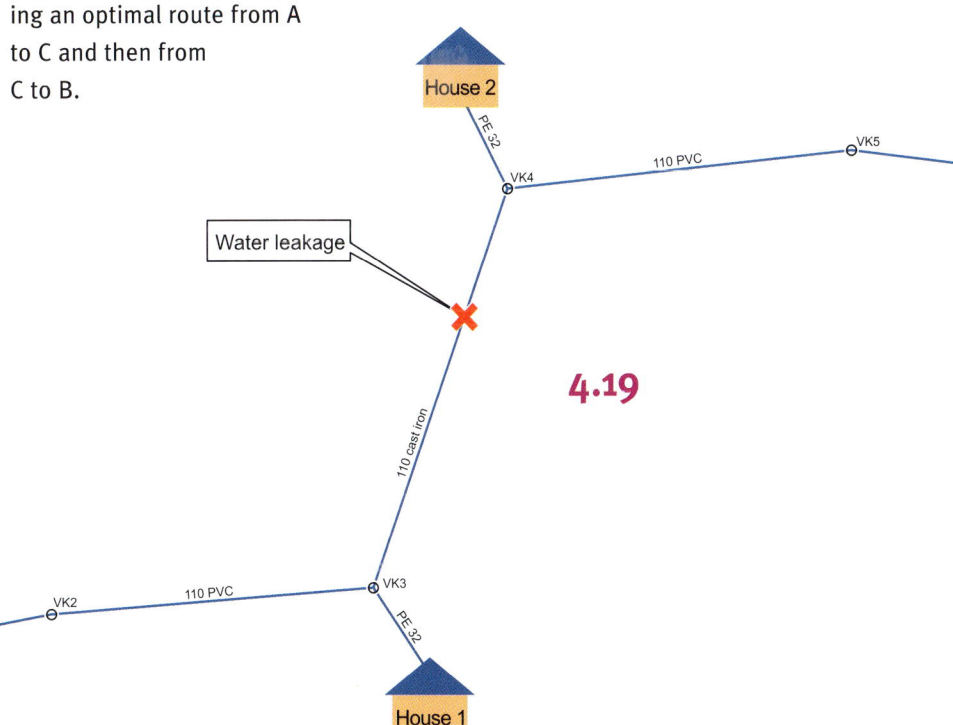

4.19 The map shows in which link and between which two nodes a water leakage has occurred.

154 Network analyses

Transport companies can compare the characteristics of one of their vehicles with the planned route's characteristics and a route planner will calculate a route that avoids obstacles. For example, a vehicle's height and weight can be compared to the heights of underpasses in road networks or a vehicle could be compared with axel weight limits on roads and bridges. A good route planner will find a route that fits the vehicle's characteristics, although the identified route might not be the fastest or shortest for all types of vehicles.

Network analysis can also be used to find the best route of one wants to visit several places without visiting them in any particular order.

Upstream and downstream analyses

From the descriptions above it is apparent that networks contain sources and end-users and that one can therefore define how something flows in a network. For example, the source of a particular type of pollution can be analysed to find out which end-users will be affected by it. One can analyse which nodes and links are located upstream and downstream from a given point. Network analyses can also be used to identify which areas will be affected by a discharge of pollutants (downstream analysis) or which areas the measured pollutants might originate from (upstream analysis). In addition, it is possible to analyse whether there are downstream or upstream connections between two selected points in the network.

Resource allocation

Another type of network analysis is the allocation of resources either to or from a central point. One example could be an analysis of which emergency calls should be assigned to which fire station in a city. For this purpose, a network calculation could be used to find the driving times from each fire station. The fire station with the quickest response times to a given area would then be assigned all of the emergency calls for that area. A further example could be which police patrol unit would be closest to a potential crime scene in terms of its response times.

COMPLEX NETWORKS

Most network analyses are carried out within a uniform network, such analyses of cars driven in a road network or the distribution of electricity in electricity networks. However, it is also possible for an analysis to cover multiple networks (e.g. multiple transport networks). This could be particularly relevant in cases when people want to travel from one place to another using a number of different modes of transport that use very different networks. Multiple transport networks could comprise combinations of the road network, bus routes, the railway network, ferry routes, tram routes, pedestrian networks, cycle paths, and flight routes. By analysing combinations of different networks, it will often be possible to identify faster routes than within one network, such as the road network (e.g. there are numerous pedestrianized routes or roads reserved solely for collective public transport), as shown in Figure 4.20.

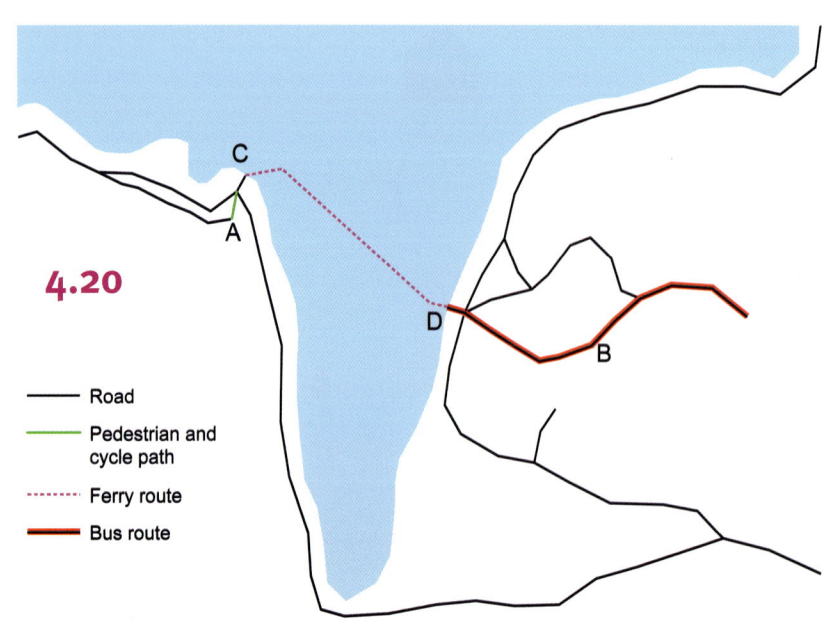

4.20 Complex network. The fastest way between A and B will be to cycle from A to C, take the ferry from C to D, and the bus from D to B. This assumes that the timetables and travel times information has been entered into the network as resistance. The online public transport planner service (www.trafikanten.no) in Oslo is a good example of the use of several such transport networks (bus, tram, underground, train, walking) for calculating the best public transport route.

Digital elevation models

Digital elevation models (DEMs) are divided into two subcategories: digital terrain models (DTMs) and digital surface models (DSMs). A DTM is a DEM that describes the ground surface, whereas a DSM is a DEM that describes a specific surface, such as vegetation and roofs. A DSM describes the situation more realistically.

A terrain model is a visualization of the surface of the terrain. Terrain models can be produced by cutting cardboard layers where cut edge of each layer corresponds to a contour. The layers are then glued into place, one on top of another, with the lowest layer representing the lowest contour. Using this method, one will end up with a physical model that can be used to visualize the surface of the terrain.

In order to create a DEM (DTM or DSM) using a computer, we have to store the coordinates X, Y, and Z of known points on the surface. We also have to decide how the z-value of any X or Y point outside the stored points should be calculated. This is called the *interpolation method* (described later in this chapter, in the section 'Storage of terrain models').
A DEM thus combines a set of known X, Y, and Z coordinates with a chosen interpolation method.

DTM applications
DTMs, which are 3D digital models of the terrain, enable advanced analyses and visualizations of the topography.

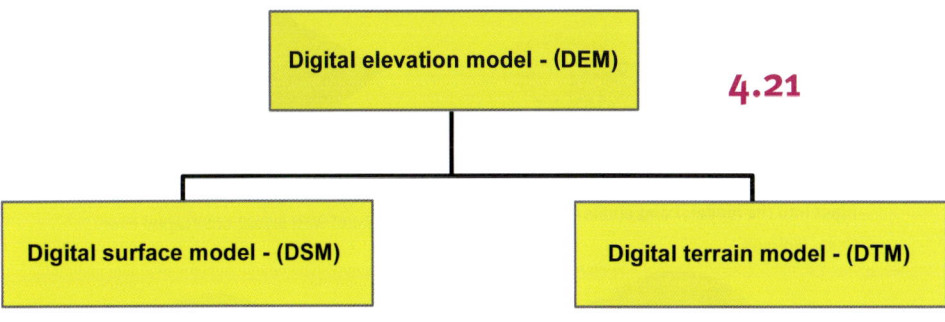

Some examples of their use include:

- Finding the elevation of the terrain
- Calculating the terrain's gradient and orientation
- Calculating what can be seen from a given point
- Calculating volumes
- Orthophoto production
- Road planning
- Calculating the coverage area of wireless communication
- Noise calculations
- Identifying areas prone to flooding
- Calculating how the air pressure varies following an explosion
- Finding exposure to sun and shade
- Reveal shading, to improve the visualization of the terrain
- Visualization of the terrain in simulators and computer games
- Virtual landscape images
- Animations.

4.21 Digital elevation models with subcategories.

4.22 The photo shows a terrain model, together with buildings, produced by gluing together thin layers of cardboard cut according to the area's contours.

4.23 The difference between a digital terrain model and digital surface model.

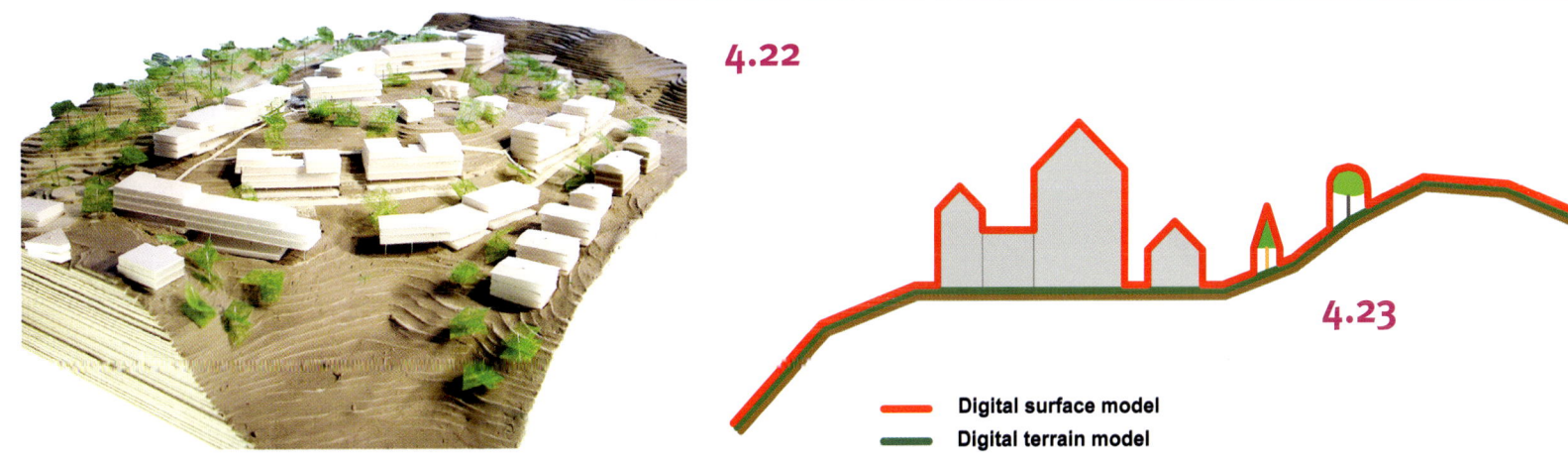

Digital elevation models

a

Input into the model. Point clusters, break lines, isolines.

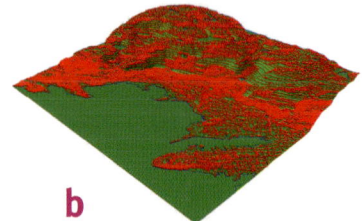

b

TIN model with "hard and soft" lines.

f

3D model with ortophoto draped over.

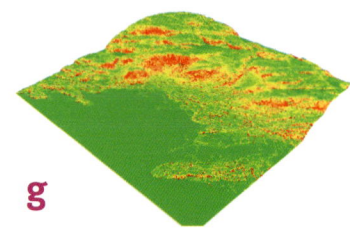

g

Slope analysis. Green symbolises flat sections, red symbolises steeper sections.

DSM applications

A DSM is a three-dimensional digital model that describes a specific surface in detail and can be used for advanced analyses and visualization of the surface. Some examples of the applications of DSMs are:

- volume calculations (e.g. for water reservoirs)
- clearance and monitoring power lines
- calculating what can be seen from a given point
- calculating how sunlight and shade falls in an area
- production of forest utilization plans and forest management
- cultural heritage mapping
- animations
- mapping erosion
- mapping obstacles affecting aviation
- urban planning

c

The actual TIN model.

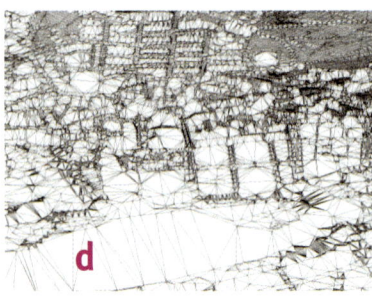

d

The triangle structure of a part of the TIN model.

h

Shadow analysis of the 3D model.

i

Sight analysis. The photo shows the expected view from the red point of the photo.

The applications of DTMs and DSMs often overlap. However, if, for example, a mass has to be calculated, DTMs must be used because the mass has to be calculated from the terrain, not from the surface. If there is to be an examination of obstacles that might affect aviation in the vicinity of an airport, DSMs must be used because the real heights of all objects above the terrain might present risks to air travel.

ESTABLISHMENT OF DEMS

Several types of data can be used to create a DEM. As in many cases, the quality of the data used as a basis will clearly determine the accuracy of the model. Today, mostly LiDAR data are used, but other more traditional ways are still used too. The following section describes the most common data types used to create a DEM, and the applications:

LiDAR and height points

Height points with known elevations can be used as a basis for making a DEM. The point layer can, for example, come from LiDAR data or from existing national reference database (Felles kartdatabase, FKB) elevation data. If it is not possible to find sufficiently good height-point data, a new round of data collection will be necessary, such as surveying with the use of a global navigation satellite system (GNSS). See also the descriptions

e

The TIN model as a completed raster model.

j

Part of the completed 3D model with ortophoto.

4.24 Example of 3D modelling and analyses of an area in Bergen.

4.24

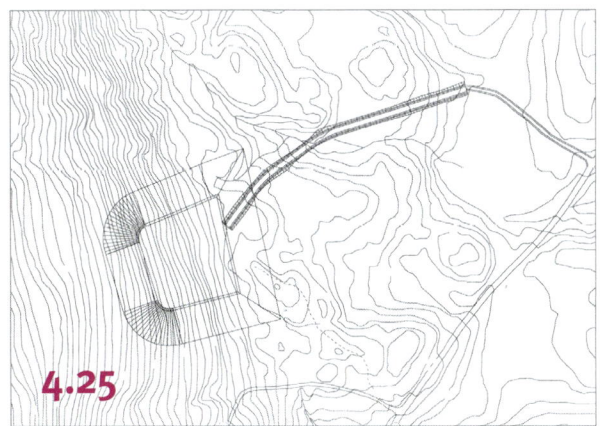

4.25

4.26

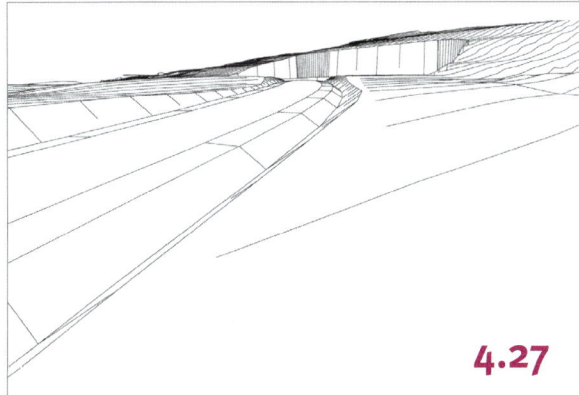

4.27

4.25 Visualisation of a new road and construction pit as a 3D map.

4.26 The same road and construction pit as in figure 5.16 but visualised with the aid of a terrain model.

4.27 Same as in figure 5.17, but from a different observation point and direction.

of data collection methods in Chapter 2 (Systems). If one measures the points oneself, the process should be planned: points should be measured at strategic locations in the terrain, such as where the terrain has natural breaks (e.g. valleys and along streams).

Grids
Grids are a commonly used method for producing a DEM. A grid is laid horizontally over the terrain surface layer and data points are selected at each of the grid intersections. By storing the z-values in a grid they can also be stored as a raster graphics image. The disadvantage of this method is that it does not take account of the variations in the terrain's surface. This can be corrected by using grid squares of different sizes: small grid squares can be used for very hilly areas and larger ones for less hilly terrain.

Contours
Contours have traditionally been the most commonly used data for generating DTMs. A contour line is a graph line in which the z-values are constant along

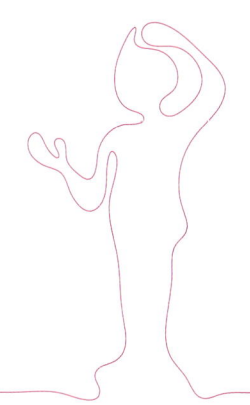

4.28

4.29

4.28 Point cloud. Elevation data from Felles kartdabase (FKB) – Norway's common map database with very detailed map data.

4.29 A point cloud with data produced using LiDAR technology.

4.30 An example of a grid. The size of the grid in the figure on the left is fixed; the size of the grid in the figure on the right varies.

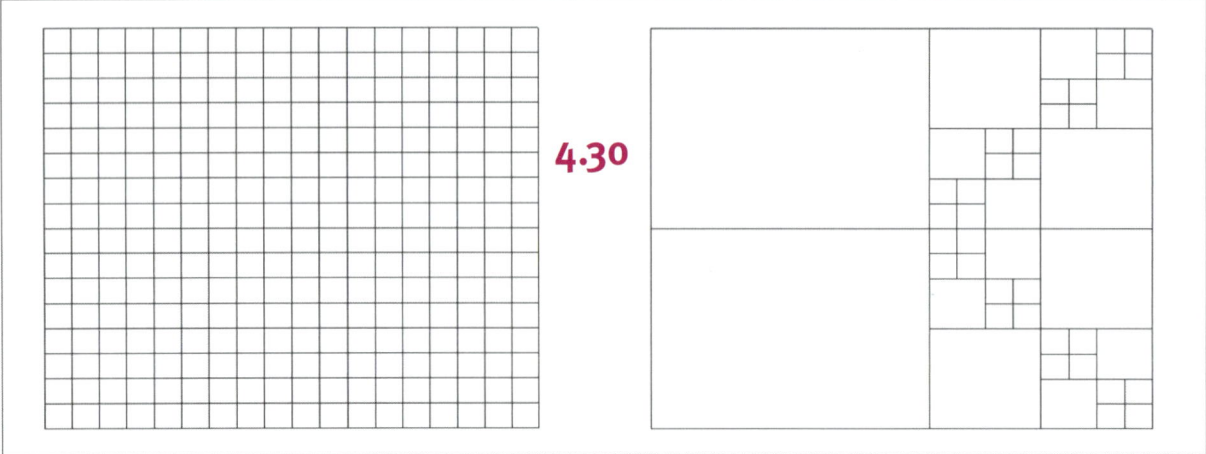

the entire curve. Contours are readily available data and are used in, for example, the national reference database (FKB) in Norway. The disadvantage of using contours is that special formations in the terrain between the contours will not be shown in the model. However, such special formations will be represented in the model if the vertical interval between the contours (i.e. the contour interval) is reduced. Hence, 1 m contour intervals will be better than 5 m contour intervals for capturing various formations in the terrain.

Parallel profiles

Parallel profiles can be used to create DTMs. The profiles consist of parallel lines along which the z-values are measured. In this case too, it is important to select points that will best reflect the physical terrain. The parallel profiles method is suitable for planning road projects.

Terrain lines

Terrain lines are used to create DTMs. In contrast to contours, where the z-value is constant along the curve, the z-value varies along the terrain line. Such lines are often

4.31 An example of isolines. Note that the elevation is the same for all the points along an isoline.

4.32 An example of parallel profiles.

4.33 An example of terrain lines. Note in particular that the elevation varies from point to point along the terrain line.

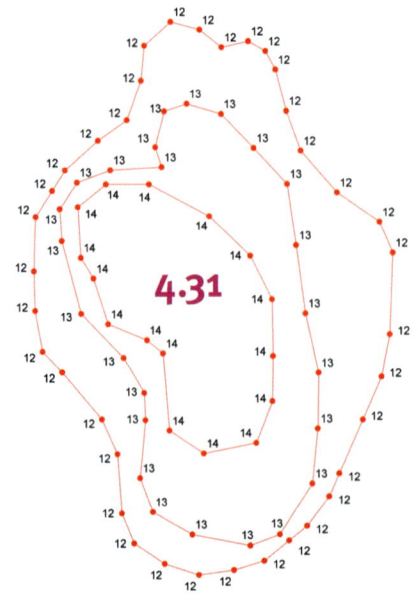

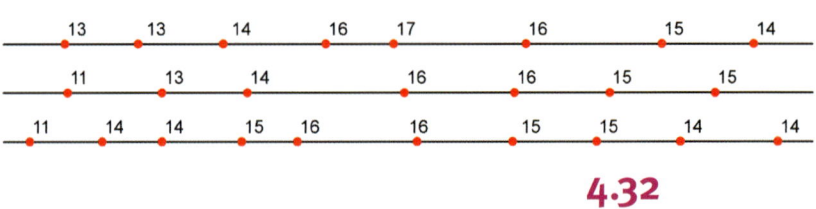

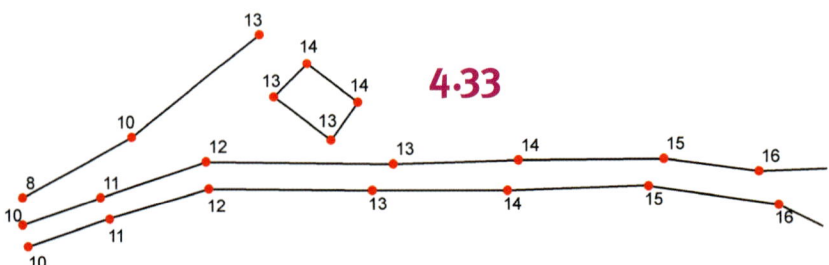

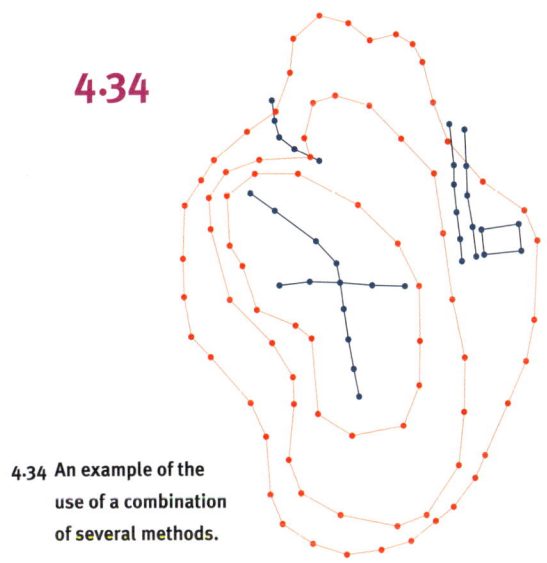

4.34 An example of the use of a combination of several methods.

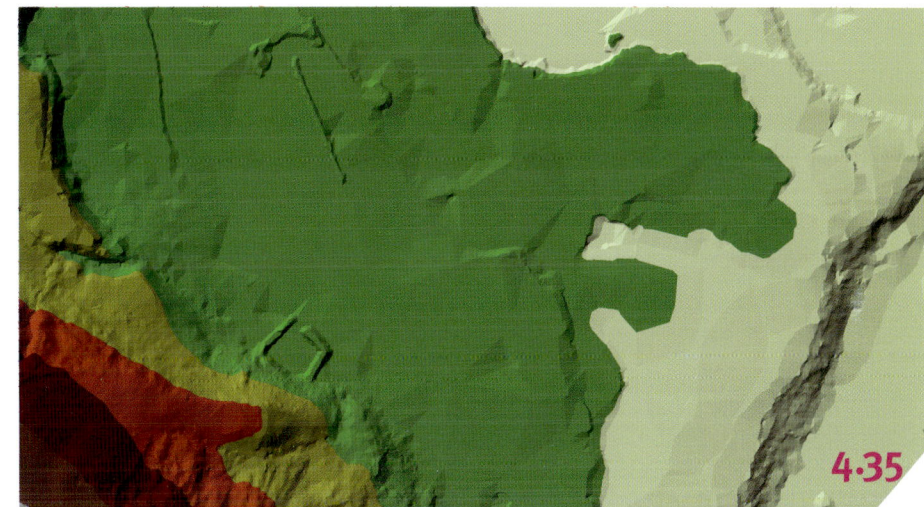

constructed along frangible lines in the terrain, such as the bottom of ditches and along roadsides.

Combined base height layers
As the subsection heading suggests, two or more of the above-mentioned methods are used in combination. In this case, the strongest components of two or more of the methods are used and the final model will be more accurate. For example, it is very common to use contours, spot heights, and terrain lines – all of which form part of the data in Norway's national reference database (FKB) – as input values for DEMs.

When using combined methods, it is also common either to use surfaces with a known elevation (e.g. lakes) as input values for heights in the model or to use a surface boundary for the area covered by the model.

4.35 Example of a DTM of Jørstadmoen military base in Lillehammer, produced from elevation contours.

4.36 Example of a DTM produced from LiDAR data classified as ground points over the same area as shown in Figure 4.35. Note that the details in the terrain show up better when using LiDAR data instead of hypsometric curves, especially on the flatter areas in the model.

4.37 Example of a DSM produced from LiDAR data for the same area as in Figures 4.35 and 4.36. Note especially that vegetation and buildings are also visible in this model.

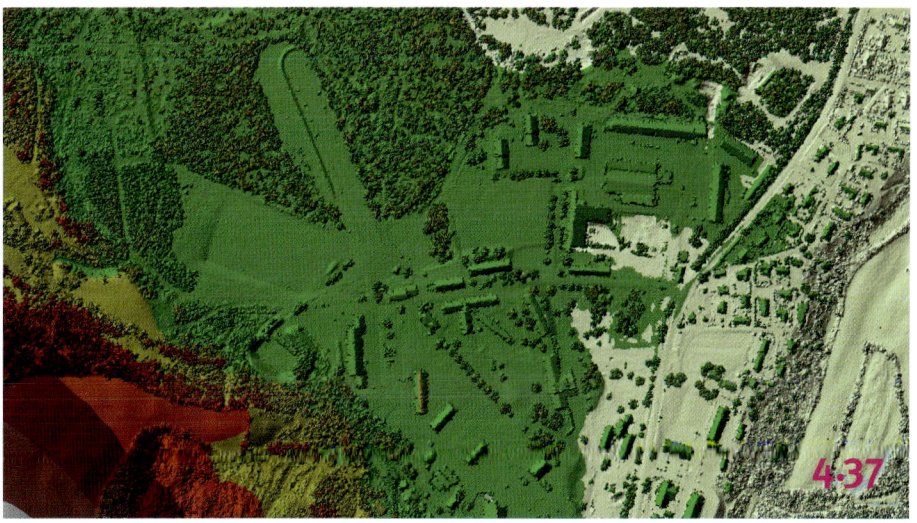

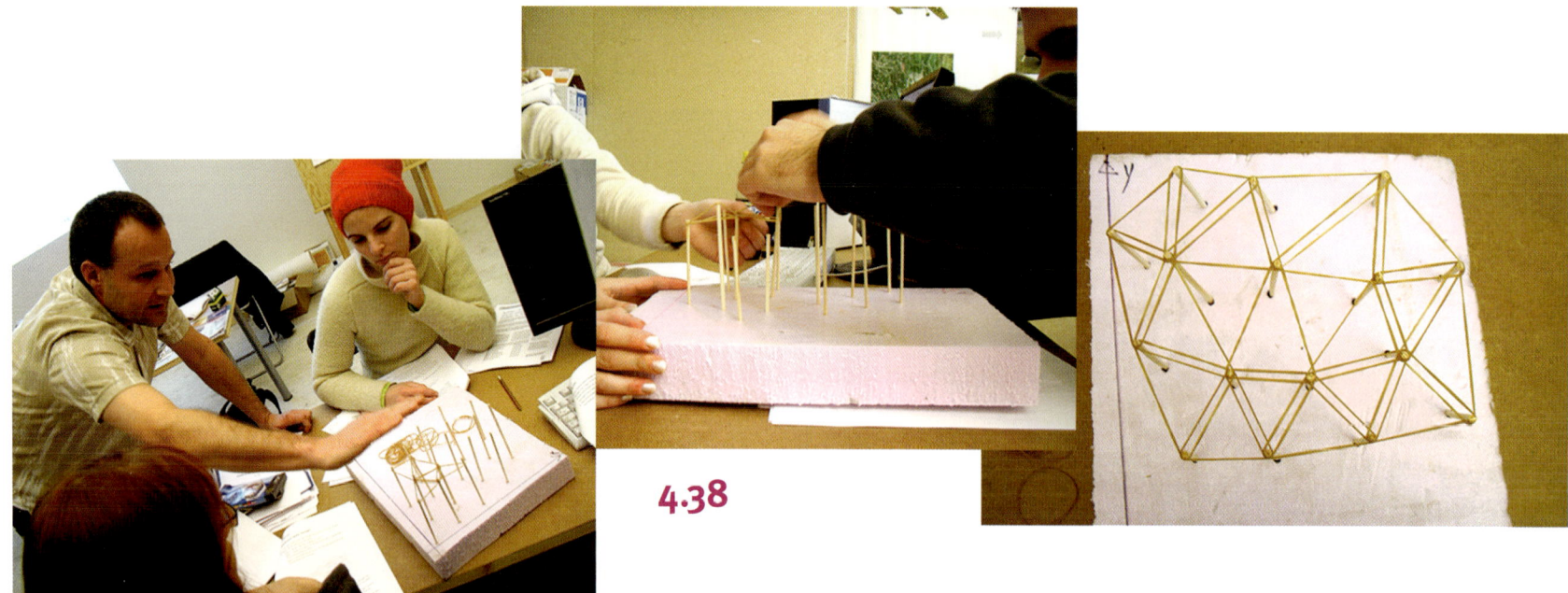

4.38

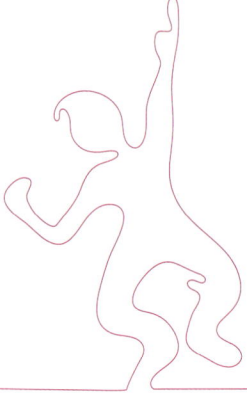

4.38 Students from the Bergen School of Architecture studying the principles of triangulation by testing physical models.

DATA STORAGE IN TERRAIN MODELS

A DEM is usually stored either as a raster model or as a *triangulated irregular network* (TIN). Raster models are described in Chapter 2 (Systems). TIN models are described in more detail later in the present chapter (in the section 'The TIN model – dividing into triangles'), with brief description of the differences between raster and TIN, and when they are normally used. However, first a general description of storage in elevation models is given.

A DEM stores information about height values of a number of points in the area represented by the model. The points may be distributed regularly and stored in a raster or stored as irregularly distributed point such as a TIN. In addition to the stored items, there also has to be a method for estimating the elevation of every other point within the area covered by the model. The method used for this purpose is *interpolation*. Interpolation is based on the assumption that spatially distributed objects also are spatially correlated. In other words, things that are close together often share the same characteristics. For example, it would be natural to assume with a high degree of certainty that if it is snowing on one side of a road it will also be snowing on the other side. However, there would be less certainty about whether it is also snowing on the other side of town. In the context of elevation models, this means that is highly probable that the height of an arbitrary point in a model will correlate with the height of the neighbouring points for which heights have been registered.

Interpolation methods

There are many interpolation methods, but all of them make assumptions about how new values can be calculated. In all cases, the more input values one has and the more evenly they will be distributed, the more reliable the interpolation will be. Interpolation methods include:

- Inverse distance weighted (IDW)
- Natural neighbours
- Spline
- Kriging
- Trend
- Linear interpolation

Inverse distance weighted interpolation means that the value of a point is estimated from the weighted average calculated from a corresponding value of a sample of other points in the data set. The sample can be the entire data set or just the points within a radius of the point to be estimated. Each point is assigned a weight equal to the inverse distance from the point to be estimated. This means that points close to the point to be estimated will make a large contribution to the point's estimated value, whereas points at a distance will make a small contribution to its value.

Natural neighbour interpolation, in common with inverse distance weighted interpolation, uses a weighted average, but the weighting and sampling are done in a different way. In this case, the group of points used for interpolating the unknown value of the new point is determined by finding the point's neighbours. This is done by splitting the model surface area into smaller surfaces so that each surface contains one point from the model, and any place that falls within the surface will be closer to this point than to any of the other points. The points in each surface adjacent to the surface with the point to be estimated are used for further selection. The area of each of these neighbouring surfaces is used as weight for each of these points contribution to the estimated value. An advantage of this method is that it is not necessary to specify any parameters for the search radius for weighting according to distance.

In *spline* interpolation, a mathematical surface is adapted to all input values. This surface is required to pass through all of the points and at the same time the curvature of the surface should be as small as possible. The method tends to overestimate values if they differ greatly over short distances.

In *kriging*, it is assumed that the distance or the direction between the sampled points reflects a spatial correlation that explains the variation in the surface of the terrain. The method approximates a mathematical function for a group of input points and thereafter the elevations of new points are calculated from this function. Kriging entails having to carry out statistical analyses of the input values. The mathematical and statistical background to kriging is not discussed here, but it is important to know that the method is most suitable when it is known that the data contain spatially correlated distances or orientations. This tends to be the case in analyses of soils and geology. Kriging is seldom used as an interpolation method for DEM models.

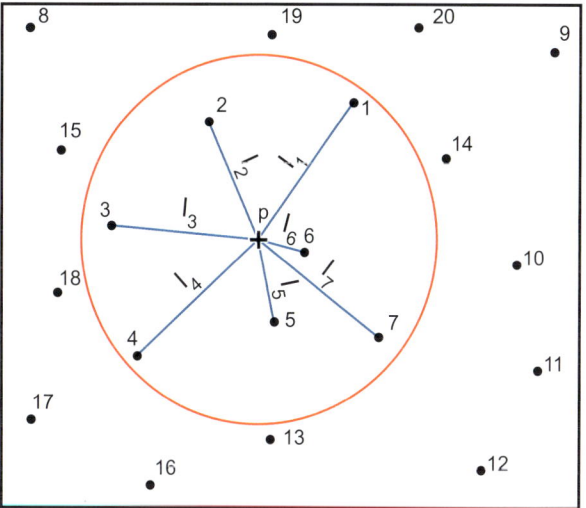

4.39 Distance weighted interpolation. The estimated height of point p is found using [equation]

$$h_p = \frac{\sum_{i=1}^{7} h_i/l_i}{\sum_{i=1}^{7} 1/l_i}$$

4.39

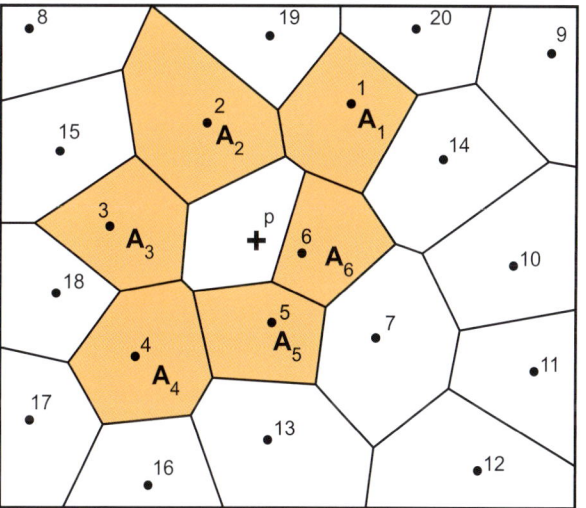

4.40 Natural neighbours: estimated height of point p is found using: [equation]

$$h_p = \frac{\sum_{i=1}^{6} h_i A_i}{\sum_{i=1}^{6} A_i}$$

4.40

Trend, in common with some of the above-mentioned interpolation methods, approximates to a mathematical function of all of the input values. The function is a polynomial of a specific order – an 'nth degree' surface. This is done by choosing the function of the nth order, in which the sum of the square of all deviations between the function and actual input values are the smallest possible. This is called the method of least squares. As a rule, this means that the estimated values will not be equal to the input values and the resulting curve will not pass through all of the input values.

In a *linear* interpolation, linear functions are used between the points. For example, it will be possible to

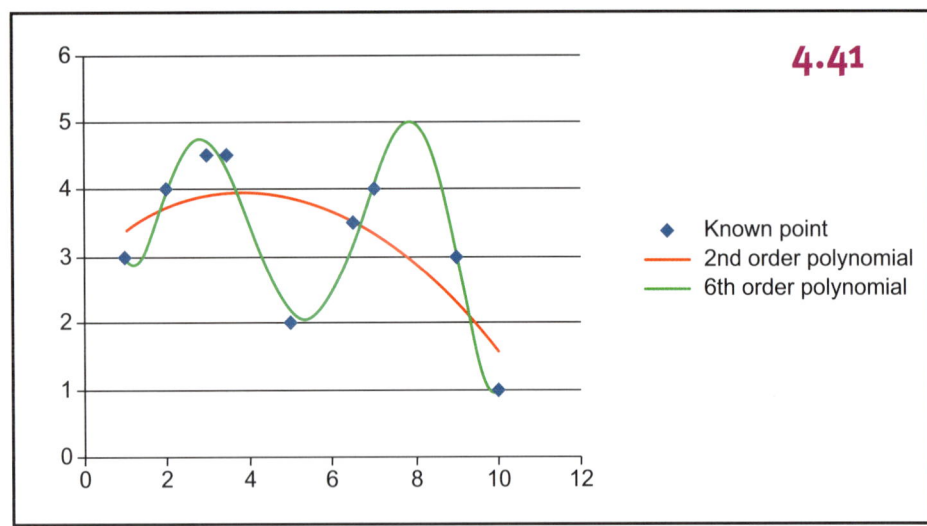

4.41 Trend interpolation with 2nd and 6th order polynomials.

add a planar linear surface between three points in a space, and then use this linear surface as an image of the surface, as surface interpolation. This interpolation method is very simple in terms of mathematics, and makes it both easy and quick to calculate the height value of a point outside the stored points. The method can result in errors if not used carefully. However, with a dense distribution of points in areas where elevation values vary widely and with less dense distributions in areas where there is less variation, it is a good and simple method and one that is widely used. Not least, the method is used in connection with storage of TINs, as described in the next section.

The choice of interpolation method will depend on the quality of the available data. If good quality data are available, a less accurate interpolation method that results in faster processing times than more accurate methods can be selected. Time is thus a factor that is taken into account when choosing an interpolation method. Also, the accuracy of the various interpolation methods should also be considered before one is selected. There is a correlation between time and accuracy: as a rule, faster methods are less accurate and slower methods tend to be more accurate. Good data in combination with an accurate interpolation method will generally yield the best results.

From any given base layer with elevation points, interpolation methods can be used to estimate the elevation of new points in a regular grid pattern and store these new elevation values in a raster grid model.

THE TIN MODEL AND TRIANGULATION

A very common method for producing a height model of irregular basis points is the TIN model. A TIN model can also be developed by using lines and surfaces. Moreover, not all of the input data will need to have an elevation value. For example, one can input lines without a height value to describe the boundary of the model. Further, one can input different types of data into the model, depending upon the software being used. Examples include break lines that typically represent a stream or a road where the terrain's gradient changes.

Polygons such as water polygons can be added at a constant height or used to refine the model. A TIN model divides a surface into triangles. A major advantage of this model is that there can be many points in areas where there are frequent changes in terrain gradient. By contrast, in areas where there are less frequent changes there can be fewer points. The triangulation model does not have problems with interpolation at *saddle points* (see Figure 4.42). Saddle points cause problems by calculating the contour lines on the basis of a grid model, and since a TIN model consists of triangles, a contour line can only cross two of the three sides of any triangle. The fact that a triangle defines a plane makes it easy to calculate both the gradient and other terrain-specific parameters for each triangle. Such information about the terrain is often sought after for analytical purposes.

However, one problem with the TIN model is that there are many different solutions when it comes to dividing a surface into triangles and consequently there are also many different approaches. The most commonly used method is Delaunay triangulation. The principle behind this method is that for each triangle a circle can be constructed that passes through

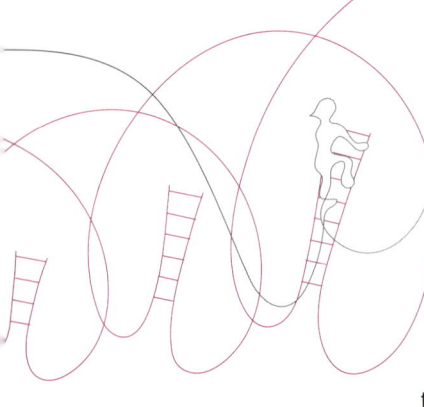

a

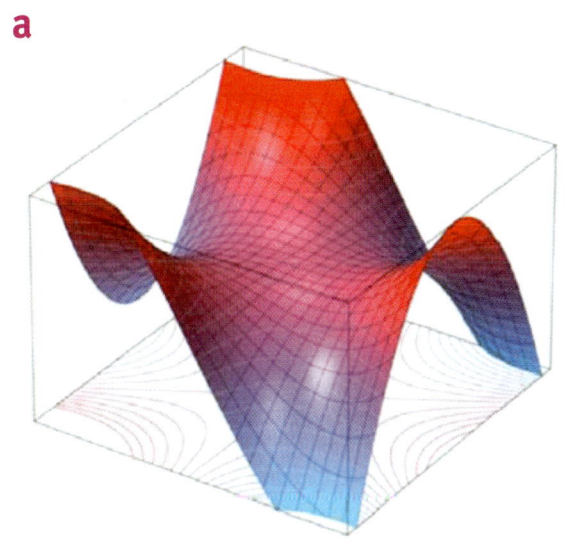

b

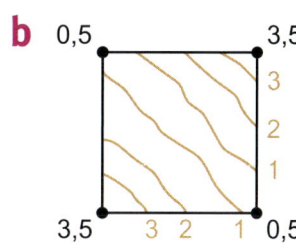 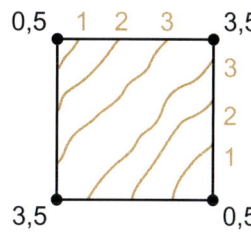 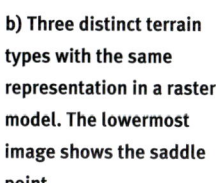

4.42

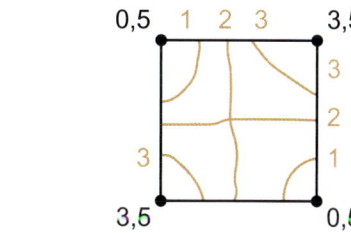

4.42 a) A saddle point.

b) Three distinct terrain types with the same representation in a raster model. The lowermost image shows the saddle point.

the three points constituting the corners (vertices) of the triangle. The circular surface that is formed should not contain any points other than the three corners.

THE RASTER MODEL VERSUS THE TIN MODEL

One advantage of the TIN model is that it is irregular. In other words, the density of the points in a TIN model can vary, such that areas in which the surface varies a lot can be full of points, but in areas where there is less variation the distance between the points can be greater. Another advantage is that the features that are used to create a TIN model are kept in the same

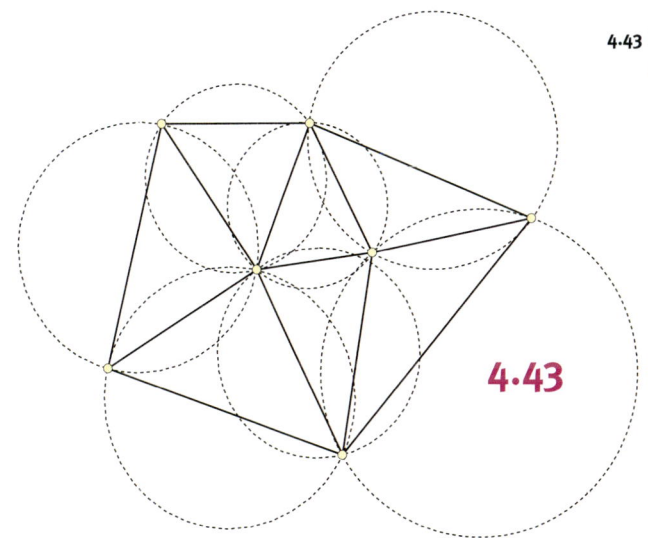

4.43 Example of Delaunay triangulation.

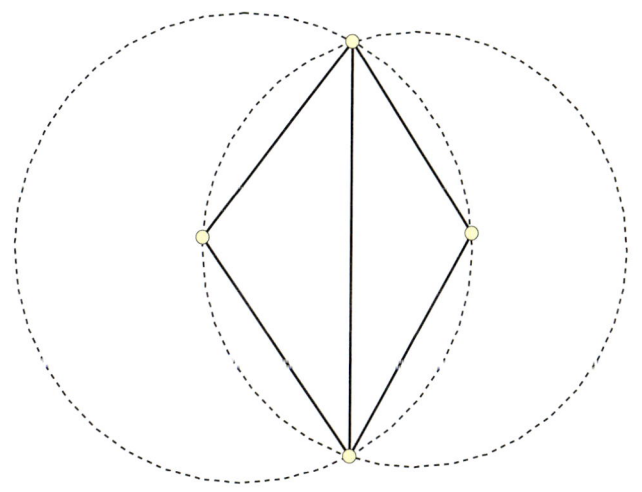

4.44

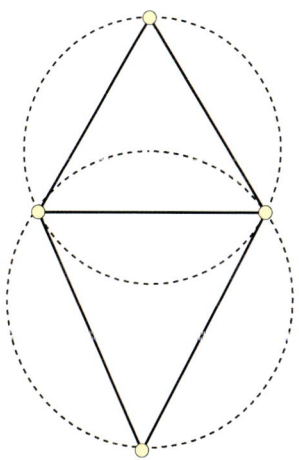

4.44 Delaunay triangulation. The triangulation to the left doesn't satisfie the Delaunay condition, the one to the right does.

4.45

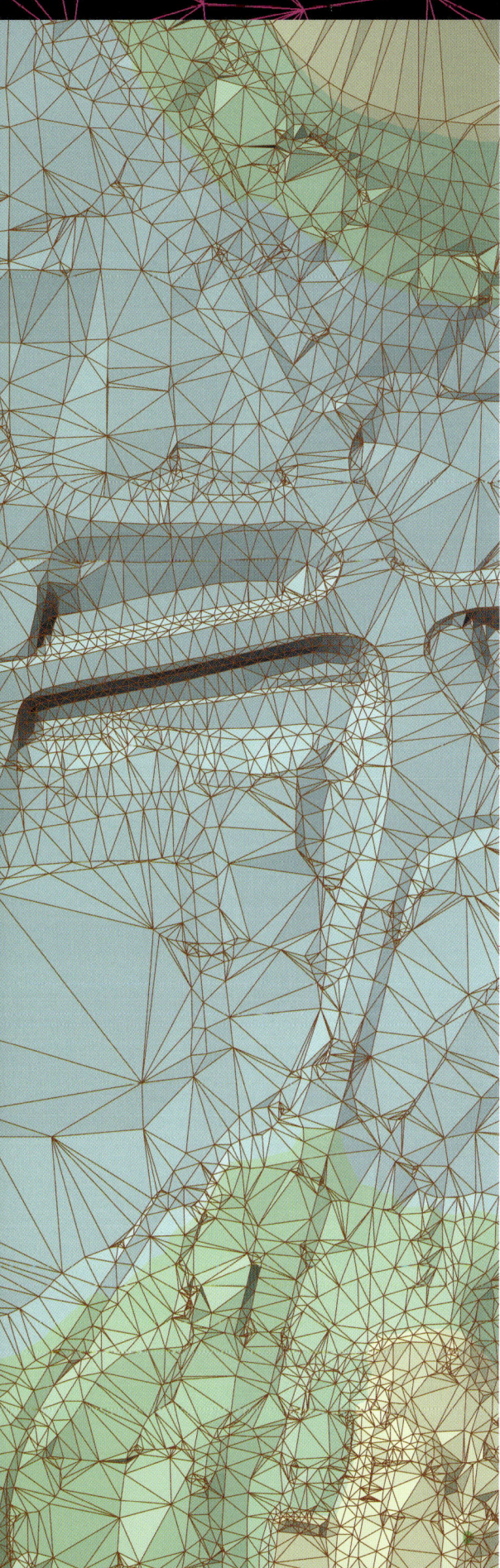

position such as the corners or edges of the triangles in the model.

Processing large TIN models (i.e. models of large areas or models based on very detailed base layers) requires a lot of computing resources and can take a long time. TIN models are therefore often used for smaller areas, whereas raster models are used for larger areas. Large TIN models are often converted to raster models for further use.

Practical analysis example

This example shows how the above-described different analytical methods can be used in practice to solve various problems. The example includes both vector and raster data, and examines spatial data manipulation, network analysis, and the use of DTMs. The example is based on a fictitious scenario that has formed the basis of an examination in engineering at the Norwegian Military Academy.

A cabin is to be constructed somewhere in Rømskog Municipality. N250 map data are available for the analysis. A greyscale map provided by Geodata AS is used as a background map for some of the results. The developer has stipulated the following criteria for the cabin's location:

- at least 1 km from other buildings
- at least 1 km from all county roads
- not be built on a marsh or water
- as close as possible to a tarn, but no closer than 100 m.

The map in Fig. 4.46 shows the actual area for the location of the cabin according to the above criteria.

The first step in the analysis is to perform a buffer analysis of 1 km around the buildings in the area. This is done by first selecting the buildings' points in Rømskog Municipality and then buffering them. The results are shown in Figure 4.47, with yellow circles representing the buffer distance from buildings. The yellow zones indicate areas that are **not suitable** for the cabin.

The second step is to add a buffer of 1 km the around county roads in Rømskog Municipality. The roads are

4.45 Part of a DTM, showing how triangles are formed to reproduce the terrain.

Practical analysis example

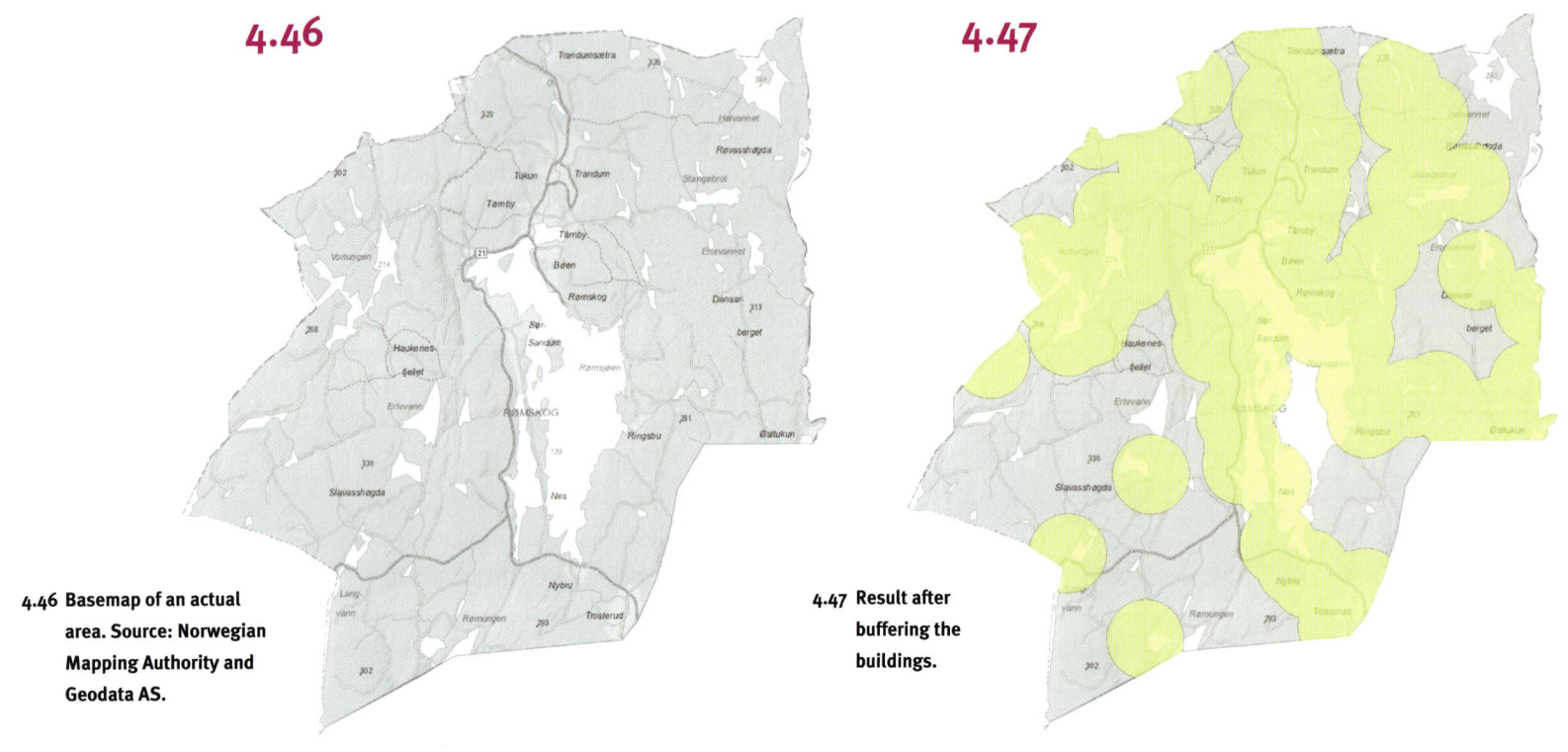

4.46 Basemap of an actual area. Source: Norwegian Mapping Authority and Geodata AS.

4.47 Result after buffering the buildings.

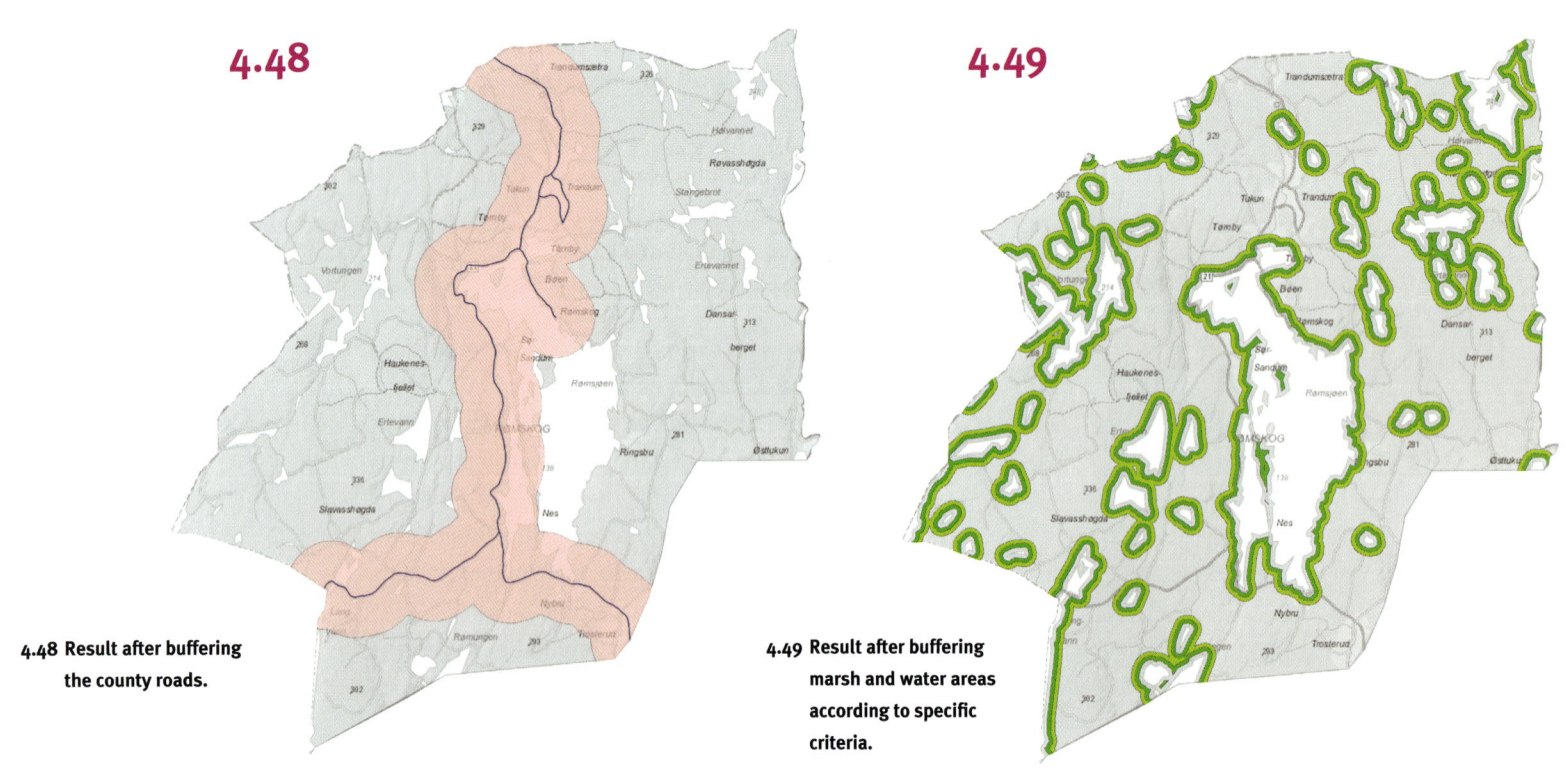

4.48 Result after buffering the county roads.

4.49 Result after buffering marsh and water areas according to specific criteria.

selected by using the attributes that indicate which road lines are located in the municipality and the type of road line they represent. Thereafter, the buffer analysis is performed for the selected roads. The results of the buffer analysis are shown as pink zone around the county roads in Figure 4.48. The pink red zone shows areas that are **unsuitable** for the cabin.

Finally, the last two criteria have to be processed. Once again, first marsh and water within the municipality are selected. Thereafter, a buffer analysis is performed on this sample. It is decided that a distance of 300 m to the marsh and water is not a relevant area. The buffer analysis is performed for three 100 m zones. The innermost zone (0–100 m) is not relevant and it is removed by sampling various buffer zones. The results of the analysis are shown in the map in Figure 4.49, where dark green areas represent a 100–200 m zone from marsh and water and light green areas represent a 200–300 m zone from marsh and water. The dark green and light green zones now show areas that **might be suitable** for the last two criteria.

So far, three buffer analyses have been performed, resulting in three different themes. Two of the theme layers show areas that are not suitable for building cabins, but one shows an area that **might be suitable**. The areas in the theme layer that **might be suitable** should not overlap the two unsuitable areas. There are several ways to proceed in order to find areas that would satisfy all of the criteria. One way would be to merge the two unsuitable areas of a theme layer and make a common theme layer for unsuitable areas. Either the union or merge method could be used, as described in the section on spatial data manipulation (earlier in this chapter). The results of merging are shown in Figure 4.50, where the light blue zone shows the merged unsuitable areas.

The next step is to find the areas that are unsuitable in one theme but suitable in another theme. This can be done using the spatial data manipulation method 'erase'. The method is based on the appropriate theme layer (Figure 4.49), and then removes the areas where this theme overlaps with the unsuitable theme layer (Figure 4.50). The result will then be a layer containing

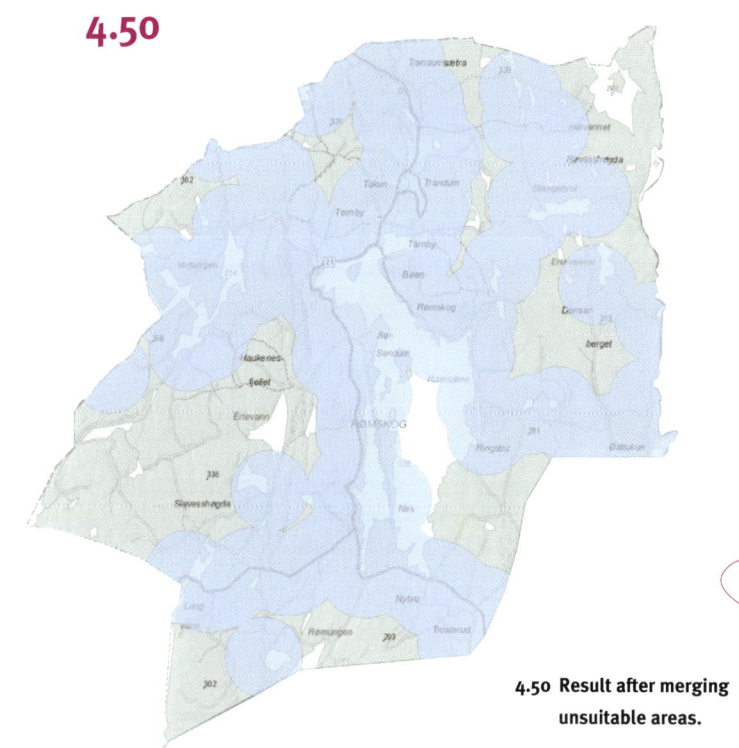

4.50 Result after merging unsuitable areas.

the areas that satisfy all of the criteria in the analysis. A simple map onto which the results have been plotted is shown in Figure 4.51. Prior to the work to find a suitable plot for the cabin, the client comes up with several new criteria for the location: a sunny place is desired and a suitable driving distance from the nearest grocery store. The new criteria assume that:

- the gradient should be south-east, south, or south-west
- the driving time to a grocery store in Rømskog should preferably be 10–15 minutes, but could be a little as 5–10 minutes.

This means that the scope of the analysis has to be widened in order to find suitable sites. The results of the analysis will depend upon the original criteria still being valid, but the theme layers uses will be converted to raster layers. The reason for this is that a later gradient analysis will result in a raster layer. The principles of overlay analysis of raster data are described in the section on spatial data manipulation.

Practical analysis example

Possible cabin areas

First, a DTM of the area is created to find the terrain gradient and the gradient orientation. Since N250 map data only contains contours at 100 m intervals (100 m equidistant), better terrain data will need to be obtained for the model. In the example, 10 m contour lines have been used. This model is not particularly accurate, but it is good enough for the example. After the DTM has been created, it can be subjected to a gradient analysis.

Thereafter, areas with a south-east, south, or south-west gradient will be merged to form suitable areas, while the remaining areas will be unsuitable. The results of the merging are shown in Figure 4.55.

The next step is to find the location of grocery stores in Rømskog. For the example, only one store is selected: Joker Rømskog.

The maps shows the result of an analysis of where the developer's criteria are met for building the new cabin
Suitable areas are colour-coded according to the distance of the nearest tarn

The following criteria were used in the analysis:
The cabin should:
- be at least 1 km from the nearest building
- be 1 m from all county roads
- not float on water or be built on a marsh
- be as close to a tarn as possible beyond a 100 m zone

Suitable areas
- 100 - 200 metres
- 200 - 300 metres

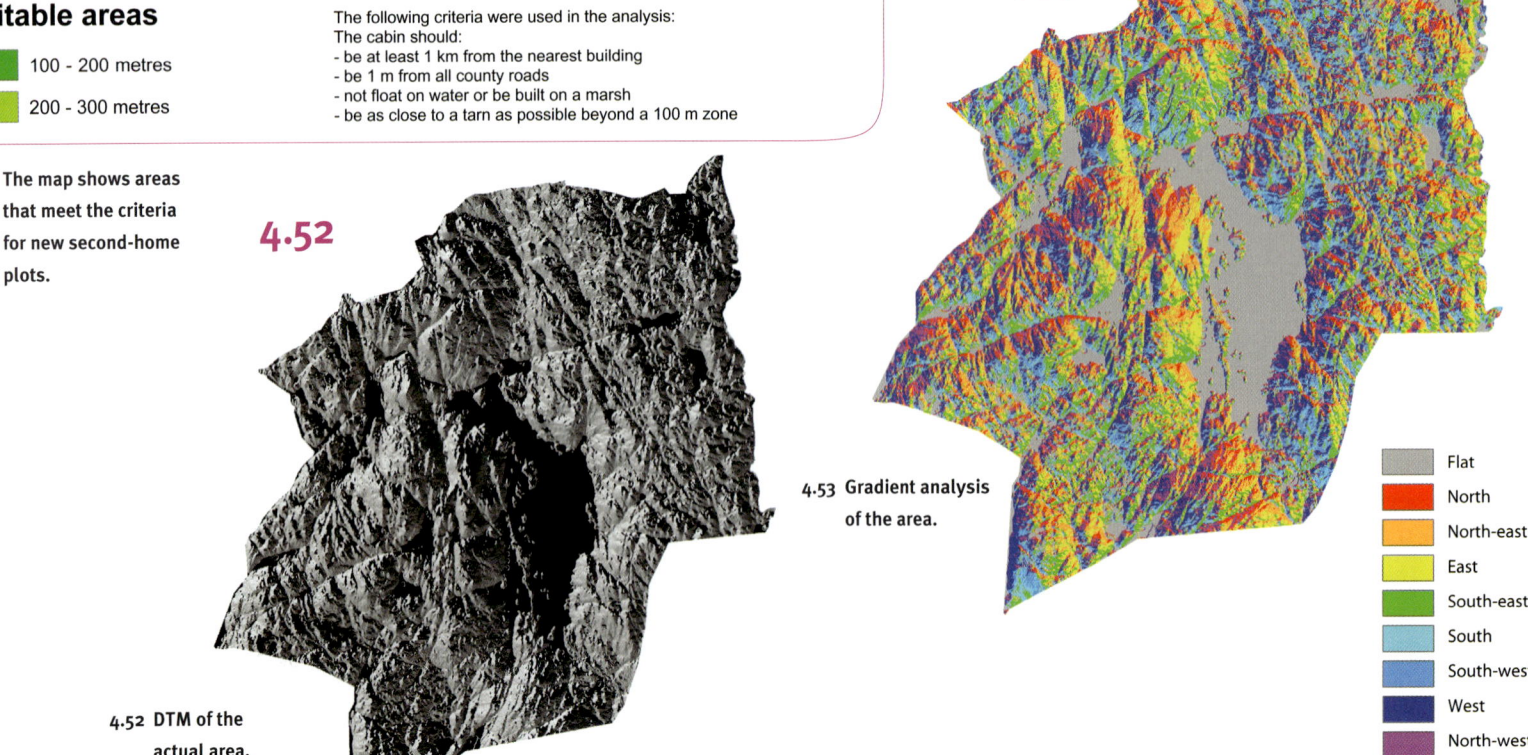

4.51 The map shows areas that meet the criteria for new second-home plots.

4.52 DTM of the actual area.

4.53 Gradient analysis of the area.

Legend: Flat, North, North-east, East, South-east, South, South-west, West, North-west

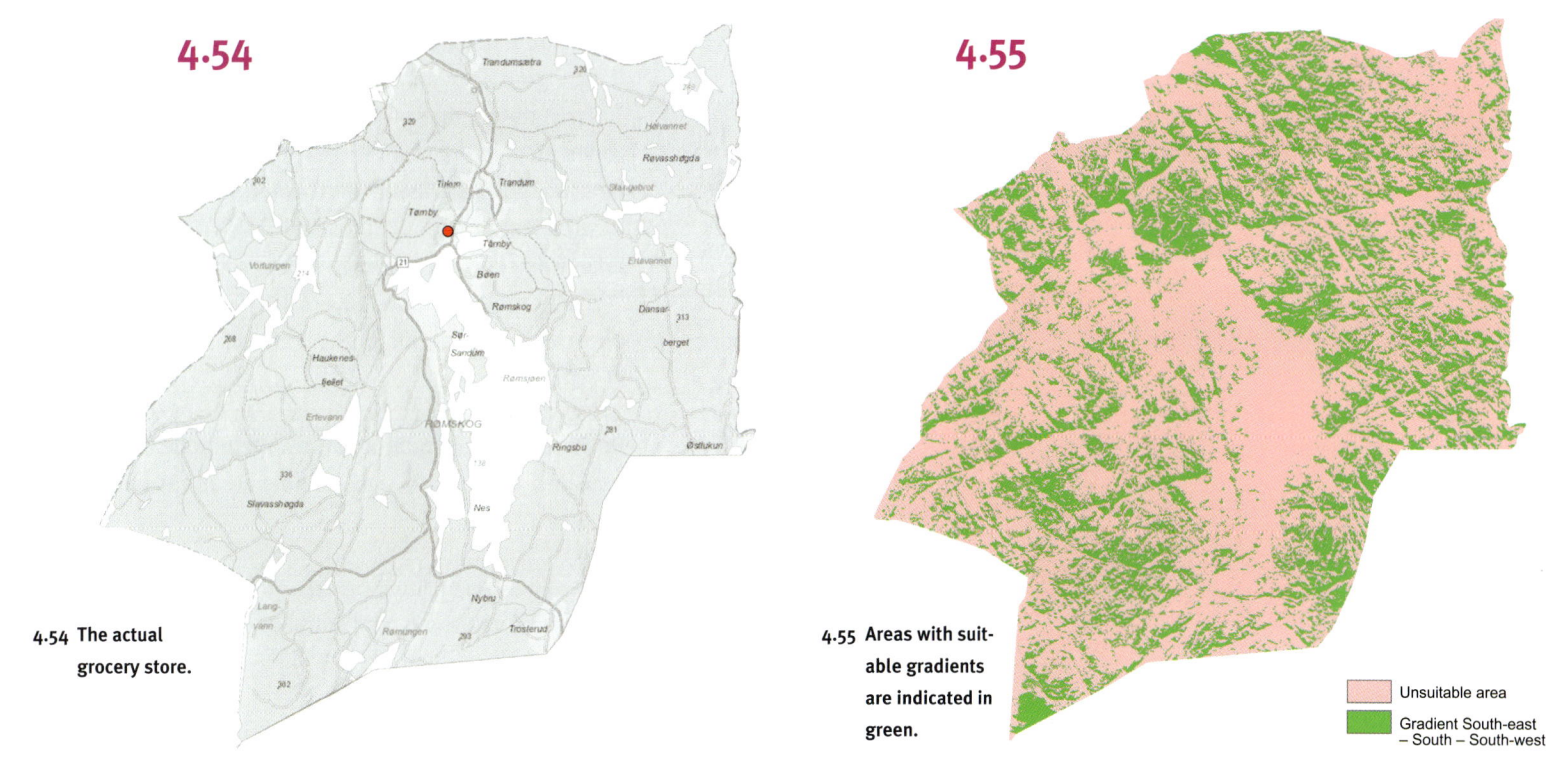

4.54 The actual grocery store.

4.55 Areas with suitable gradients are indicated in green.

Unsuitable area

Gradient South-east – South – South-west

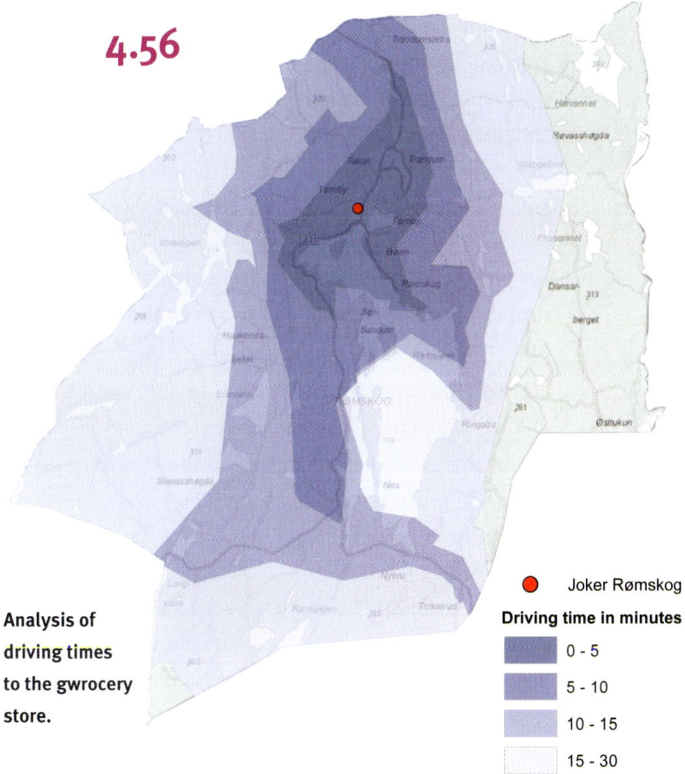

4.56 Analysis of driving times to the gwrocery store.

● Joker Rømskog
Driving time in minutes
- 0 - 5
- 5 - 10
- 10 - 15
- 15 - 30

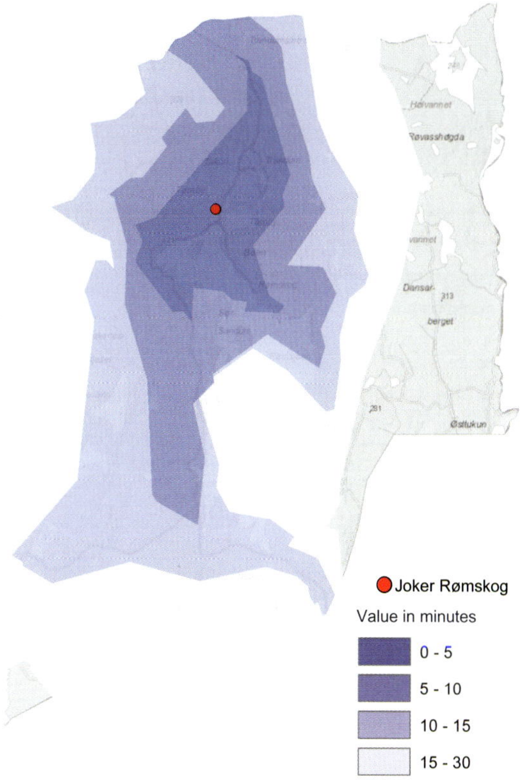

4.57 Figure 4.56 in raster format.

● Joker Rømskog
Value in minutes
- 0 - 5
- 5 - 10
- 10 - 15
- 15 - 30

Practical analysis example

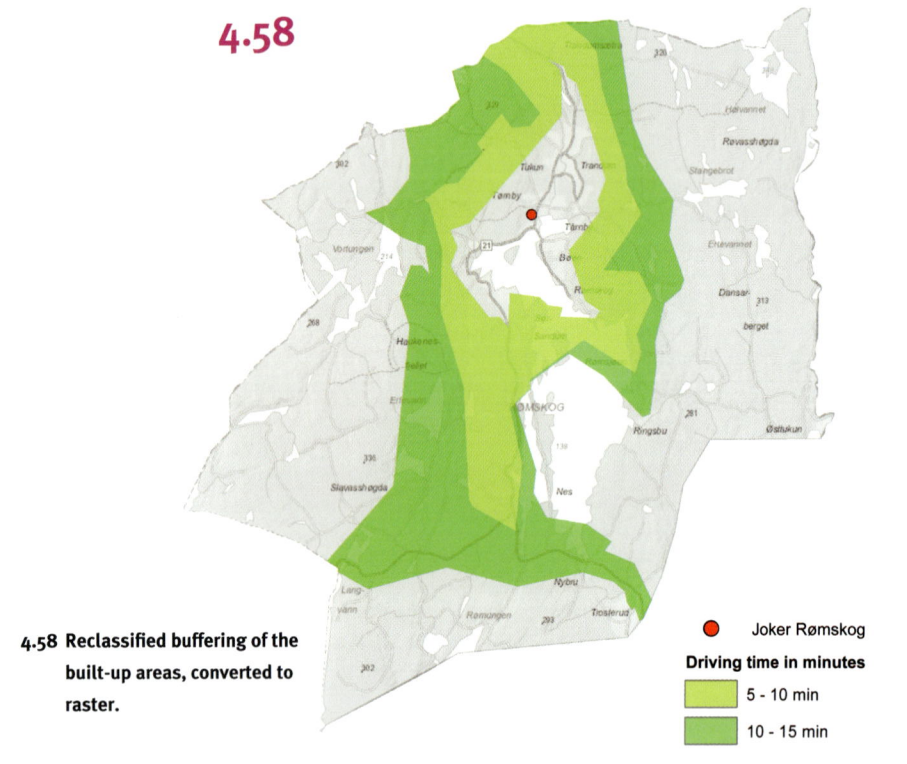

4.58 Reclassified buffering of the built-up areas, converted to raster.

In Figure 4.56, the driving times are shown as vector data. Figure 4.57 shows the same results as raster data.

From the theme layer showing driving times, the zones located in accordance with the given criteria can be selected. This results in a layer that shows areas located within 5–10 minutes and 10–15 minutes from Joker Rømskog, as shown in Figure 4.58.

Thereafter, the raster layers are reclassified. A simple schematic diagram of the reclassification is shown in Figure 4.59:

> Buffer analysis of built-up areas. Unsuitable areas assigned a value = 0, and suitable areas assigned a value = 1.
>
> Buffer analysis of county roads. Unsuitable areas assigned a value = 0, and suitable areas assigned a value = 1.
>
> Buffer analysis of water and marsh. 100–200 m from water and marsh assigned a value = 2, and 200–300 m assigned a value = 1.
>
> Gradient. Unsuitable areas assigned a value = 0, and suitable areas assigned a value = 1.
>
> Network analysis of driving times. 10–15 minutes assigned a value = 2, and 5–10 minutes assigned a value = 1.

Then, a network analysis is performed to find the driving time to the grocery store. For this step, road network data from the National Road DataBase (nasjonal veidatabank, NVDB) are used. In the example, other driving times are shown than those in the stipulated criteria, to enable a wider context to be seen.

4.59 Basic principles of classification.

For these raster layers, a raster calculator is used to find the final suitable areas in which the cabin could be built. The raster calculator is used to perform mathematical operations on the raster layers, such as to sum or multiply pixel values in the raster.

For the example, each pixel in the resulting raster is assigned one of the following values when the raster calculator is used for the five raster layers:

Areas with value = 1 → 1 x 1 x 1 x 1 x 1 = 1
Areas with value = 2 → 1 x 1 x 2 x 1 x 1 = 2
Areas with value = 4 → 1 x 1 x 2 x 1 x 2 = 4
Areas with value = 0 → All areas where one of the criteria indicates that the area is unsuitable.

4.60

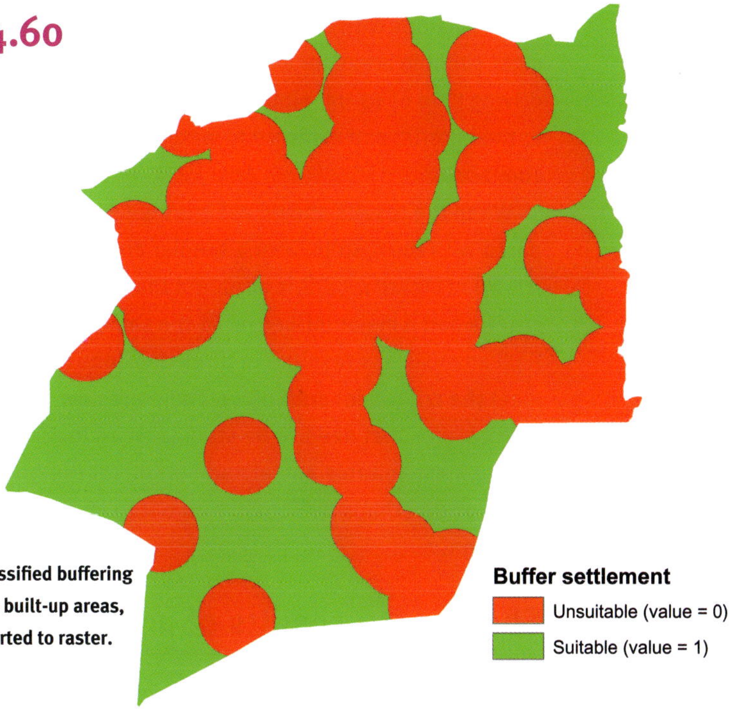

4.60 Reclassified buffering of the built-up areas, converted to raster.

Buffer settlement
- Unsuitable (value = 0)
- Suitable (value = 1)

4.61

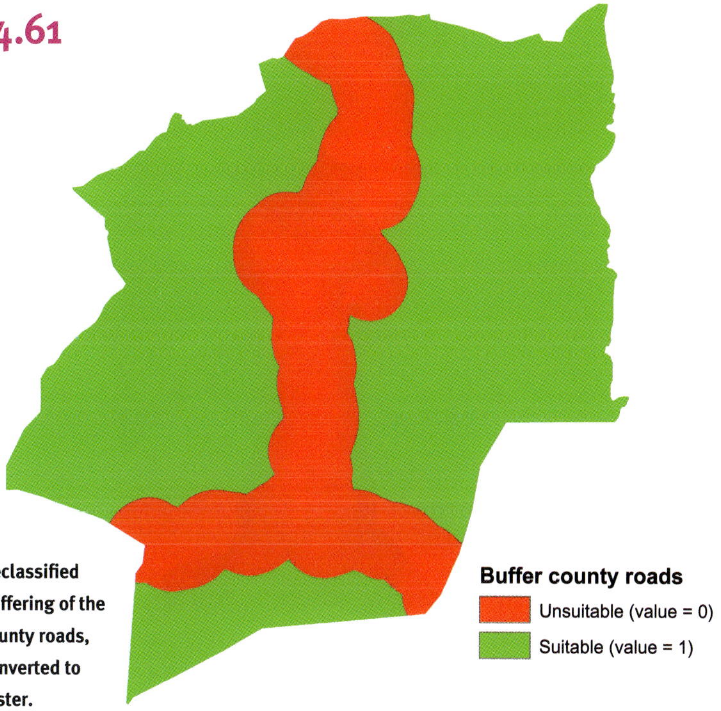

4.61 Reclassified buffering of the county roads, converted to raster.

Buffer county roads
- Unsuitable (value = 0)
- Suitable (value = 1)

4.62

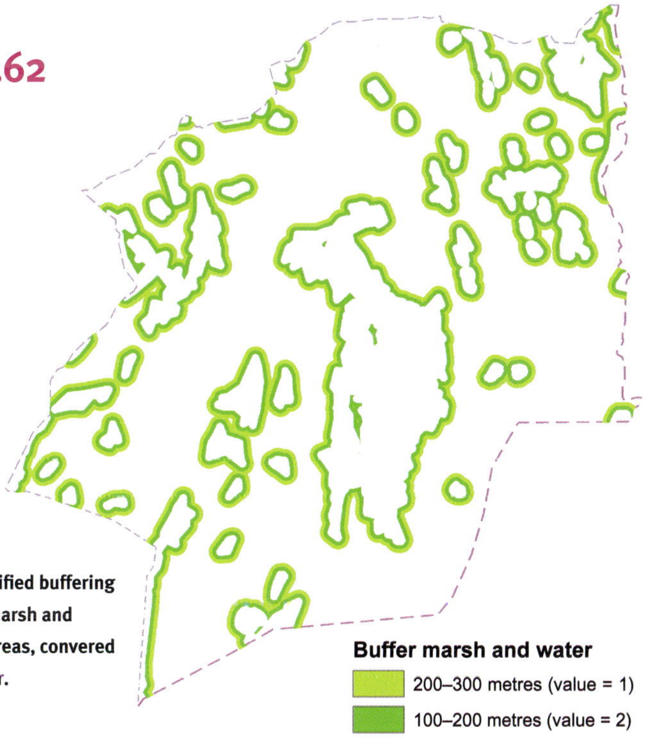

4.62 Reclassified buffering of the marsh and water areas, convered to raster.

Buffer marsh and water
- 200–300 metres (value = 1)
- 100–200 metres (value = 2)

4.63

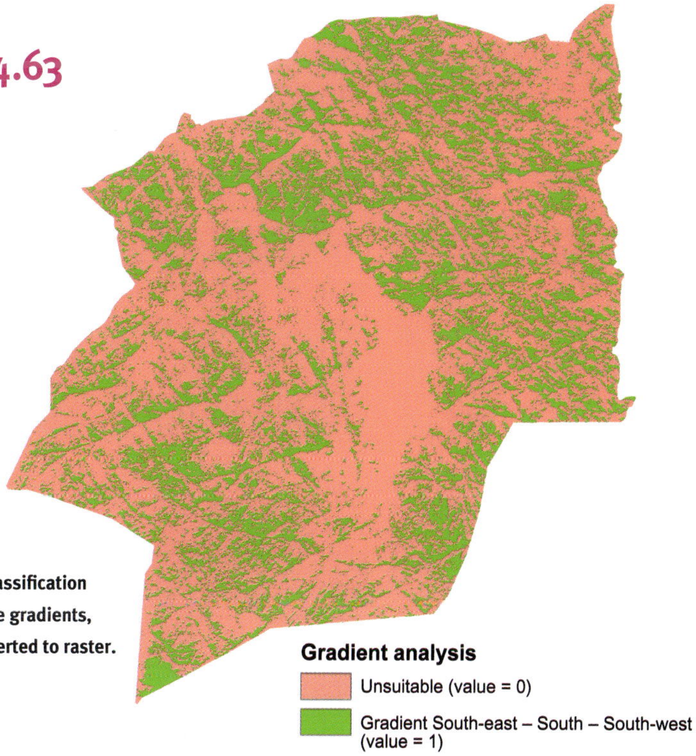

4.63 Reclassification of the gradients, converted to raster.

Gradient analysis
- Unsuitable (value = 0)
- Gradient South-east – South – South-west (value = 1)

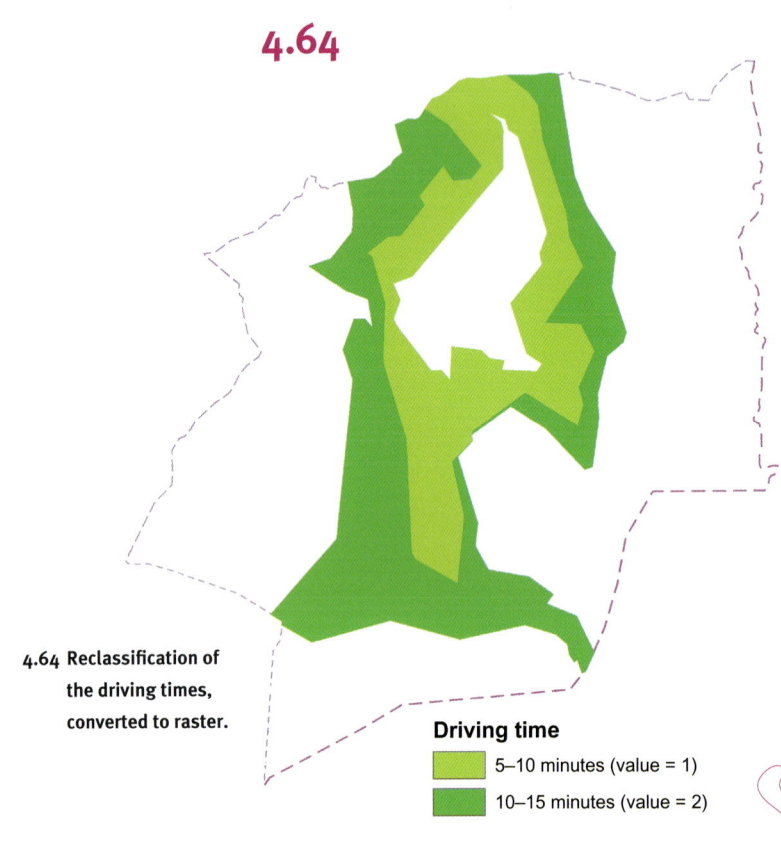

4.64 Reclassification of the driving times, converted to raster.

Driving time
- 5–10 minutes (value = 1)
- 10–15 minutes (value = 2)

4.65 Basic principles of the raster calculator.

Thus, areas with a value = 4 are considered the most suitable areas for a cabin. In this case, the driving time of 10–15 minutes is the best according to the criteria, and distance to water and marsh is 100–200 m.

When the results shown in Figure 4.66 are compared with the results from a simpler analysis, shown in Figure 4.51, it is apparent that the last two criteria had a major influence. At the same time, the final analysis has resulted in three divisions that meet the criteria for suitability. There are now far fewer places that need to be studied when looking for the most suitable and attractive site for the cabin, and perhaps this could turn out to be near Ertevann.

Sharing the results of analyses

The practical example of finding a suitable plot for a cabin is a relatively complex geographic analysis. Generally, the end users of such an analysis would not have the necessary software or expertise to know which analytical tools would need to be used to obtain accurate results. Consequently, either a GIS expert might have to perform all of the analyses or GIS would not be used to achieve the result. It is therefore important that GIS experts not only publish the results of their analyses, but also share their analytical tools, so that end users can perform the analyses themselves.

Today, GIS software provides GIS experts with opportunities to make analyses available via the Internet, such as via a map portal, so that all users of that portal can carry out their own analyses. The end user only specifies the parameters that are relevant and then receives the results of the analysis based on these parameters, without knowing anything about the tools and processes that run in the background. For example, in the practical example presented above, the end user specified the distances from a county road, the desired distance from water, and proximity to a grocery store in order to find a suitable site for their cabin.

Another example could be an analysis of driving times when the client wants to know the fastest route to a given destination. The client may not be interested in the method, but is concerned about the outcome.

One benefit of sharing the results of analyses is access to powerful data processing and GIS software on other servers. Moreover, necessary data for an analysis will generally already be available and accessible.

Since analytical tools are used by persons other than those who have established them, it is important that the tools are well-documented. For users of an analysis, it is similarly important to read the documentation for an analytical tool before it is used, in order to take into account any assumptions or simplifications relating to that tool. It is also important to examine any information about the quality of specialist data before using such tools. For example, it may be the case that an analysis of shortest driving route has taken into account that the original client has claustrophobia and therefore will want to drive through the fewest number of tunnels possible.

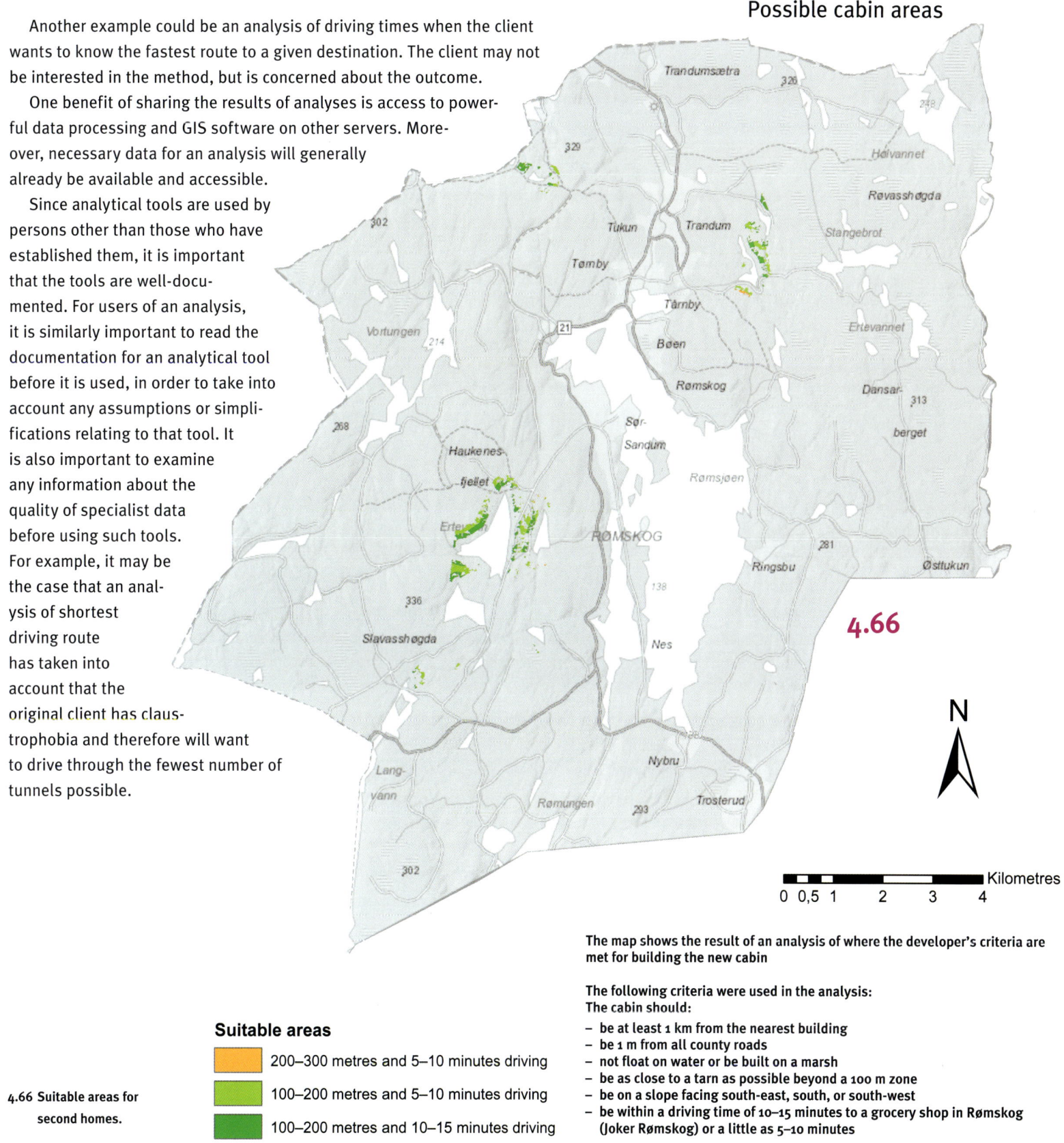

4.66 Suitable areas for second homes.

Suitable areas
- 200–300 metres and 5–10 minutes driving
- 100–200 metres and 5–10 minutes driving
- 100–200 metres and 10–15 minutes driving

The map shows the result of an analysis of where the developer's criteria are met for building the new cabin

The following criteria were used in the analysis:
The cabin should:
- be at least 1 km from the nearest building
- be 1 m from all county roads
- not float on water or be built on a marsh
- be as close to a tarn as possible beyond a 100 m zone
- be on a slope facing south-east, south, or south-west
- be within a driving time of 10–15 minutes to a grocery shop in Rømskog (Joker Rømskog) or a little as 5–10 minutes

Steinar Nilsen

holds a master's degree in land consolidation and spatial planning from the former Norwegian College of Agriculture (now the Norwegian University of Life Sciences). He works as a senior architect in spatial planning at the Norwegian Defence Estates Agency and has previously worked with GIS and thematic map production for the same agency and for the Norwegian Mapping Authority.

Presentation 5

Cartography	*176*
Perception level	*178*
Information variables	*179*
Graphic elements	*180*
Visual variables	*180*
The perceptual properties of the visual variables	*184*
Cartographic method	*185*
Point symbolization	*186*
Line symbolization	*187*
Area symbolization – choropleth maps	*189*
Other maps with area symbolization	*192*
Cartographic work process	*196*
Map design	*199*
Web cartography	*201*

5.1 An aeronautical chart from the M517 Air series for low flying which combines a topographic base map with information for safe flying (in blue). Published by the Norwegian Military Geographic Service.

Cartography

TOPOGRAPHIC MAPS AND THEMATIC MAPS

Maps are often categorized as either topographic maps or thematic maps. A topographic map presents a general overview of the reality without any emphasis on particular elements. Examples of topographic maps are those in Norway's main map series at a scale of 1:50 000.

By contrast, a thematic map highlights a particular theme, and the other themes found in a topographic basemap are either completely removed or are considerably toned down. Examples of thematic maps include geological maps and weather maps. Thematic maps are easier to understand than topographic maps and are therefore better suited for teaching cartography. Hence, in this chapter emphasis is on thematic maps, with the use of examples.

Some maps are hybrid forms that use a topographic map as a basemap, to which particular themes are added. Such maps are usually made for specific purposes when there is a need for a complete topographic basemap, such as hiking maps and aeronautical charts.

GIS AND CARTOGRAPHY

Maps are the most effective way of communicating and visualizing geographic information. A good map can convey all of the information in an area much faster than text or a table listing numbers. Maps that convey information visually can be likened to graphs that present data from numbers tables. A large amount of visual information can be instantly perceived, as shown in Figure 5.2, where the settlement patterns in densely populated areas, cities and towns, and unoccupied areas are all clearly visible.

In a geographic information system, map presentation is one of the final stages of data collection and processing. The amount of work that has gone into the collection of data and/or GIS analyses is irrelevant if the end-product in the form of a map fails to convey information correctly. A map should present a result, communicate a message, or serve as a research tool. It can be produced either on paper or on a computer screen that serves as a presentation medium.

Before the time when databases first appeared, maps served as archives for geographic information, in addition to having to communicate that information. Today, the role of maps and cartographers is to extract, present, and communicate information, and the storage function is now entrusted to geodatabases.

5.1

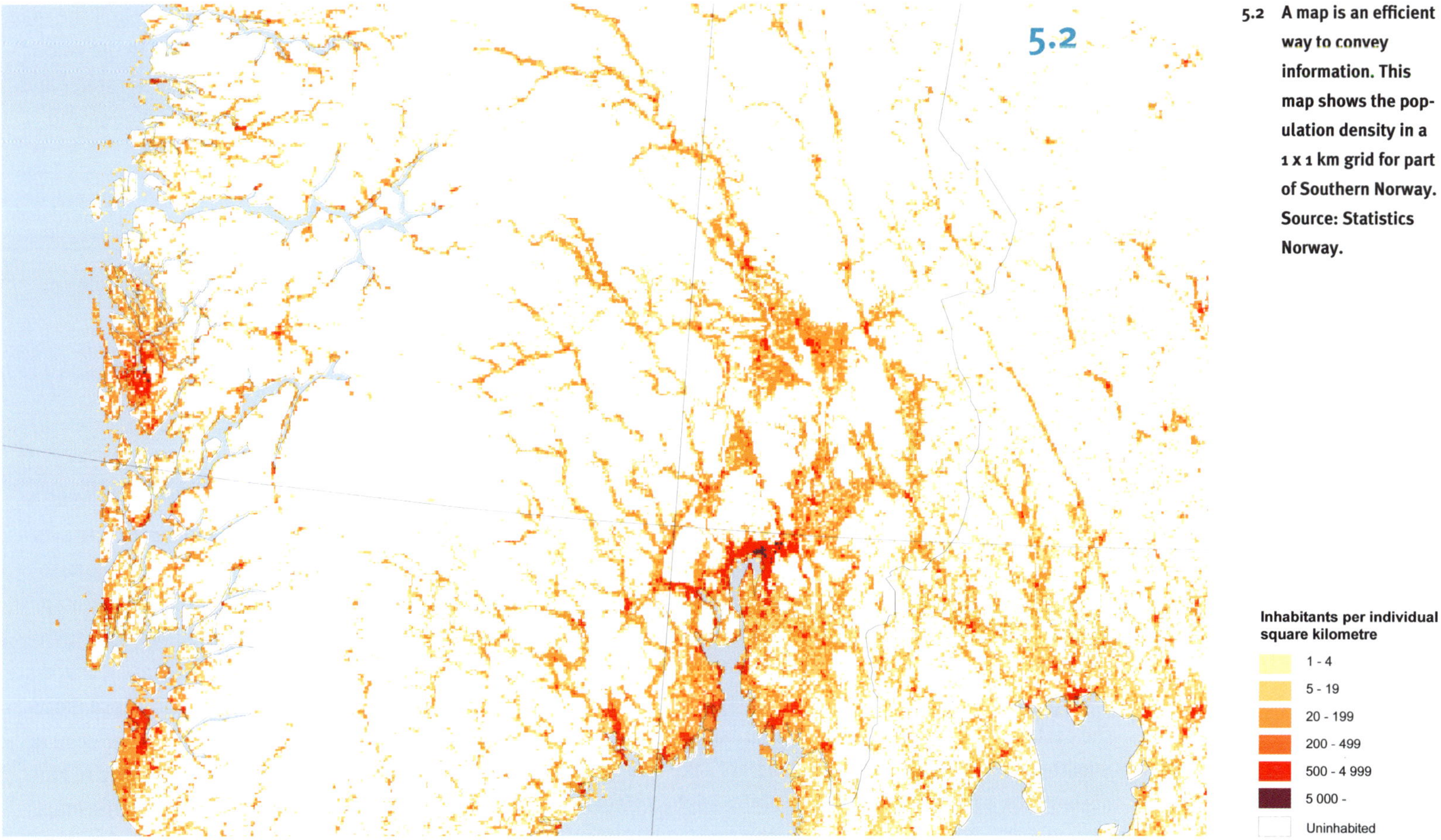

5.2 A map is an efficient way to convey information. This map shows the population density in a 1 x 1 km grid for part of Southern Norway. Source: Statistics Norway.

Inhabitants per individual square kilometre
- 1 - 4
- 5 - 19
- 20 - 199
- 200 - 499
- 500 - 4 999
- 5 000 -
- Uninhabited

Maps are symbolic abstractions of the reality on the ground, and the question of what themes are included or left out depends on the value attributed to them by the cartographers. For example, on topographic maps, churches are marked with large symbols and text that reflect both their prominence in the landscape and their importance in society 100–200 years ago. Today, petroleum stations – car drivers' 'temples' – are important elements in the landscape, but are less visible on maps compared to churches.

CARTOGRAPHIC COMMUNICATION

Cartography is the application of graphic symbols to communicate geographic information. It is related to other forms of visual communication such as graphic design, but also uses characteristic features such as coordinate systems, map projections, and scales. Both cartography and graphic design have to take into account fundamental rules concerning how people perceive their surroundings.

A good map is one that communicates information effectively. In order to communicate information correctly and effectively with the aid of a map, the visual variables have to be appropriate. The visual variables are the differences in the shape, orientation, colour, texture, value, or size of map symbols. Visual variables have different perceptual properties. In principle, cartography is a matter of choosing the correct visual variable for how geographic information is to be communicated. If a cartographer is not aware of this fact, his or her maps will fail to convey information effectively and instead they will provide misleading or even erroneous information. The relationship between geographic features may be hidden from view or non-existent relationships may be

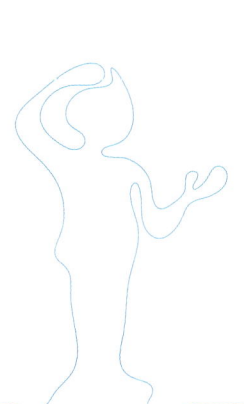

Perception level

introduced. It is therefore important to be aware of what a map is intended to convey or, in other words, what questions the map should provide answers to.

In principle, cartography involves an assessment of the level of information of the data to be presented on a map and the associated graphical elements. Thereafter, the appropriate visual variable can be selected.

The map of snow depths shown in Figure 5.3 has been reproduced from the daily newspaper *Aftenposten*. At first glance, it appears that a tall bar indicates a high value for the theme and a short bar indicates a low value, where the theme is the amount of snow. This is in accordance with human's capacity for visual perception. However, a closer look at the numbers on the bars will reveal that the cartographer intended to convey the opposite: a short bar indicates that most of the bar is covered by snow; in other words, the snow depth has a high value. Thus, a map reader (the recipient of the information) may perceive information differently from that intended by the cartographer (the data provider). In the case of Figure 5.3, the cartographer has allowed the information to be misinterpreted because he or she did not take into account the perceptual properties of the applied visual variables. Such misinterpretations and 'noise' in cartographic communication may also occur if the map reader is colour-blind, if there are errors in the colour reproduction (i.e. printing), or if the cartographer and map reader have different levels of knowledge and therefore interpret the reality differently.

Perception level

A map can communicate on three different levels of perception:

- elementary or local perception level
- group level
- global perception level.

Maps that communicate on element level are *read*: a map reader can only understand (perceive) one symbol at a time and then has to shift his or her gaze to the next symbol. It is only possible to read the map for a single location at a glance, and therefore such maps are called *readable maps*. Topographic maps are designed to function as readable maps. These maps should show the reader *what* features are present at a specific location. The topographic basemap in Figure 5.1 was designed to be used as a readable map.

Maps created on the global perception level are called *viewable maps*. With a viewable map, it is possible to understand the message at a glance. Viewable maps are intended to show *where* features are located. Figure 5.2 shows a viewable map that clearly communicates the population density in part of southern Norway.

Map symbols that communicate at group level form an intermediate category, where more symbols can be understood (perceived) simultaneously. On the aeronautical chart for low flying map show in Figure 5.1 it is possible to view the blue symbols as a group. The rectangular white symbols with black crosses representing churches can be viewed as another group.

Communicable maps combine the overall perception of a map's content (on global perception level) with information about occurrences at individual locations (local perception level). The result is an oversimplification of the underlying data. Figure 5.46 shows an example of a communicable map.

5.3 Map showing snow depths in the southern part of Norway. The user can find out whether there is little snow or a lot of snow at a specific location. Source: *Aftenposten*.

Information variables

When presenting data on a map, it is important to be aware of what level of information of the data visualization is needed. The attribute data can be categorized into three types: qualitative, ordered, and quantitative. Quantitative data can be subdivided into interval data and ratio data, and should not be presented at a higher level than the level of information in the data set. Ordered data should not be presented as quantitative data. By contrast, data with a high level of information can be presented at a lower level of information.

QUALITATIVE
Qualitative data are characterized by objects that are different but of equal value. Examples include the land classes coniferous forest, deciduous forest, wasteland, and farmland, or counties with, for example, county numbers 01, 12, and 16. The objects are either the same or not the same – they are qualitatively different. The objects cannot be ordered.

ORDERED
Ordered data values can be ordered or ranked, but do not provide any information about their relative values (i.e. low, medium, or high suitability).

QUANTITATIVE
Quantitative data are characterized by the fact the intervals between the data values are known and specific measurement are used, such as temperature, altitude, area, people's age, business turnover (NOK), or precipitation. A distinction is made between ratio data and interval data.

Interval data, such as temperature measured in degrees Celsius, lack a zero value to signify the absence of the attribute. It is not possible to say that 40 °C is twice as hot as 20 °C, and it is not possible to say anything about the absolute differences in magnitude between two values.

By contrast, *ratio data* can be measured on the basis of a zero value. The data values are therefore directly comparable and it is thus possible to calculate the intervals between values. Ratios can be subdivided in absolute data and relative data. Example: the number of votes cast for a political party in one municipality during a general election was 1570 votes, which is an absolute number. The number of votes cast for the party in relation to the total number of votes cast was 1570/7049 = 0,223 (i.e. 22.3%), which is a relative number.

5.4

QUALITATIVE	SPRUCE	PINE	OAK
ORDERED	Small	Medium	Large
INTERVAL DATA	<---- 30 years old	30 – 90 years old	----> 90 years old
RATIO NUMBERS	0.4 m³	1.0 m³	1.6 m³

5.4 Data about objects can be categorised according to whether they are qualitative, ordered, interval or ratio data. The categorisation of forest into different types of forest according to tree type: spruce, pine or oak, will be a qualitative categorisation. The classification of trees into small, medium and large is an ordered categorisation. Forest can be categorised into felling classes, which depend on the age of the trees. Felling class categorisation is interval data. The amount of timber is measured in m³, a ratio number.

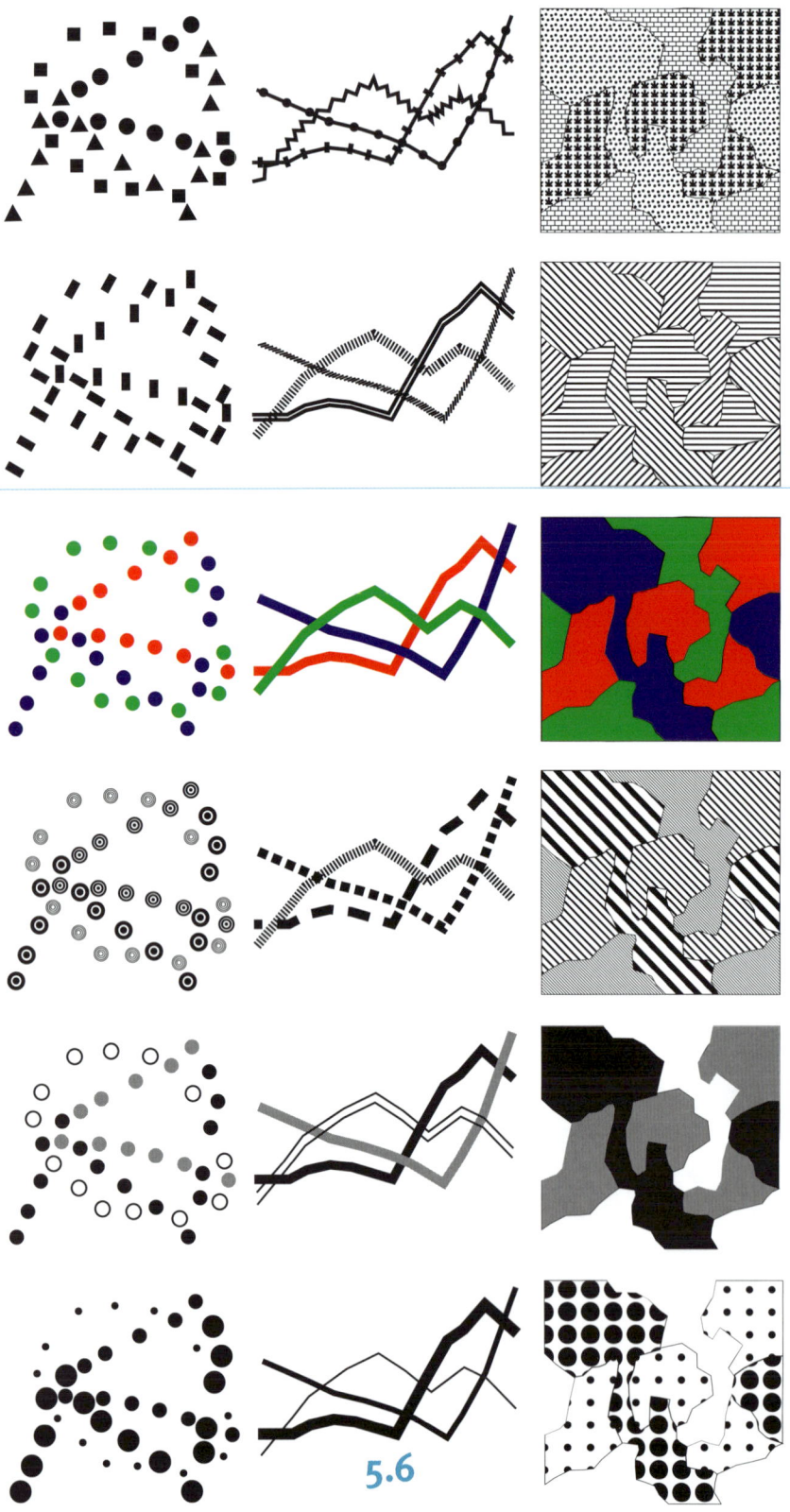

Graphic elements

Geographic features (objects) are presented on a map by the graphic elements points, lines, and areas. Their locations are fixed by the variables x and y (coordinates).

Attribute data are linked to the geographic objects represented as points, lines, or areas. These data describe the various types of attributes (information variables) for each object. Attribute data are represented on a map with the use of the following visual variables: shape, orientation, colour, texture, value, or size.

Visual variables

VISUAL PERCEPTION AND GRAPHIC COMMUNICATION

The current rules for cartographic symbolization were originally analysed and systematized by Jacques Bertin (Bertin 1983). Figure 5.6 shows the six visual variables: shape, orientation, colour, texture, value, and size visualized as points, lines, and areas.

Each of the visual variables will immediately generate specific perceptual stimuli. For example, differences in *value* communicate the order of objects from a little (light) to a lot (dark). Bertin introduced four different perceptual properties: *associative, constant visibility, selective, ordered,* and *quantitative*.

Choosing the appropriate visual variable to visualize geographic objects so that their immediate perceptual properties correspond with the information level of the data (qualitative, ordered, and quantitative) is of fundamental importance in cartography.

5.6 The 6 visual variables visualised as points, lines and areas. Bertin (1983).

SIZE

Variation in *size* is the only visual variable that can accurately communicate quantitative data. Figure 5.7 shows a visual representation of the values 1, 3 and 9. Maps with proportional circles are examples of the use of size as a visual variable.

VALUE

The variable *value* states how much of the area is filled. Value varies from completely empty (white) to 100% filled (black). A scale with variations in value is well suited to visualizing ordered information. A high value communicates 'a lot' and a low value communicates 'a little'. 'Value' is the variable most often used to communicate ordered data.

TEXTURE

Texture communicates ranking from coarse-grained (a lot) to fine-grained (a little). Variations in texture can be compared with photographic enlargements or reductions in which the relationship between black and white is constant. Texture is the only visual variable that can convey both ranking and constant visibility.

COLOUR

Colours are defined using different models. The CMYK model is used for printing and reproduction. A colour defined as 100% cyan (ice blue), 100% magenta (purple), and 100% yellow will result in a maximum dark colour. This maximum dark colour cannot be used to produce black and therefore the CMY model is often supplemented with the pure black colour named K (meaning 'key'). Printing inks are usually defined in CMYK.

The RGB colour model is based on the primary colours of light and is used to define the colours that appear on computer monitors and television screens. A colour that is 100% red, 100% green, and 100% blue will result in white. The primary colours in the CMY model are complementary to the primary colours in the RGB model: red – cyan, green – magenta, blue – yellow. Mixing two primary colours in the CMY model

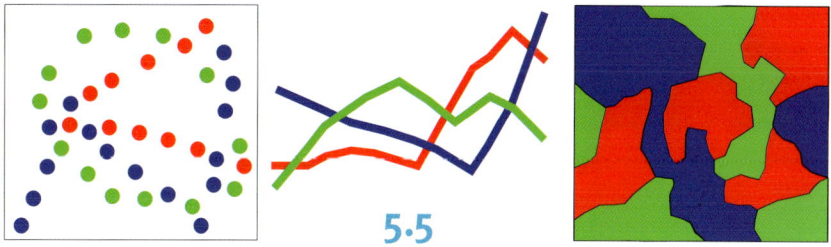

5.5 Geographic objects represented as points, lines, or areas. After Bertin (1983).

5.7 Variations in size. The largest square is 3 times bigger than the square in the middle, which is 3 times bigger than the smallest square.

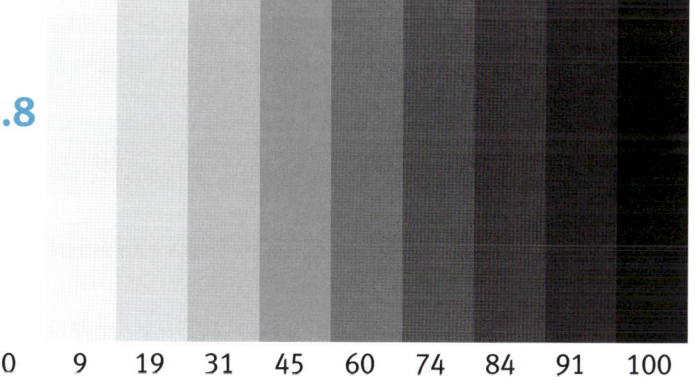

5.8 Equidistant intervals of value from 0% black to 100% black. By equidistant steps we mean that the visual interval between the steps is constant. Bertin (1983).

5.9 Point symbols with variation in texture.

Visual variables

5.10 Hues and values: the rows have the same values, but different colours. After Bertin (1983).

results in a primary colour in the RGB model: cyan and yellow produce green, magenta and yellow produce red, and cyan and magenta produce blue.

The HSB colour model (also known as HSV – hue, saturation, and value) is based on human perceptions of colours. In the HSB model all colours are defined on the basis of three fundamental properties:

- *Hue* (H) is the wavelength of the light reflected by an object. Hue is more commonly identified using names such as red, orange, or green. In the HSB model, hue is expressed in degrees in the range of 0–360°.
- *Saturation* (S) is the strength or purity of a colour. Saturation represents the amount of grey in proportion to the hue and is measured as a percentage between 0% (grey) to 100% (fully saturated colour).
- *Brightness* (B) (or value) is the relative lightness or darkness of a colour, usually measured as a percentage ranging from 0% (black) to 100% (white).

In this book, the visual variable colour is defined as the variation in hue and saturation, while brightness is the same as the variable *value*. Light red and dark red are the same colour, but vary in value.

Colour is a very selective variable. It is easy to combine it with other variables and it is easy to read. However, at the same time, it should be remember that colour cannot be used on its own for an ordered component; rather, it must be combined with value to communicate ranking. Colour is very attractive and will catch the eye much more easily than black and white.

Figure 5.10 shows the relationship between the visual variables *colour* and *value*. Each row has the same *value* but varies in *colour*. The pure colours (100% saturation) are indicated by white dots. If only the pure tones are used, the result will be a series with maximum saturation. This series will also vary in brightness (value), ordered from light to dark. The parts of the spectrum on either side of yellow result in an ordered series, with yellow as the lightest colour. The cold series

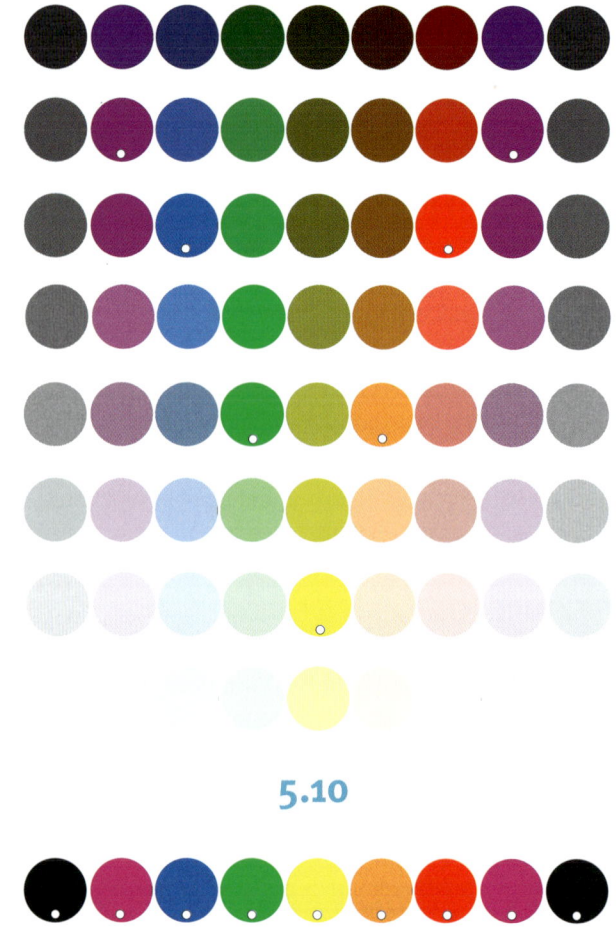

Colour

5.10

Saturation

is yellow-green-blue-violet-black, and the warm series is yellow-orange-red-purple-black. These two series, shown in Figure 5.11, are very suitable for use in cartography since they combine variations in *value* for the visualization of ranking with the selective properties of *colour*. Combining the warm and cold series results in a 'traffic light' scale that can be used for data series on both sides of a zero point.

It is important to differentiate between the general perceptual laws for colours, which cannot be departed from, and the culturally determined associations relating to colours, which vary according to different

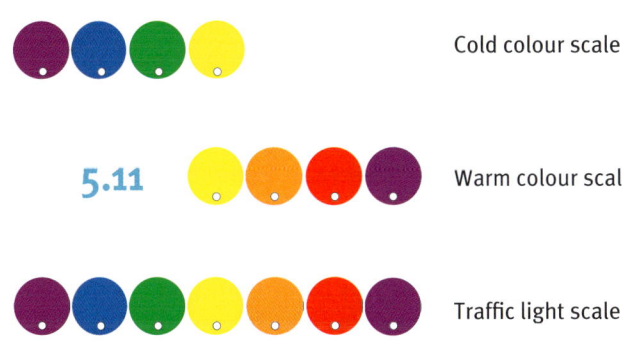

Cold colour scale

Warm colour scale

Traffic light scale

5.11

SHAPE

The visual variable shape is used for qualitative data. Map objects can exist in numerous shapes and it may therefore be tempting to use the variable shape when making maps. In such cases, it is important to know the limitations of the perceptual properties of shape as a visual variable. Shape is the only visual variable that is not selective.

Shape variations for point symbols can be divided into three different types: pictograms, mimetic symbols, and geometric symbols.

A *pictogram* is a stylized drawing of the object it represents. The drawing should be very simplified but no more than necessary for a user to be able recognize the object it represents. For maximum utilization, the symbols should be sufficiently logical that a map key is not needed. The pictograms in Figure 5.13 are taken from the standard *Symbolfonter for friliftsliv og sport* (Symbol fonts for outdoor activities and sports) (Kartverket, 1997).

Mimetic symbols fall somewhere between pictograms and geometric symbols. Many geometric symbols have been designed to form a shape

5.11 The warm and the cold colour scales can in combination be used for data series on either side of a zero point.

cultures and traditions. The associations that we relate to colours should be exploited in cartographic contexts: blue is associated with cold and water; red is associated with danger, heat, and socialism; and green is associated with fertility, whereas brown is associated with infertility. In the Western world, death is associated with the colour black, while in some African societies death is associated with the colour white. Furthermore, some associations can be determined by purely professional conventions (e.g. the colours on a land use planning map) or the colours used on geological maps.

5.12 Four selective orientations. Note that the angles of the oblique lines are 30° and 60°, not 45°.

5.13 Variations in shape: pictograms, mimetic symbols, and geometric symbols.

ORIENTATION

Orientation can convey qualitative characteristics. It can distinguish between two orientations represented as lines (parallel and perpendicular) and between four orientations represented as points. Representations of orientation are less selective on an area representation (see the representation of lines and areas in Figure 5.6).

Pictograms	Mimetic symbols	Geometric symbols
	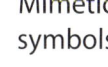	●
		▲

5.12

5.13

that can be associated with the theme it is intended to represent. The symbols in Figure 5.13 indicate churches and mines and have been taken from the general topographic map series for Norway, which is at a scale of 1:50,000. The mimetic symbol for a church can be compared with the pictogram for a church.

Circles, triangles, and squares are all examples of *geometric symbols*. One cannot see from a geometric symbol alone what it is intended to represent. Rather, one has to consult the map key.

The perceptual properties of the visual variables

QUANTITATIVE – Q
The general rule for the presentation of quantitative data is that there should be equal distances between the data values and the visual presentation of the intervals. With graphical presentation, only the visual variable size can reproduce the variations in distance. By using, for example, the variable value, the information presented will be reduced from a quantitative level to an ordered level.

ORDERED – O
In order for map objects to be arranged relative to each other (ordered data), the variables size, value, or texture have to be used. These variables are the only ones that can express ranking. Colour cannot express ranking: red is not greater or less than green. Shape cannot express ranking either: a circle is not greater or less than a triangle. An ordered series is distinguished from a quantitative series by the fact that the distances between each class or category are equal. The graphic presentation therefore has to maintain this state; the visual distances between the increments in the data scale must be constant (see Figure 5.8).

SELECTIVE – ≠
It is easy to distinguish map features that differ in colour or size (e.g. one can see at a glance all of the red objects on a map), whereas it is difficult to distinguish between objects that only vary in shape. Colour and size are selective map symbols. This is evident from Figure 5.6, where the < in the red points and the 'N' in the blue points can easily be distinguished from the other colours. The same effect can be achieved with the use of texture, value, and size, but the variable shape would fail to show up the < and 'N'. Furthermore, all triangular objects have to be identified one by one. The variable shape cannot convey selective properties.

ASSOCIATIVE (CONSTANT VISIBILITY) – ≡
The visual variable texture has the same visual weight and hence constant visibility. The variables value and size have varying visual weight. Conveying that objects are equally important, but have qualitative variations requires that size and value are kept constant, while shape, orientation, colour, or texture may vary. In Figure 5.6, the area presentation of the variables size and value fails to form a square, unlike the other four variables. This can also be seen in point representation, where the points of shape, orientation, colour, and texture are equally weighted. The visual variable texture is the only one of the ordered visual variables that can also express constant visibility.

5.14 The visual variables' properties. The figure illustrates which properties the six visual variables can communicate. Associative map symbols have the same visible weight, selective symbols are perceived as different, ordered symbols are perceived as having a ranked order and quantitative symbols have proportional size in relation to each other.

SIZE		≠	O	Q
VALUE		≠	O	
TEXTURE	≡	≠	O	
COLOUR	≡	≠		
ORIENTATION	≡	≠		
SHAPE	≡			

5.14

Cartographic method

In essence, the cartographic method involves the cartographer making conscious decisions regarding the types of problems described in the sections above:

- *Perception level* – readable or viewable maps
- *Information level* – qualitative, ordered, or quantitative
- *Graphic element* – point, line, or area
- *Visual variables* – to convey the information as intended

First, it should be determined whether the map should provide answers to questions at local perception level (what features exist at a specific location) or whether it should provide answers to questions about *where* features are located, In the former case, one should create a readable (topographic) map, and in the latter case a viewable thematic map. Thereafter, the level of information of the data and the graphic elements should be evaluated and the visual variable that can convey the information level (qualitative, ordered, or quantitative) of the data in accordance with desired perceptual properties (quantitative, ordered, selective, or constant visibility) should be chosen.

Figure 5.15 shows a matrix of nine types of pure thematic maps that result from a combination of the graphic means of presentation (point, line, area), the level of information (qualitative, ordered, quantitative), and a visual variable that conveys the level of information accurately.

Quantitative data are very rarely visualized using the size of an area shape (as shown), and then only for absolute values of the ratio data. Quantitative data are normally presented using the variable value, and thus reduces the information variable to the ordered level.

During the preparation of a topographic map, the above-described process has to be done for each theme, such as roads, rivers and waterways, buildings and built-up areas, and land surfaces.

The following section examines the symbolization of the individual graphic elements that result in the different types of thematic maps.

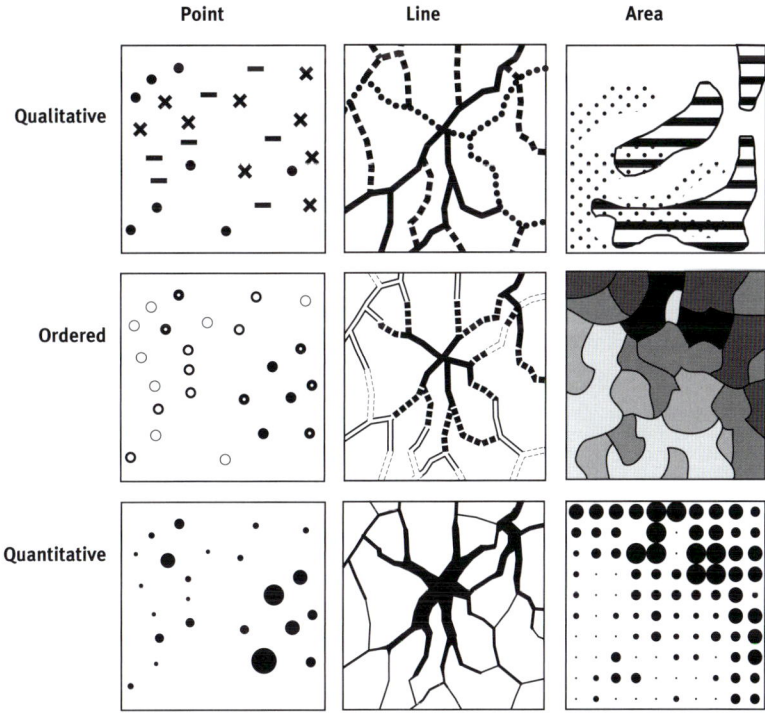

5.15 The type thematic of map is determined by whether the data is qualitative, ordered or quantitative in nature and whether it is visualised as points, lines or areas.

Point symbolization

SELECTIVE POINT SYMBOLS

Variation in shape does not convey selectivity. One exception is the variation between the three point symbols point, line, and cross. These three symbols can be combined with two variations in texture and six variations in colour: grey, violet, blue, green, brown, and red. In addition, lines can have four variations in orientation. Other colours have different values (e.g. yellow and black) and will not give a sense of constant visibility. These combinations are only useful when the map is viewed at global perception level and with constant visibility and selectivity as desired perceptual properties. All other symbols will appear equal at global level, and it will not be possible to group them visually.

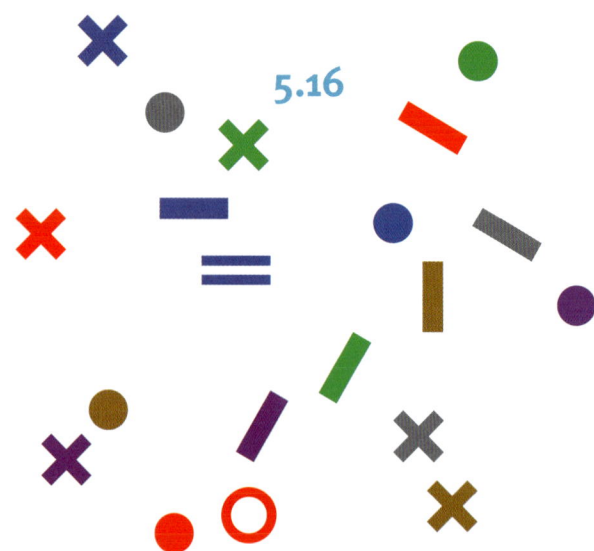

5.16 Selective point symbols with equal visibility that have been derived from combinations of three shapes and six colours. The red point symbol and blue line symbol differ in texture and the lines have four different orientations.

QUANTITATIVE POINT SYMBOLIZATION

Figure 5.17 shows a map with *area proportional circles* that represent quantitative data associated with a point or area. The area of each circle is proportional to the data value it represents. Area proportional circles are particularly well suited when a wide range of data values is to be visualized.

The human eye will have difficulties perceiving that a circle with an area of 20 measurement units is twice as large as a circle with an area of 10 units. The brain has a tendency to underestimate the size of large symbols relative to small symbols. One can correct for this by using a logarithmic scale that exaggerates size.

The size of the circles should be calculated in such way that the area covered by all of the circles together covers between 5% and 10% of the meaningful part of the map, as this will optimize the readability of the map. In other words, a circle that represents all values should cover just under 10% of the map's area.

Classification should be avoided on maps with area proportional circles, since it reduces the value of the information provided by the data. However, when the range of data values is very large, it may be necessary to classify parts of the data set.

The *abacus method* shown in Figure 5.18 shows the quantitative attributes of the data associated with an area as a collection of identical point symbols. Each point symbol represents a specific data value. Unlike a dot density map, no attempt is made to position the symbols accurately in spatial terms; rather, they are grouped in a geometrically shaped cluster within the area they represent. Squares are often used as point symbols.

A *dot density map* shows the quantitative data associated with an area by the density of dots. The dots are the same size and each dot represents a specific amount of data. The choice of size and unit value of the dots should ensure that the dots almost merge together in the densest areas. The dots are located evenly within the area unit they represent. If one knows how the data are distributed within each area unit, the dots can be positioned such that most of them are visible where there is the greatest quantity of data.

A dot density map is visually perceived in two ways: as the density of the dots (the ratio between black and white, i.e. dot/no dot) and as the number of dots (as long as the dots do not blend together). The presentation form of a dot density map is not unlike that of a spatial positioning map, but the visual perception of the density and quantity of the dots is more important than their exact spatial position.

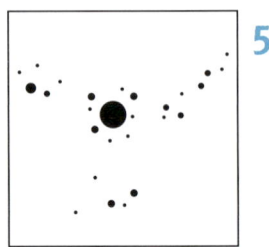

5.17 Quantitative variations shown as proportional circles.

5.18 Quantitative data presented using the abacus method.

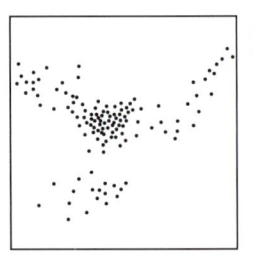

5.19 Dot clusters on a dot density map give a clear representation of quantitative variations.

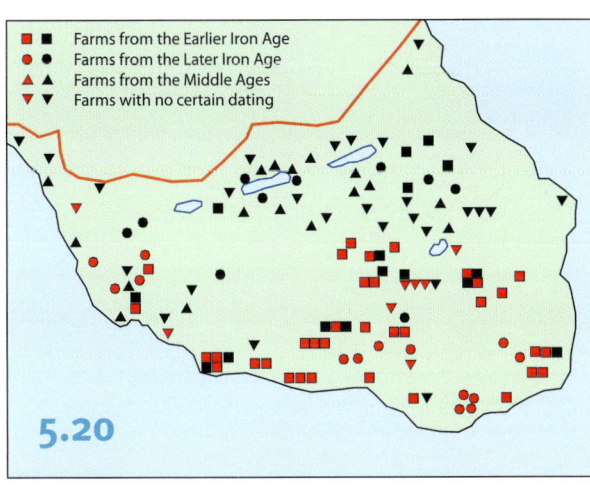

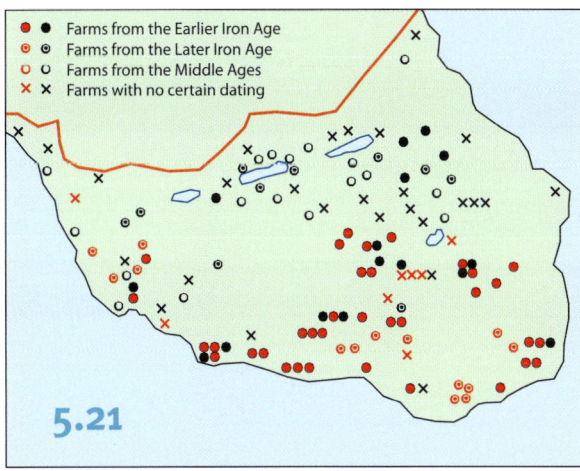

5.20 Deserted farms after the Black Death. The map symbols' variation in shape does not communicate the data set's ranking.

5.21 Deserted farms after the Black Death. The points' symbolisation corresponds with the data's message.

EXAMPLE: CHOOSING POINT SYMBOLIZATION – DESERTED FARMS AFTER THE BLACK DEATH

Figure 5 shows farms in Nes, Hedmarken, represented as point symbols. The time periods in which the farms were established are shown by variations in the point symbols' *shape*, whereas the status of 'deserted farm' and 'continued farm' after the Black Death (14th century) are differentiated by *colour* (red and black respectively).

Time periods are an ordered series and therefore when using symbols one has to use a visual variable that communicates order; *shape* cannot do this. Figure 5.21 shows the use of the variable *value*, which is a variable that conveys order. The variable *shape* is used to indicate farms of unknown dates with using a cross instead of a circle. The symbol 'X' also has associations with 'unknown'. The use of *colour* (red and black) is kept, even though black has a greater value than red. The point symbols therefore do not have constant visibility. However, red is the strongest colour in terms of the effect of symbols, and black results in an association with 'death' (deserted farms).

Line symbolization

The background topography of a thematic map primarily contains general line elements such as coastlines, rivers, and administrative boundaries. In visualization of lines, it is very important from a graphics point of view that the basemap's lines can be differentiated from the thematic data lines. The base map should therefore be drawn in a toned-down form, whereas the thematic data should be strongly visualized. The visual variables colour and texture have to be used to create selective lines. The colours grey, violet, blue, green, brown, and red, which have the same value, have to be used to ensure that the lines are visually perceived as equal (constant visibility). Typical examples of ordered lines are administrative boundaries (national, county, and municipality boundaries) and roads (highways and county roads).

Figure 5.23 shows quantitative data in linear form represented as lines of varying width. The width of each line is proportional to the data value at that point along the line.

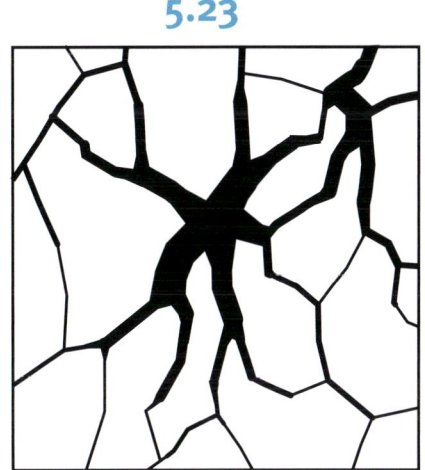

5.23 Width proportional bands communicate quantitative line data.

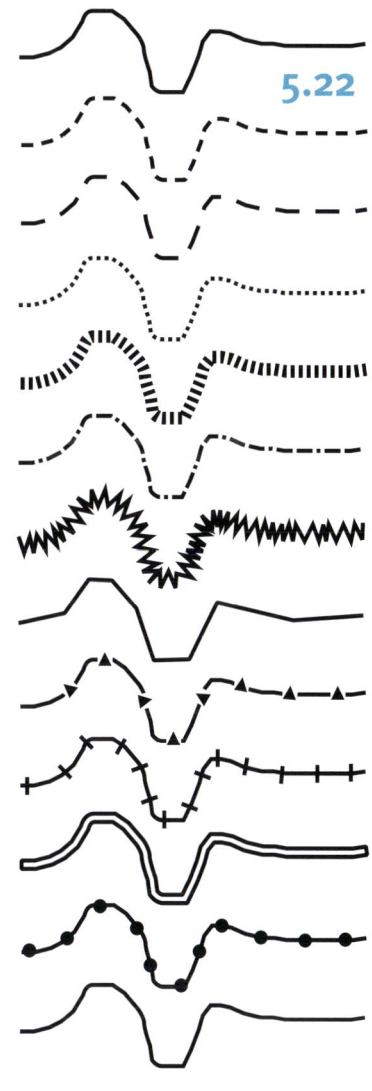

5.22 Lines that vary in shape.

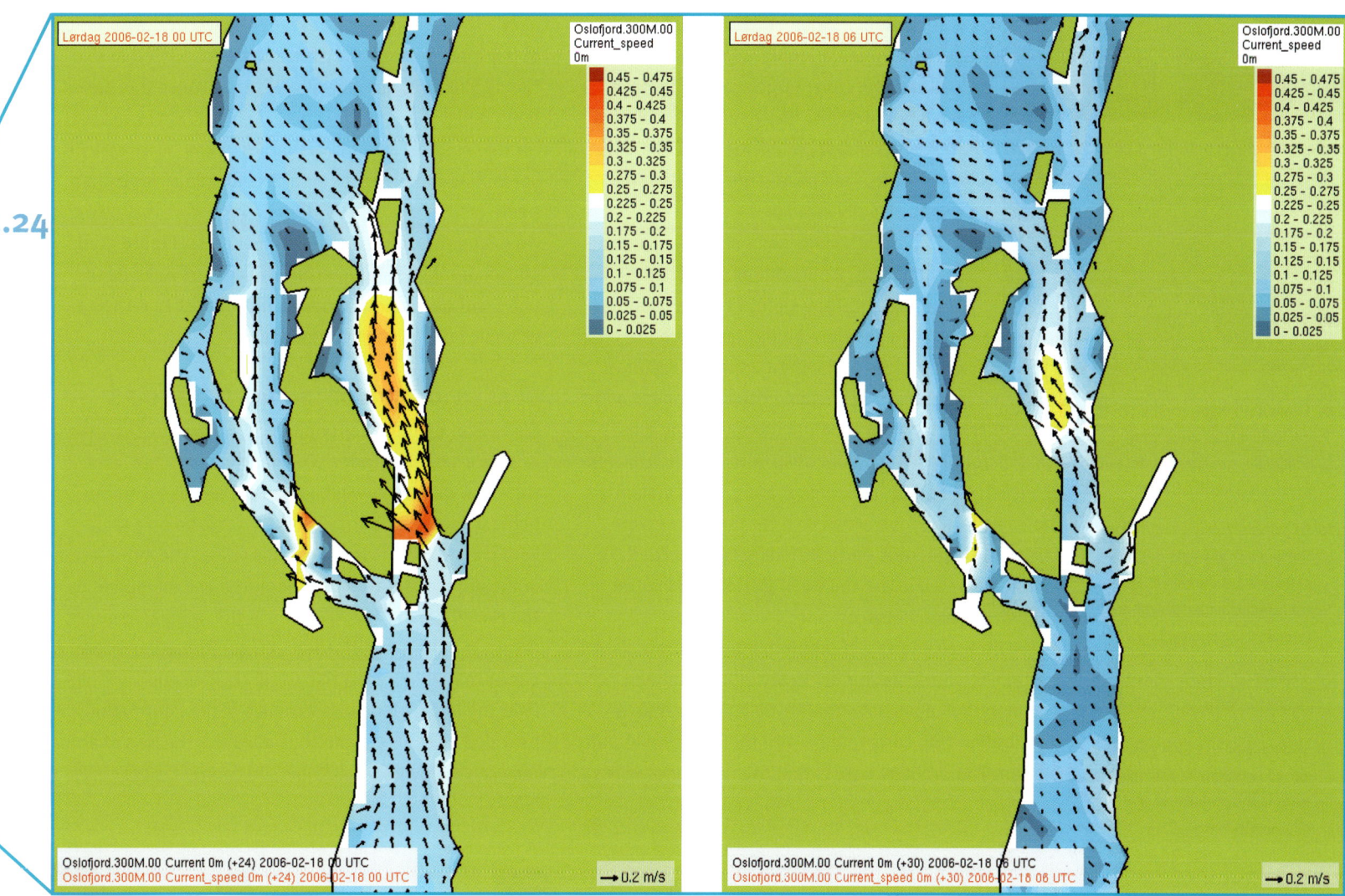

Area symbolization – choropleth maps

In choropleth maps, value is used to visualize ordered data associated with areas. Most choropleth maps visualize quantitative data with the aid of the ordered variable value, and in practice reduce the data to an ordered level.

Koro is Greek for 'place', and *plet* (Greek: pletos) means 'quantity'. A choropleth map is a map where each unit area is given a value depending on the class data values. The readability of a map with variation in value depends on the tension between the smallest (brightest) and largest (darkest) values.

Choropleth maps are one of the most frequently used map types and the weaknesses inherent in the method have therefore been extensively researched. Choropleth maps visualize the average value within each area unit and class, and any variations within an area unit or class will be hidden from view. This will give the impression that the data change suddenly from one area unit or class to another.

The area units ought to be equal in size. Area units that differ in size will have great impact on visual perception and might result in an erroneous perception

5.24 The map of the currents in the Drøbak Sound is essential information for participants in the annual Færder Sailing Regatta. The speed of the current is visualised by a combination of the length of the arrows (size) and colours in the traffic light scale (value).

190 Area symbolization – choropleth maps

5.25

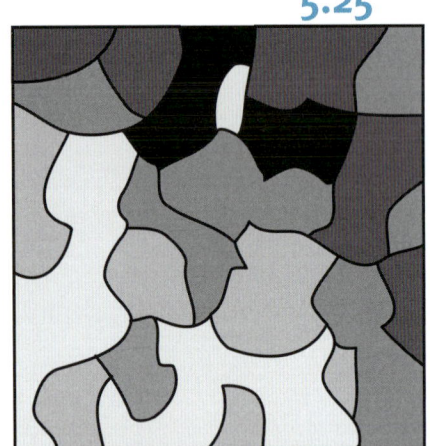

5.25 Choropleth maps are the most common way of presenting ordered or quantitative data in an area representation.

5.26 A choropleth map produced on the basis of absolute data results in a different message to the one the map producer really intended to communicate.

5.27 A choropleth map produced with the correct variation in value.

5.28 Using colour in which value is varied randomly is a common error in the production of choropleth maps.

of the average value of a larger area. For example, population data for rural municipalities that cover large areas (and have small populations) will visually dominate a map, whereas city municipalities that cover small areas (and have large populations) will be visually swamped.

Choropleth maps can only visualize rates or ratios (e.g. a percentage of, per km², or per capita. They should not be used to visualize absolute quantities, since the different sizes of the area units will result in erroneous visual perceptions. If one wants to produce a map with absolute numbers, one must make a dot density map or a map with area proportional circles. The maps in Figures 5.26–5.28 all visualize the same data series: votes cast for the Norwegian Christian Democratic Party (Kristelig Folkeparti, KrF) per municipality in the county of Østfold during the general election in 1997.

Figure 5.26 shows a choropleth map of a data series containing votes cast for KrF in absolute numbers. The map thus does not show anything other than a distribution that corresponds with the variation in population. The problem here is that the number of votes needs to be represented independently of the variation in population size – they have to be calculated in relation to the total number of votes cast. Once this has been done, the map will show the variation in support for KrF. Figure 5.27 shows that Rømskog, with 34% KrF votes, has the highest value. A cold colour series has been used in this map, with pale yellow as the lowest class and dark blue as the highest class. Figure 5.28 shows another serious error: the colours have been used without taking into account the fact that the variation in value is the active variable.

CLASSIFICATION OF CHOROPLETH MAPS

Classification means grouping the quantitative values in a data set into groups. The groups can range from the class with the lowest values to the class with the highest values, and they can be visualized on a map with the aid of the visual variable *value*. The information in the original data set is generalized and simplified by the classification. The accuracy in the original numbers is lost, but the simplification of the information makes it easier to read the material. Good classification will highlight the important characteristics of the information provided by the data and thus minimize the loss of accuracy caused by generalization. By contrast, poor classi-

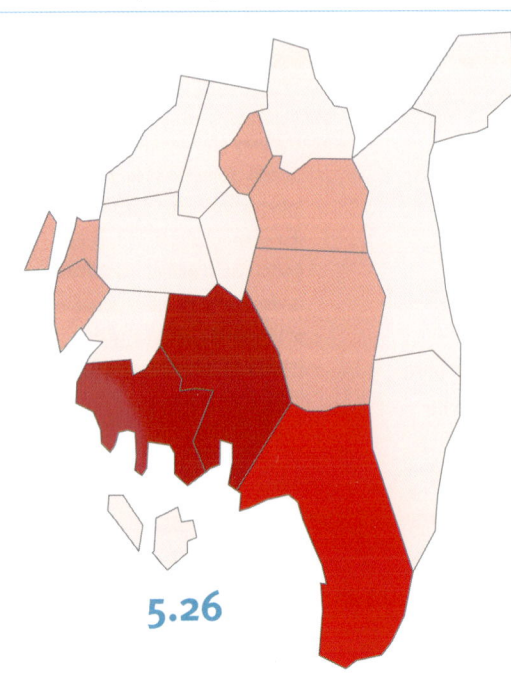

5.26

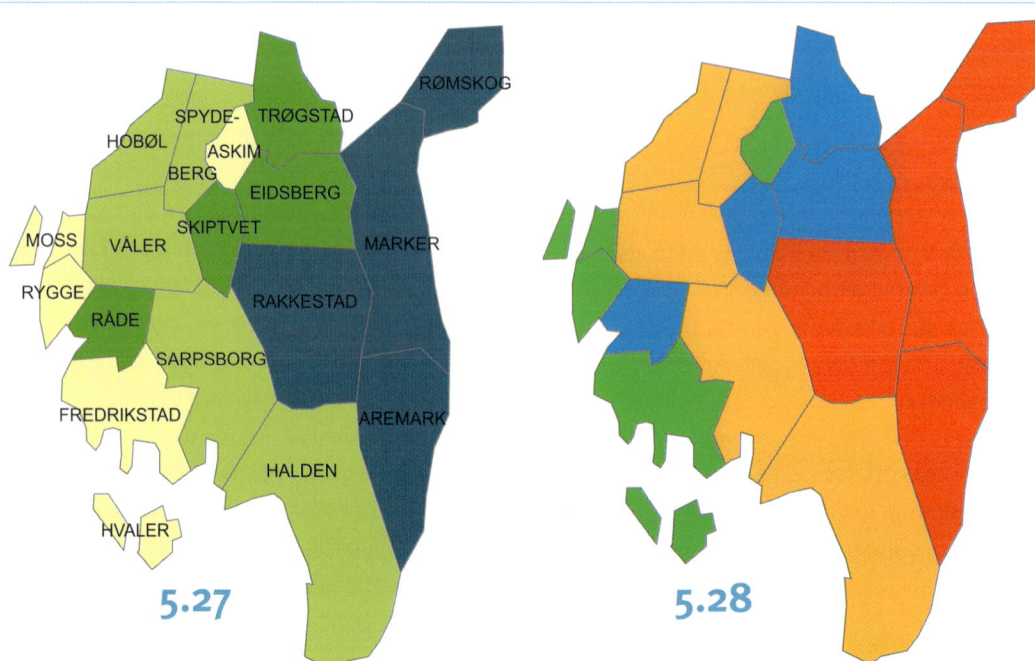

5.27   5.28

fication can distort information in a data set or remove it. The challenge presented by classification is thus to simplify without losing the original information.

Choosing the number of classes involves a compromise between the desire for detailed information (requiring many classes) and the map user's ability to differentiate between the classes (requiring few classes). The human eye cannot differentiate between more than eight classes if a black and white scale is used. The more classes there are, the harder it is to differentiate between them. Hence, one should not have more than eight classes and the message will often be clearer if there are fewer classes.

The distribution of the data values is often decisive for the choice of classification method. An overview of the distribution of the data can be gained by sorting the data values from the lowest to the highest and then creating a bar chart of the distribution. Many classification methods have been designed to take into account the different distribution patterns that can occur in data. None of the classification methods satisfies all stipulated requirements. The good methods provide good solutions to some of the requirements, but they will necessarily provide poorer solutions for other stipulated requirements. Therefore, the choice of classification method should be based on an assessment of the requirements. The class thresholds should, if possible, be round numbers, as this would result in a readily understandable classification that is easy to remember.

Figures 5.29–5.32 show classification methods in a data set of the votes cast for the KrF per municipality in the county of Østfold in the 1997 general election. The value for each municipality is shown in a bar chart where the lowest value (Moss, 8.1%) is to the left and the highest value (Rømskog, 34%) is to the right. The classification values are shown on the Y-axis.

For *equidistant classification* (Figure 5.29), the class thresholds are set with equally large intervals between them. The range of the data is calculated (here the maximum is 34.0% (Rømskog) and the minimum is 8.1% (Moss)) and the number is then divided by the number of classes one wants to use (in this case four). This gives the class width (in this case 6.5), and the exact thresholds of the classes are found by repeatedly adding the class width ('equidistant') first to the lowest value, then adding the result of this, and so forth. This method is best suited for data with a regular, linear distribution. If the method is used for data that do not have a linear distribution, some of the classes will contain many of the data observations, while other classes will contain few or no data observations.

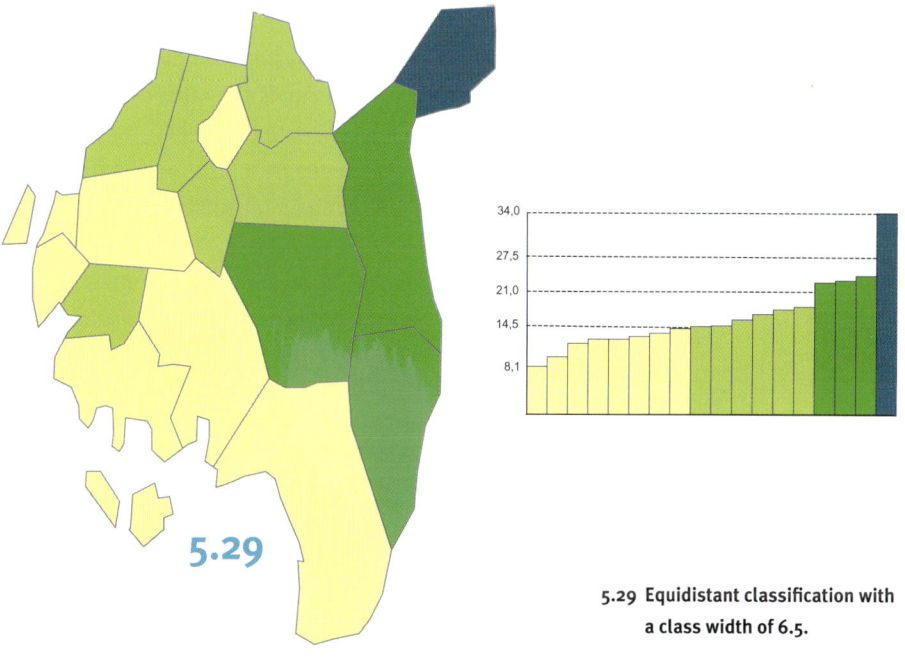

5.29 Equidistant classification with a class width of 6.5.

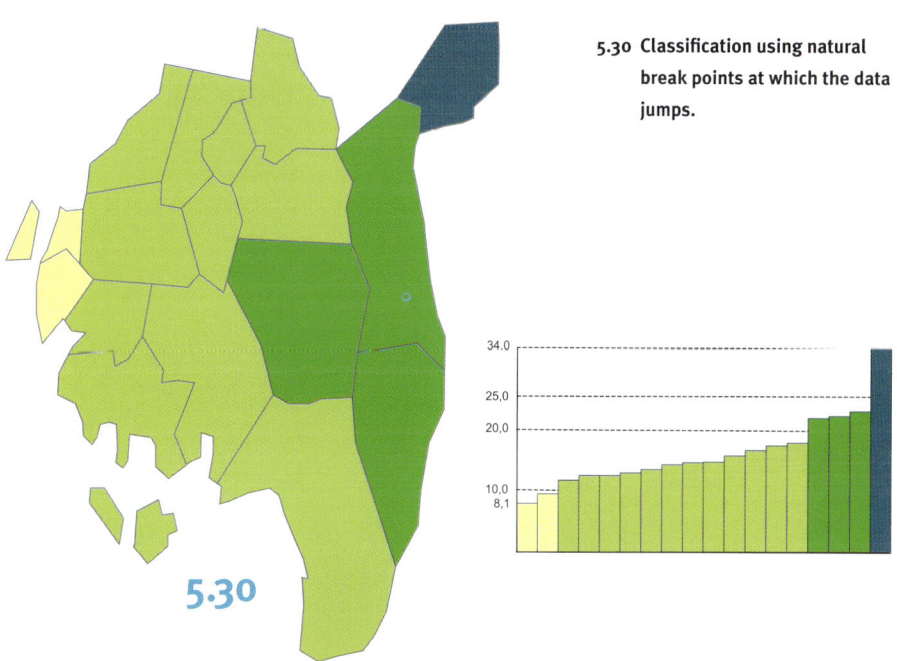

5.30 Classification using natural break points at which the data jumps.

Area symbolization – choropleth maps

5.31 Quintile classification with an (almost) equal number of units in each class.

5.32 In the classless method the symbolisation of value accurately corresponds with the amount of basic data and annuls the need for classification.

For *classification using natural break points* (Figure 5.30), the class thresholds are natural breaks in the data set. The strength of this method is that the classification maximizes the information content since it is tailored to the map user's perception of the map. The map reader will perceive that area units in the same class have the same data value. The class thresholds should be set with this in mind, such that data values of approximately the same value end up in one and the same class. The method provides a true visual image of the data values. In the other methods, the class thresholds can often cut through clusters of data that belong together.

Quantile classification (Figure 5.31) results in an equal number of data observations in each class. With four classes (quartiles), each quartile contains 25% of the data observations, with five classes (quintiles) each quintile contains 20%, and so forth. This method is highly appropriate when one data set is to be combined with the data sets of other themes. If the data strongly deviate from a linear distribution, the absolute class width could display great variations. In the example, five observations are used in the two lowest classes and four in the highest classes.

A variant of quantile classification is *percentile classification*, in which each class contains a specific percentage of the data observations. Although the classes each contain a proportion of the quantity of data (like the quantile method), they are adapted for other types of data distribution than linear distribution. The relative scales, quintiles, and percentiles enable comparisons to be made with other themes. For example, the spatial positioning of the highest 20% of values in theme 'X' can be compared with the spatial positioning of the highest 20% of values in theme 'Y'.

The classless method (Figure 5.32) resolves the classification problem by quite simply removing the classes, which is why the method can be called the classless classification method. The visual variable *value* ranges from 0% to 100% black. The data span in the example is 34.0 – 8.1 = 25.9. The area unit with the value 8.1 is assigned a value of 0% (white) and the area unit with the value 34.0 is assigned a value of 100% (black). The method results in a map image that in terms of its variation in value accurately corresponds to the real distribution. This method provides the most accurate picture of the data distribution. The method does not result in any simplification of the quantity of data. The map image is thus less comprehensible and less easy to read than a map produced by a generalization with the aid of classification. The method is difficult to use with data distributions containing a few extreme values.

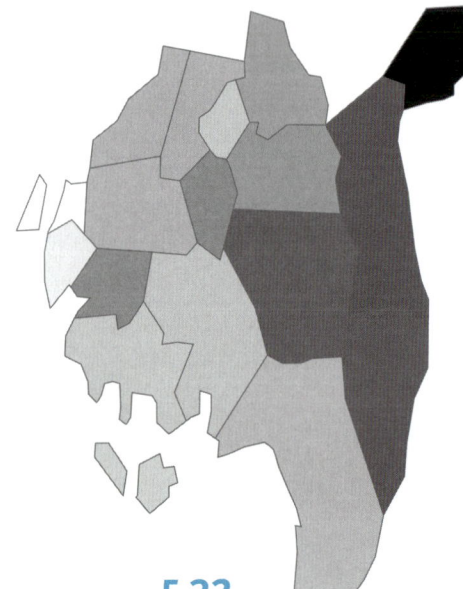

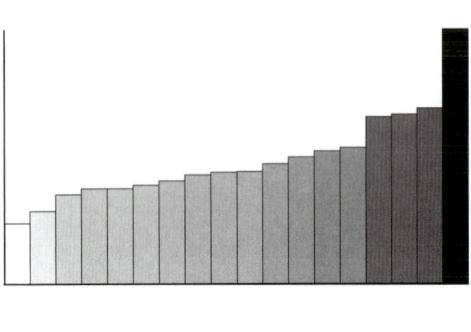

Other maps with area symbolization

DASYMETRIC MAP

A *dasymetric map* is a choropleth map in which the area units are processed on the basis of geographic knowledge. The element *dasy* is Greek for 'density', and *metric* is derived from 'metron', which is Greek for 'measure'. Ordinary choropleth maps provide an average value for the whole area unit. Dasymetric maps have an advantage over choropleth maps in that they contain additional information. Many themes show data relating to populations. If we know which parts of the country are unpopulated (geographic knowledge) we can leave the unpopulated areas out of the choropleth map. The map will then show variations in the theme in only those areas (the populated areas) in which the theme actually exists. In the example (Figure 5.33), the dasymetric method reduces the dominance of the municipalities that cover large areas, which would be a weakness in a choropleth map.

GRID MAP

A *grid map* is a choropleth map in which the area units (grid squares) are equal in shape and size. Since the areas are equally large, the map can visualize data containing absolute numbers. By using the same grid, different themes can easily be compared. The grid can also easily be aggregated into larger grid units, if privacy considerations or data processing requirements make this desirable. Figure 5.2 shows the population density with use of a 1 x 1 km grid.

ISOLINE MAP

An *isoline map* is one in which themes with continuous geographic distribution are visualized with the aid of isolines. Isolines are lines that pass through points of equal value. The word element *iso* is Greek for 'equal' or the 'same'. On a topographic map, isolines that pass through points of equal elevation above sea level are called *contours*. Isolines are used for phenomena that are continuously distributed. The production of such a map requires many geographically dispersed observation points, and the accuracy of the isolines depends on the value of the data points. The intervals between the values of the isolines should be constant (equidistant). A pure isoline map, such as the map in Figure 5.35, does not contain any visual variables. In order to perceive the isolines as ordered, one has to apply the visual variable *value* in the area between the isolines, as shown in Figure 5.36.

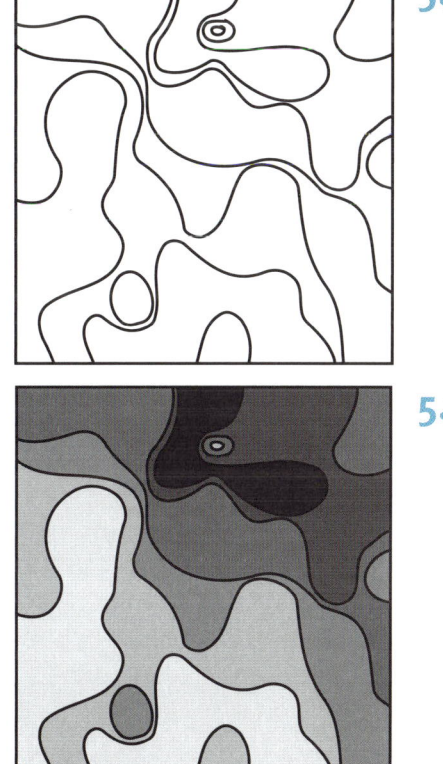

5.33 A dasymetric map is a refinement of the choropleth map's message based on geographic knowledge.

5.34 A grid system map is another name for a raster map.

5.35 Isolines pass through points with the same value.

5.36 Isolines with the visual variable value.

194 Other maps with area symbolization

5.37 A visualisation of the total amount of forest, and a selective visualisation of different types of forest are achieved by using shape and texture vis-à-vis the area symbolisation.

5.38 Sámi and Finnish speaking areas in Sweden. From the National Atlas of Sweden: 'Population'.

5.39 Visualisation of Sámi and Finnish speaking areas in Sweden in which the symbolisation corresponds with the basic data.

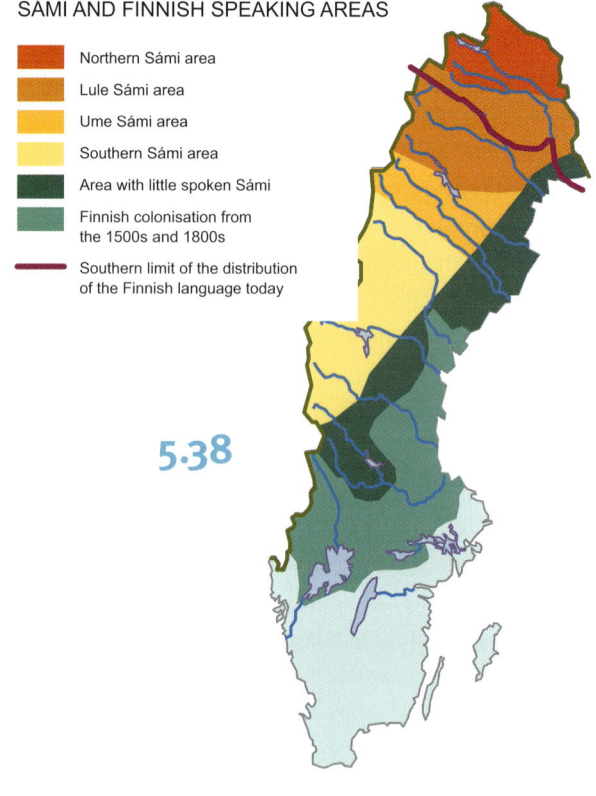

5.38

DISTRIBUTION MAPS

Distribution maps show the distribution of one or more (qualitative) themes within an area. In this case too, it has to be decided whether the whole area should be visible at a glance or whether the relevant data should be ordered. The purpose of Figure 5.37 is to show the total area covered by forest (i.e. the totals for all types of forest). In this case, it is necessary to use associative symbols with constant visibility (i.e. texture, shape, or colour).

EXAMPLE: CHOOSING AREA SYMBOLIZATION – SÁMI-SPEAKING AND FINNISH-SPEAKING AREAS IN SWEDEN

Figure 5.38 has been reproduced from *Sveriges nationalatlas* (national atlas of Sweden). The texts in the map key indicate that the map shows two different language groups. The visual distinction between them should therefore be as high as possible, as in Figure 5.39. In this map, variations in colour have been kept for the

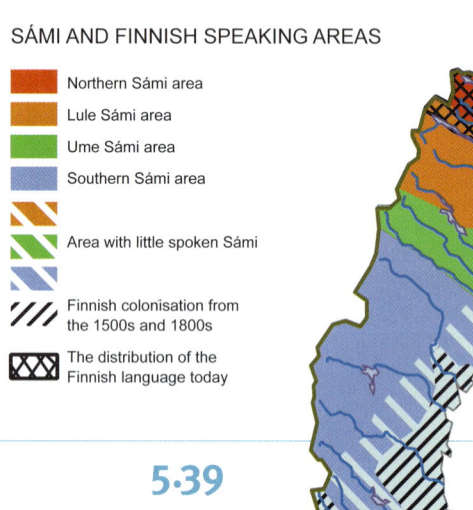

5.39

Sámi-speaking areas, whereas the Finnish-speaking areas are visually separated from the Sámi-speaking areas through the use of shape (a raster pattern) in black colour. Area representation has been used to delineate the 'Present-day distribution of the Finnish language', not a difficult-to-read line as is used in Figure 5.38. The highest value (cross-hatching) is used where Finnish is still spoken, compared to the area where it was previously spoken. In Figure 5.38 the Sámi-speaking area is divided into subgroups by variations in hue and value. The variation in value is incorrect in relation to the differentiation between qualitative area data, and therefore the colours in Figure 5.39 have the same value. It is completely wrong to use the colour green for areas where little Sámi spoken, since this colour is also used for Finnish-speaking areas. Rather, it would be appropriate to use the same colour as for the Sámi group to which the area relates, but a lower value should be used, in this case the striped pattern, since the area where little Sámi is spoken is an ordered perception in relation to the main area where the Sámi language is spoken. Variations in value are reflected in the relations between the stripes with colour and the stripes without colour.

ONE DATA SET AND MANY MAP TYPES – CHOLERA IN LONDON

Figure 5.40 is taken from the *Atlas of Disease Distributions* (Cliff & Haggett, 1988). It shows the different types of map that can be produced from the same data set. In this case, the data values represent the death rates for cholera in London in 1854. The basemap consists of an undulating line that depicts the River Thames, and the locations of St Paul's Cathedral (SP) and Westminster Abbey (W) are visualized as black squares. ND is an abbreviation of 'No data'. On the dot density map, each dot represents 10 deaths. The flow-line map shows the times of the cholera epidemic outbreaks.

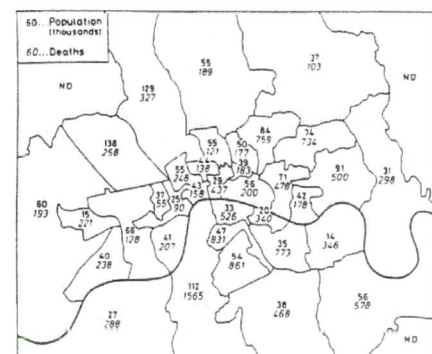
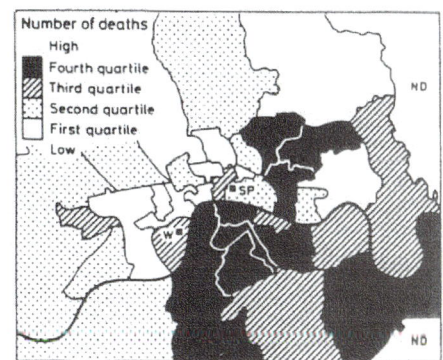
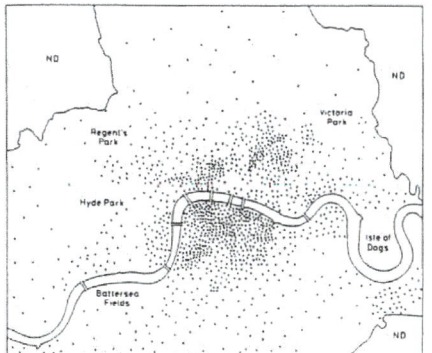
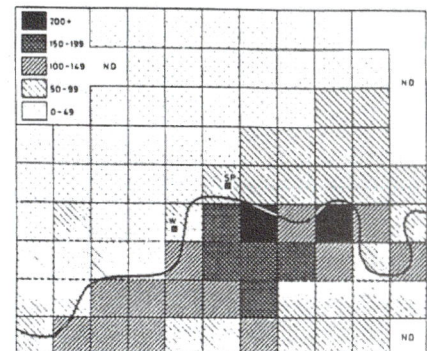

5.40

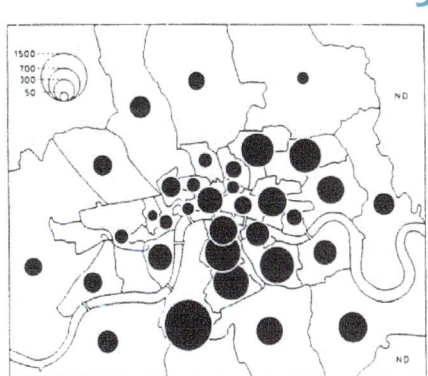
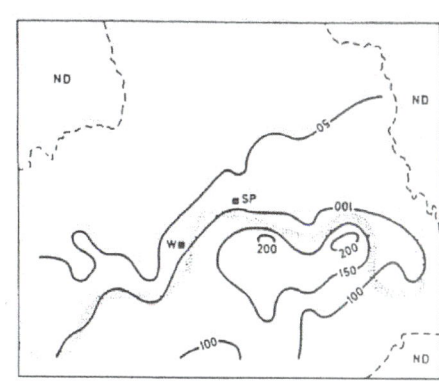
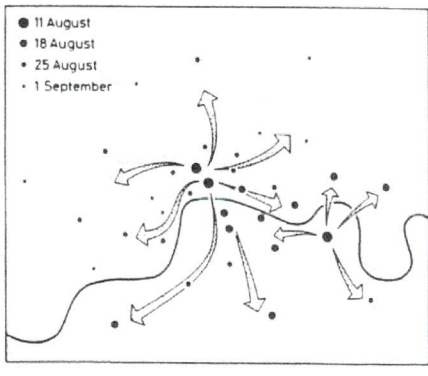
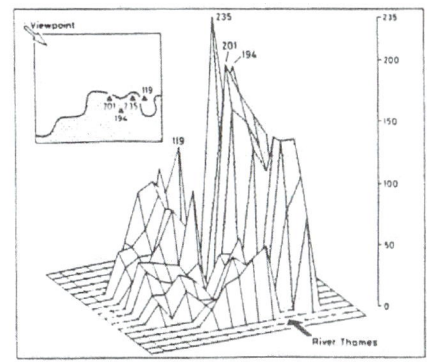

5.40 Cholera in London. A.D. Cliff & P. Haggett: 'Atlas of Disease Distribution'. The maps were all produced from the same data set: mortality and death rates from cholera.

Cartographic work process

Figure 5.41 shows the votes cast per municipality in the county of Østfold in the general election in 1997. Using these data as a starting point, this section describes the work process in the production of a thematic map, from raw data, through grouping and ordering, and generalization, to the final map that provides new knowledge.

READABLE MAP
Figure 5.42 shows a *pie chart map*. In diagram maps, charts are used as point symbols. Each point symbol contains more than one data variable. A graph has to be prepared for each unit area, in this case a municipality. The most common diagram maps are bar graphs (or bar charts) and pie charts. A pie chart with only one theme is a map with area proportional circles. It is not uncommon to see a pie chart map as cartographic presentation of a data set with many variables, such as listed in Figure 5.41.

Figure 5.42 shows nine theme variables: the number of votes for eight parties and the total number of votes per municipality shown as the size of the circles. The map provides answers to questions related to a single location regarding the characteristics of the party composition in a municipality. The map has to be read symbol by symbol, and it fails to provide

5.41 The table shows the number of votes cast in each municipality in Østfold County during the parliamentary elections in 1997.

5.41

Municipality	SUM	Frp	H	V	KrF	Sp	Ap	SV
0101 Halden	15 071	2 189	1 528	363	2 147	884	7 279	681
0104 Moss	14 335	2 502	2 671	387	1 227	332	6 379	837
0105 Sarpsborg	26 867	4 393	2 802	856	3 685	1 407	12 461	1 263
0106 Fredrikstad	37 569	6 174	4 903	1 530	5 002	1 663	16 105	2 192
0111 Hvaler	1 986	418	318	108	244	100	695	103
0118 Aremark	911	85	68	23	211	195	297	32
0119 Marker	1 991	183	158	26	470	374	730	50
0121 Rømskog	434	35	23	4	149	67	151	5
0122 Trøgstad	2 886	538	277	93	465	560	875	78
0123 Spydeberg	2 525	419	345	88	380	388	788	117
0124 Askim	7 397	1 547	1 051	272	973	423	2 804	327
0125 Eidsberg	5 359	915	614	190	916	768	1 765	191
0127 Skiptvet	1 794	302	138	50	332	344	575	53
0128 Rakkestad	4 159	427	364	108	946	919	1 300	95
0135 Råde	3 651	614	481	82	659	497	1 214	104
0136 Rygge	7 239	1 296	1 389	195	729	447	2 839	344
0137 Våler	2 217	379	368	56	325	312	696	81
0138 Hobøl	2 375	425	308	95	358	333	760	96
Total	138 766	22 841	17 806	4 526	19 218	10 013	57 713	6 649

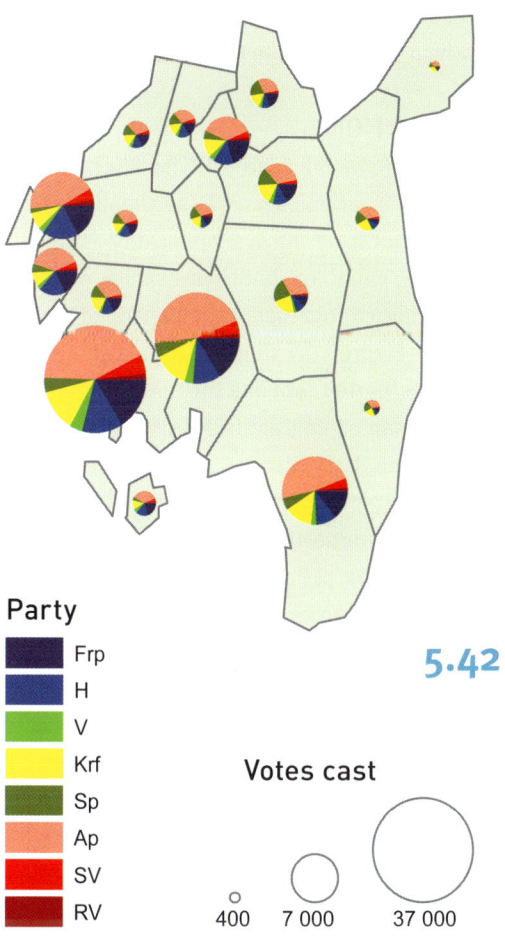

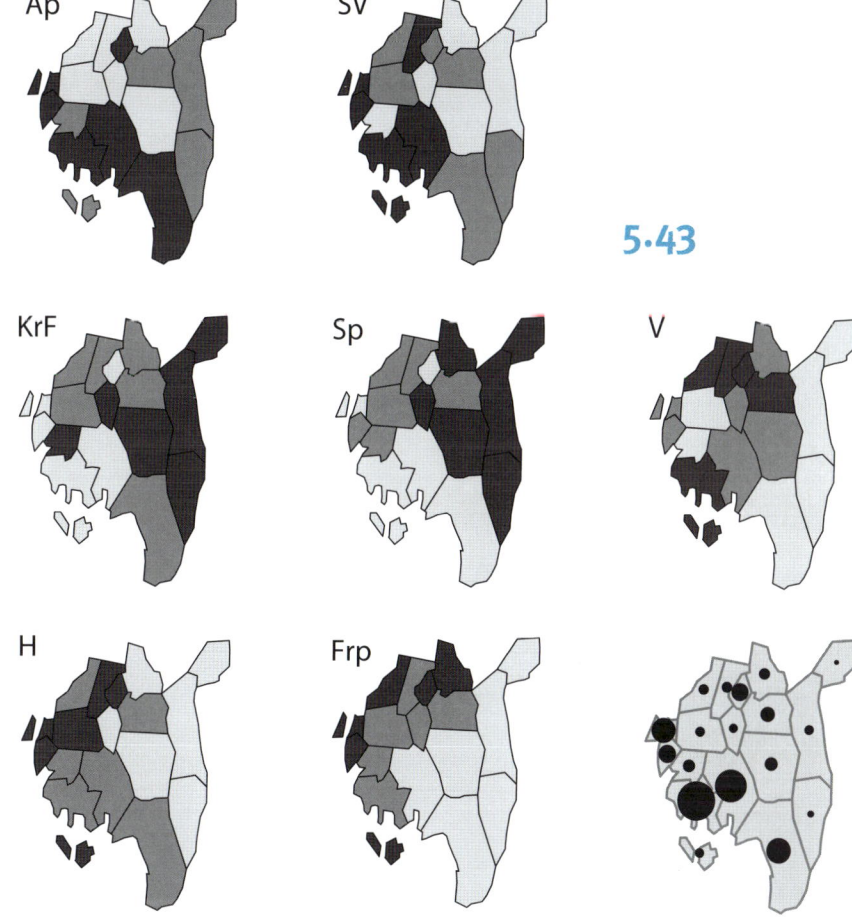

an immediate answer to the question of where the KrF party dominates. Maps with many variables can only be read at local perception level. A readable map does not allow for comprehensive perception, and a table provides an equally rapid and simultaneously more precise answer than a map. Thematic pie chart maps thus fail as a communication medium and tables would be a more effective means. Such maps may work to some extent if only two or three variables are presented.

VIEWABLE MAP

Figure 5.43 shows a series of choropleth maps. Each map contains only one theme variable. All of the maps are based on quantile classification with three classes. At a glance, the maps quickly reveal information about where each party is strongest – the map is viewable. The map with area proportional circles show the total number of votes cast per municipality. Viewable maps allow one to find geographic patterns. Answers to the question of where the KrF party is dominant are readily apparent from the map, and the map can immediately be perceived as a whole. In a viewable map it is possible to perceive a defined theme in the whole image – the map is read at a global perception level.

Each choropleth map shows the distribution of a particular party and is the appropriate cartographic solution to answer questions regarding the degree of dominance for a given party. However, in order to compare the distribution for one party with the distribution for the other parties, one has to shift one's gaze from one map to the next. The information resulting from the comparison of

5.42 The pie chart map fails to communicate information.

5.43 A series of choropleth maps convey the geographic variations in the data. The total numbers of votes cast in each municipality are represented by proportional circles.

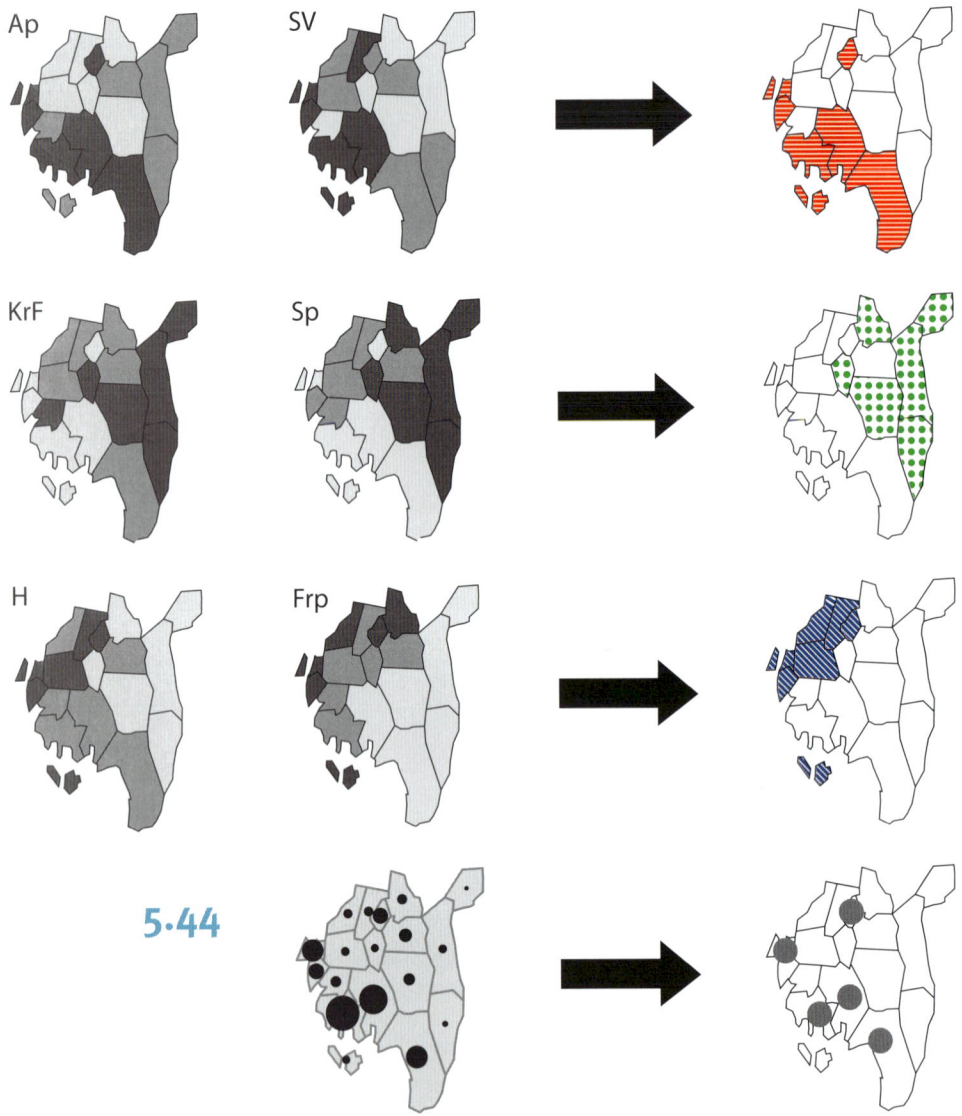

the eight maps is new (a synthesis of the individual topics), which is not possible to maintain visually.

GENERALIZATION

In order to understand a synthesis visually, the individual theme variables have to be simplified considerably. In cartography, simplification is called generalization. Variables that have a similar distribution are grouped together. A comparison of the eight maps in Figure 5.43 shows that the Labour Party (Arbeiderpartiet, Ap) and the Socialist Party (SV) have the same geographic distribution, as do the Conservative Party (Høyre, H) and Progress Party (Fremskrittspartiet, Frp), and the KrF and Centre Party (Senterpartiet, Sp). The distribution of votes for the Liberal Party (Venstre, V) shows a more irregular pattern and is therefore omitted. The distributions for Ap and SV (both left-wing parties) have been simplified to a binary map showing where this group dominates. The same has been done for H and Frp (both right-wing parties) and for Sp and KrF (both centre parties). The map showing the number of voters has been simplified to show the five municipalities with the highest number of voters (the towns of Moss, Fredrikstad, Sarpsborg, Halden, and Askim) compared to other municipalities (rural municipalities). The result is four binary maps. This method can also be performed with bar graphs instead of choropleth maps.

5.44 Choropleth maps, with similar distribution are grouped together and simplified to binary maps.

5.45 The binary maps are compiled.

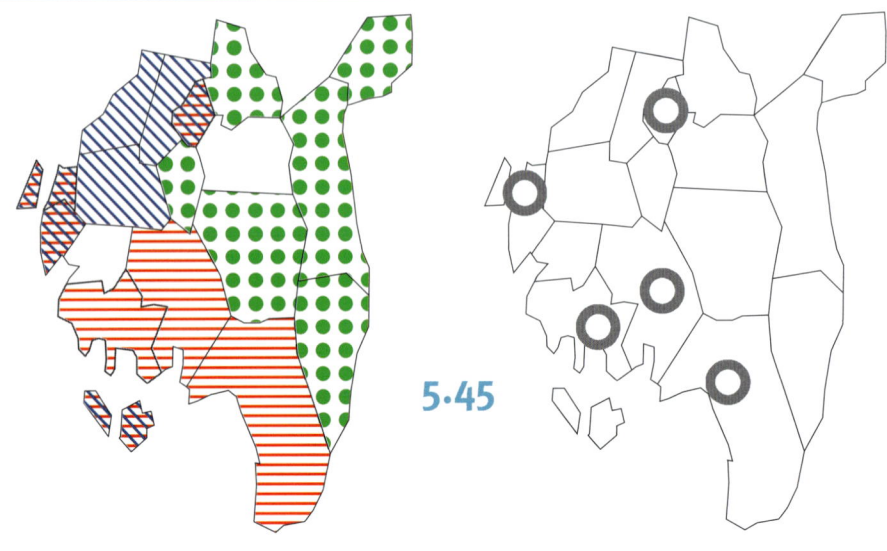

COMMUNICABLE MAPS

When binary maps such as those in Figure 5.45 are combined, the result is a map like the one shown in Figure 5.46.

The map includes all nine theme variables, but in a greatly simplified form. The information content of the map is reduced. In contrast to the pie chart map in Figure 5.42, the map in Figure 4.56 enables the essence of the data to be perceived immediately. Compared with viewable maps, a communicable map provides (simplified) answers to the question of political party distribution in, for example, Rakkestad (the centre parties dominate). As already mentioned, the disadvantage is that the information is highly simplified. It is important that a map is memorable, and the details of the information are less important. If a map communicates the essence of a geographic analysis, it is a communicable map.

Communicable maps combine the global perception of the whole image with information about occurrences at each location.

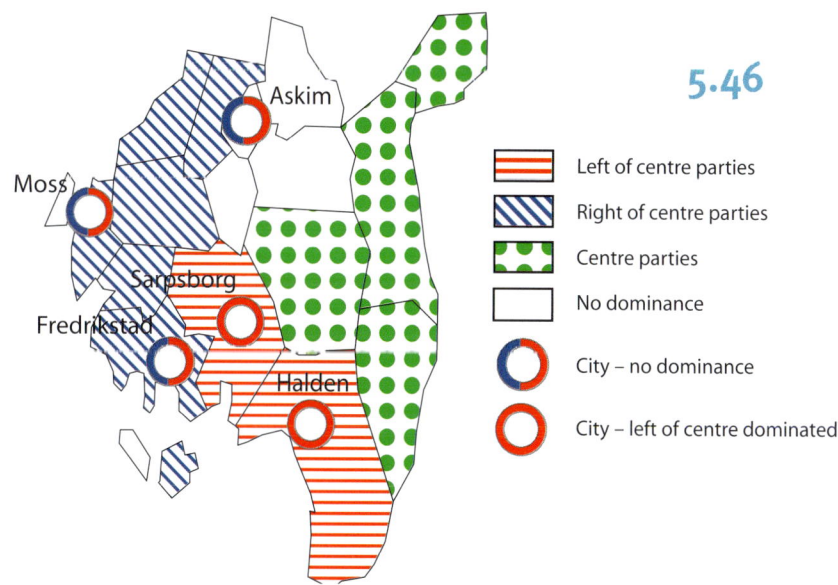

Map design

FOREGROUND AND BACKGROUND

Some elements will highlight a theme and should therefore be in the foreground, while other elements (basemap elements) should be in the background. Colour, value, and shape can be used to create the effect of depth so that some elements are more prominent.

Pale yellow or grey (possibly including some hints of blue, green, or brown) should be used as the background colour. Yellow and grey are often in harmony with the primary colours. Complementary colours go well together in terms of contrast and aesthetics; complementary colours are red-green, blue-orange, and yellow-violet. For example, a complementary colour to the background colour can be selected for the map symbols that need to be highlighted.

Contour lines around map objects should only be used when necessary. Contour lines are only required when two adjoining map objects have the same value. The use of contour lines can thus be avoided by increasing the value or the contrast in brightness between two adjacent map objects, such as a light tone for sea and a darker tone for land.

TITLE AND KEY

In addition to the map itself, the following elements should normally be included:

- A *title*, which should identify the theme, what the map shows. The theme may imply the need for a rather long title, but a short, simple title can be created that differentiates the map from the other elements in it, followed by a concise description of the theme as a subtitle. The subtitle should identify the geographic area covered by the map. However, if the area is widely known, such identification can be omitted (e.g. maps of Norway).

- A *key* (or *legend* in American English), which explains the symbols used. The key should identify the symbolization used for each theme. A good map should be self-explanatory to the extent that a map user can understand what the map is about at a glance, without necessarily having to consult the key.

5.46 The communicable map provides a simplified and immediate visual perception of distribution of voting patterns for political parties in Østfold.

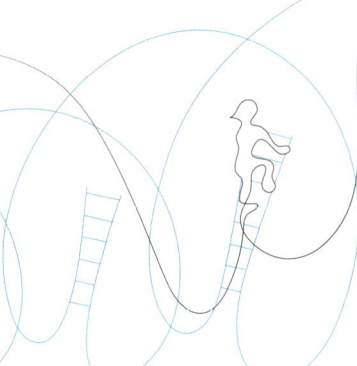

200 Map design

5.47 County names, region names, and place names in the southern part of Norway, with good and bad examples of the positioning of the names and with most of the elements normally included on maps, except for overview maps. Map data: © Norwegian Mapping Authority.

5.47

This can be ensured when the map information conforms to the visual rules and there is correspondence between the thematic data and the visual variables.

- A *scale*, which is shown graphically as a ruler, often with scale numbers. A scale may also appear as scale numbers, but cannot replace the linear scale.
- A *north arrow*, which shows the direction in which north is located
- A *coordinate system*, which is illustrated by a grid or some other form of marking
- An *overview map*, which shows the location of the actual area
- *Source information*, which provides information about who produced the map, the date it was last revised, the quality or source of the data, and the map's serial number.

TEXT AND PLACE NAMES

Antikva fonts, in which the thickness of the individual letters varies, such as Times New Roman, are the best font types for continuous text on a map. In titles and other individual text elements such as place names, fonts with letters of the same thickness will be better, for example grotesque fonts such as Arial or Helvetica. Text can be made more readable by adding a white background around text (to mask out information under the letters).

Place names on maps identify the geographic objects presented on the map. Ideally, the place name of a point object should be positioned slightly above and to the right of the point, as in the case of the name *Hamar* in Figure 5.47 (north of Oslo). The place name *Koppang* (Hedmark County) is an example of a name that is poorly positioned to the right of the point, resulting in a flat and lifeless appearance. The position should be adjusted in relation to other place names and other map elements, just as *Gjøvik* has third priority in placement, *Flisa* fourth, and *Kongsvinger* sixth. Place names should be shown either over land (e.g. *Mandal*) or over sea (e.g. *Egersund*), but not across coastlines, as in the case of *Sandnes*.

Place names can refer to a point object such as a village (e.g. *Tynset*), describe the length and orientation of a line object (e.g. *Rendalen*), or define the extent of an area object (e.g. *Østerdalen*). If the area object is so small that the name does not fit the available space, it will be treated as a point (e.g. *Smøla*).

The names of line objects should follow the curvature of the lines and placed above the object (e.g. *Lågen*), not below them (e.g. *Glåma*). The extended parts of the letters force a word's baseline to be positioned farther away from the object, and therefore in the case of a place name that has more letters with ascenders (e.g. b, d, f, h, k, l, and t) than descenders (e.g. j, g, p, and y) it will be preferable to position the name above the object's location.

5.48 The order of priority for locating the place names of point objects.

Web cartography

DATA PRESENTATION

Old paper maps functioned both as a form of storage for spatial data and as a presentation form. Today, geodatabases and presentation are increasingly becoming two separate processes. In order to create standardized map presentations (map products), first a selection of data sets in the geodatabase has to be made (often, not all of the data will be included on the map). For each selected data set, rules will be establishing specifying how different geographic objects are to be presented on maps at a specified scale. There will be different selection rules and sign rules for each scale and for each purpose. Examples of such illustration rules include the Norwegian Mapping Authority's map standard *Grafisk utforming av kart i målestokk 1:500 – 1:10 000* (Graphic design of maps at scale 1:500 – 1:10 000) for the presentation of **Felles kartdatabase** (FKB) (national database) geodata and the Ministry of Local Government and Modernisation's *Nasjonal produktspesifikasjon for arealplan og digitalt planregister: Del 2 – Spesifikasjon for tegneregler* (National product specifications for land-use and digital plan registry: Part 2 – Specification for signs rules) for presentation of land use plans.

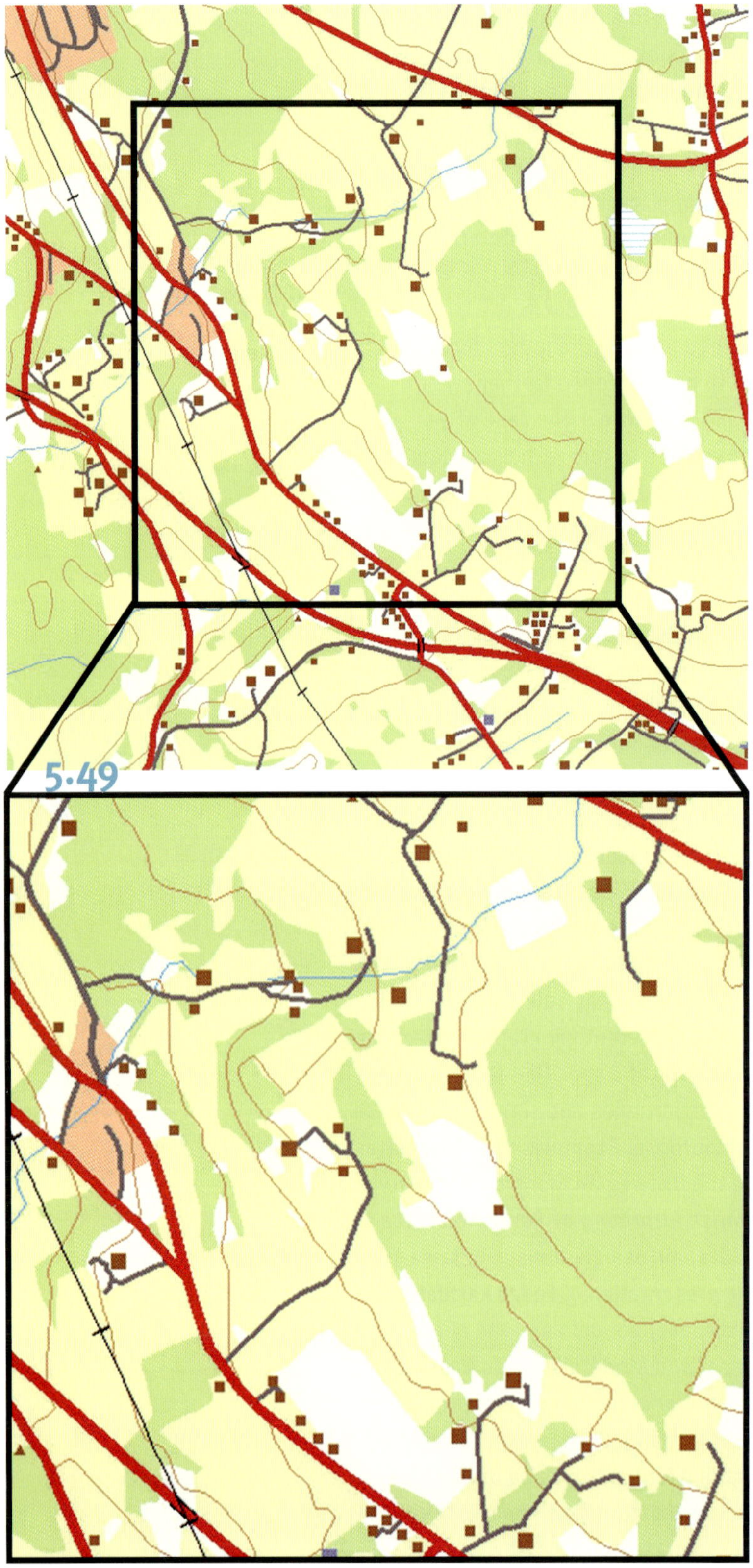

202 Web cartography

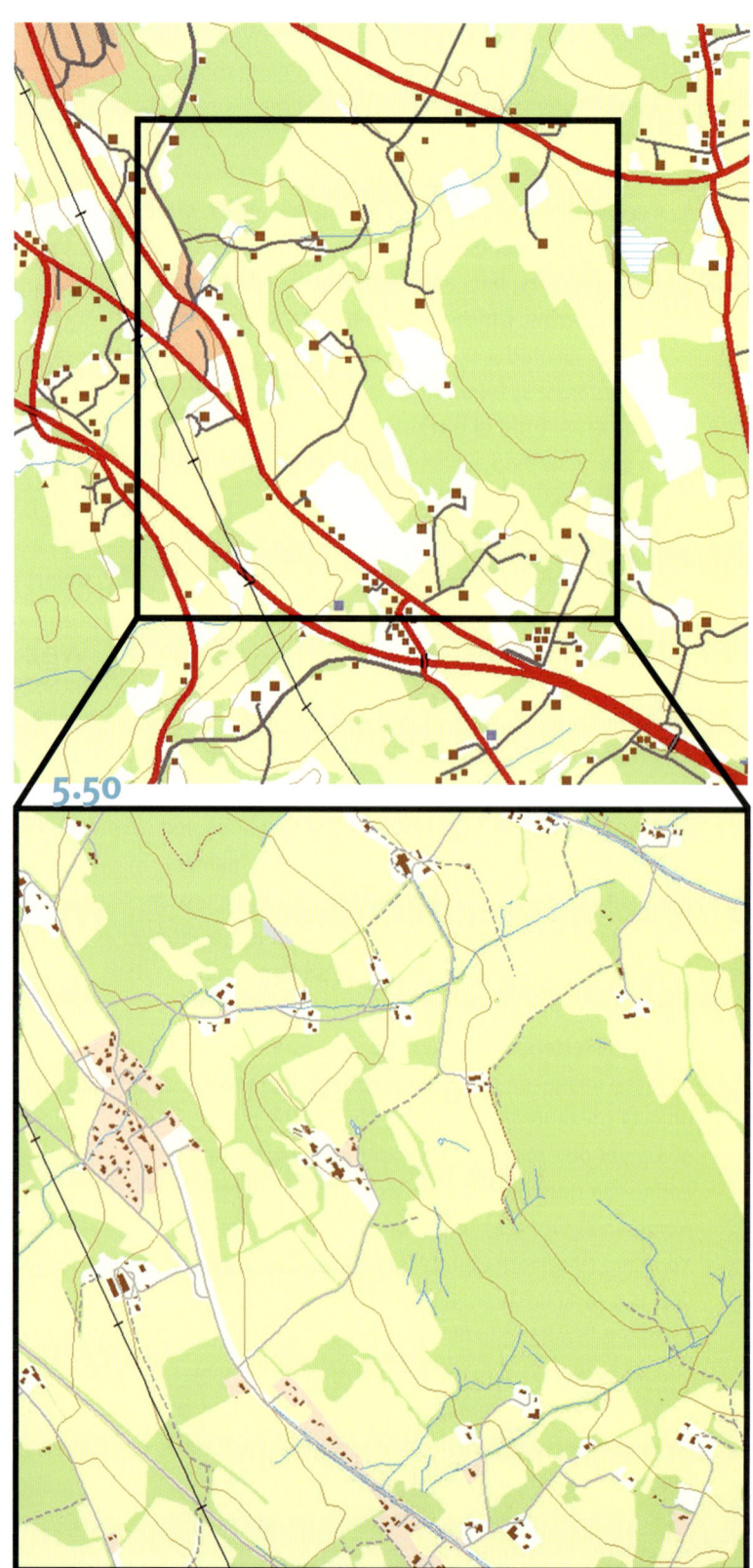

5.49

5.50

Ideally, computers should generate good map images from the relevant selection rules and sign rules. This is possible today, but computer-generated images of maps are normally complex. Hence, it is often necessary to establish auxiliary data in the form of presentation data suitable for different series of maps and map presentations. The most important element in the presentation data is text. In addition, false polygons (masked polygons) are used to remove information from areas where the map image is unclear.

SCALE AND DETAILING

Printed analogue maps are designed with a defined scale, whereas web maps can tolerate presentation at a larger scale. Zooming between the detail and the whole map presents new challenges for web cartographers. Should it be possible to enlarge the map without any changes to the content (*static linear zoom*) or is it desirable that the map content becomes more detailed as one zooms in and, if so, what scale intervals should be used? With *static non-linear zoom*, a finished map is laid over the same area with different generalizations for defined scale intervals. With *dynamic zoom*, the image is presented with selected data from the same database, for example where the database polygon objects (e.g. house) is presented as a polygon at a large scale (e.g. 1:5000), whereas in a smaller scale (e.g. 1:50 000) it is visualized with a point symbol.

ANIMATION

Screen media provide opportunities for maps to be presented as animations in which the images change, usually with time as a variable. The American geographer Alan M. MacEachren has defined four dynamic visual variables to describe time as a visual variable (MacEachren 1995):

- Duration
- rate of change
- order
- phase – rhythmic repetition of events.

Duration refers to how long a frame or image is displayed in the animation. *Rate of change* refers to the degree of change in position and attribute values for theme objects from one image to the next. The variable *order* refers to whether the still images are sorted chronologically or by attribute value. The variable *phase* can be used when it is desirable that the theme object should flash (rhythmic repetition of an event).

5.49 In the case of static zooming the map image is enlarged without the map content becoming more detailed.

5.50 In the case of dynamic zooming the map content becomes more detailed. Note how the square point symbols for buildings are visualised with a building outline when zooming in.

5.51 Still images from an animation showing areas with precipitation and clouds on 26 November 2015 at 11, 12, 13, and 14 hours (from the weather portal yr.no). Forecasts from Yr are provided by the Norwegian Meteorological Institute and the broadcasting company NRK.

5.51

Time 11.00

Time 12.00

Time 13.00

Time 14.00

Magnus Fjetland graduated from the Norwegian Military Academy in 2011. At the academy he studied Military Geography and got a bachelor degree in civil and military engineering. He holds the rank captain and works as an instructor in engineering at the Norwegian Military Academy. Captain Fjetland has a background from both national and international service in the Norwegian Armed Forces.

Alexander Steffensen graduated from the Norwegian Military Academy in 2010. At the academy he studied Military Geography and got a bachelor degree in civil and military engineering. He holds the rank captain and works as an instructor in engineering at the Norwegian Military Academy. Captain Steffensen has a background from both national and international service in the Norwegian Armed Forces.

Military geography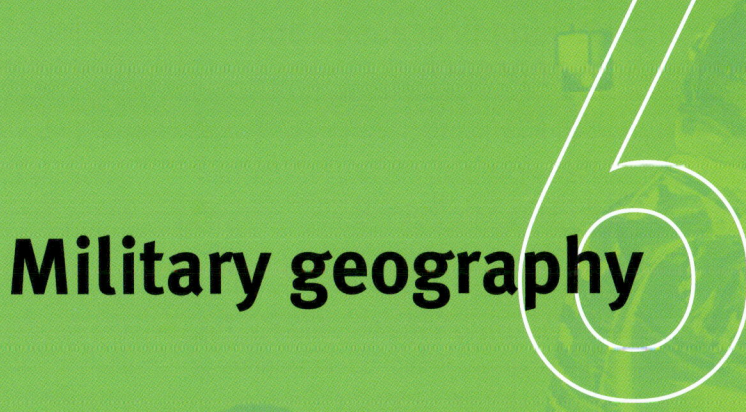

Military working processes and methodical approach   *206*
Contribution categories   *209*
Military methodology example – planning and decision-making process   *210*
Military use of GIS   *215*

Military working processes and methodical approach

Various regulations have been established, doctrines formulated, and books published in order for the Norwegian Armed Forces to fulfil their missions as effectively as possible. Through such literature, continuity and consistency have been maintained in work across departments, both nationally and internationally. The field of military geography (MilGeo) is no exception in this regard. Years of experiences have resulted in more or less standardized methods for creating a common understanding of how analyses should be performed and how products should appear and be presented. The use of GIS is central in this respect, and so examples of the application in military planning procedures are covered in the following sections, without giving detail on how the GIS analyses themselves, were performed.

THE GIS CYCLE

Chapter 4 (Analysis) describes a working method that can be used in all types of analyses. The steps in the work methodology described in Chapter 4 largely correspond to the work methodology used by most MilGeo analysts in the Norwegian Armed Forces. The GIS cycle depicted in Figure 6.1 is an imagined systematized working method adapted for the MilGeo field.

The method starts with a type of mission analysis (Point 1 in the cycle, Requirement), where the purpose is to define the core objective of the mission. At this stage, *user competence* can present a challenge. What questions does the user want to answer and to what extent does the person concerned have the appropriate qualities to meet this need? Users' expertise and knowledge of the field can lead to misunderstandings and therefore high demands are placed on the MilGeo analyst's ability to communicate effectively with different users in such ways that their needs are met. After an initial mission analysis, the MilGeo analyst will be responsible for ensuring that there is a common understanding of what is to be produced.

Point 2 in the GIS cycle shown in Figure 6.1 identifies what type of geographic information is required to perform the analysis. If there are any deficiencies in existing databases, measures will have to be taken to obtain the necessary information. Geographic or positional information is described in detail in Chapter 3 (Geographic data).

The Norwegian Armed Forces use several methods and means to obtain geographic information, such as through national and international cooperation agreements or the physical collection of data. During the implementation of a military operation, information will also be made available through, for example, reporting. For all such retrieval methods the data have to be organized and managed in digital databases before they can be used in a GIS. This process is particularly important in order to maintain the best possible data quality.

Analysis start at point 3 in the GIS cycle. At this stage, it is important to be aware of the importance of the following concepts: analysis and evaluation, conclusion, and product. As part of their analytical work, MilGeo analysts evaluate the results of their analyses. Their conclusions are based on these evaluations and are distributed as products. The results of an analysis can be part of the product, but they can often provide too much information, to the extent that the results will be weakened. What constitutes the final product will be largely determined at point 1 in the GIS cycle,

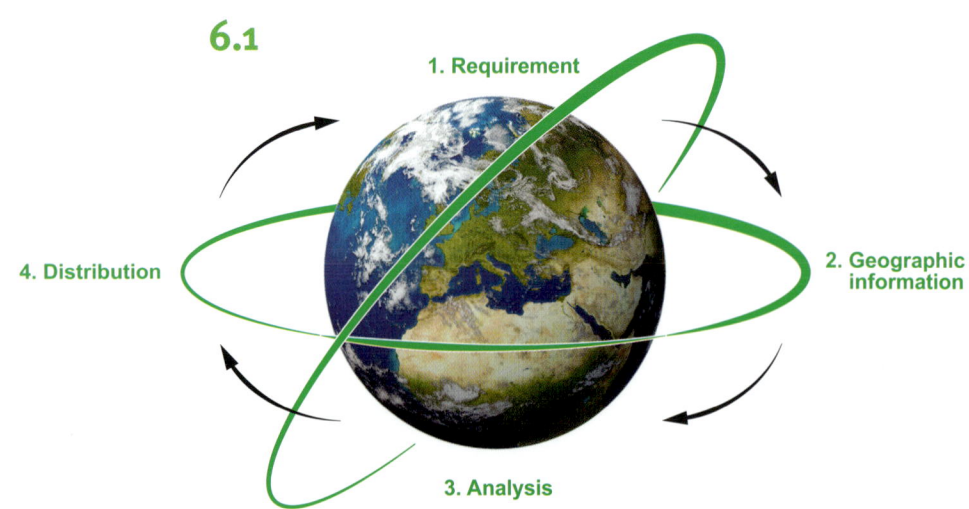

6.1 The GIS cycle. Source: Lightspring/Shutterstock.

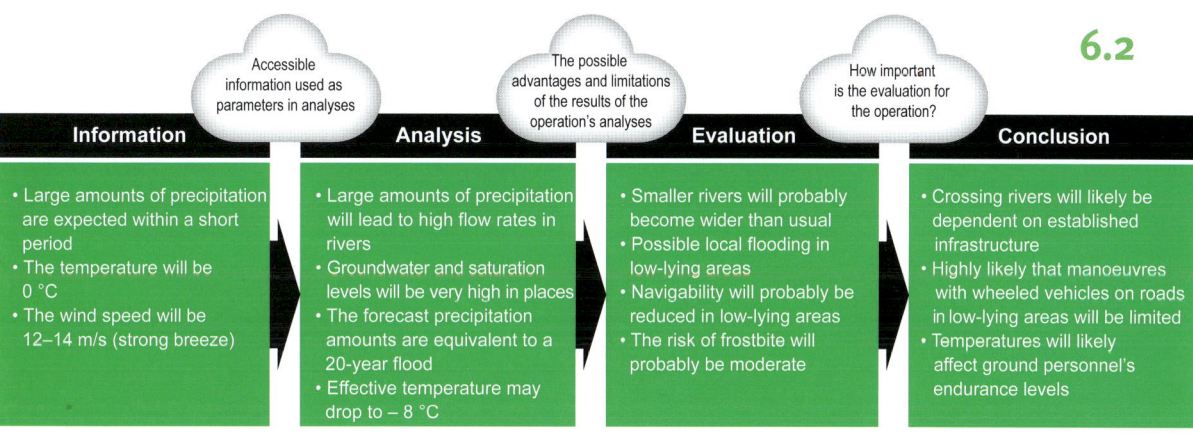

6.2 Progressing from information to conclusions.

where the need is defined. Hence, to ensure product relevance, the analyses and conclusions have to be continually evaluated against the original need, often in consultation with the user.

In common with practitioners from other specialist fields within the Norwegian Armed Forces, MilGeo analysts often work under the pressure of time constraints with tight deadlines. The delivery deadlines reflect the high pace of the work during operations. This may mean that products supplied after the deadline will be irrelevant. If it takes too long to produce the product, the users will no longer have any need for it. The analytical work should therefore be adjusted so that the assessments and conclusions are made during the production process. This would ensure a degree of relevance in cases where unfinished products have to be distributed, and would provide a considerably better decision basis as opposed to one based only on parts of the analysis.

Figure 6.2 shows a hypothetical example of a military decision-making process starting with information and ending with a conclusion. Available information constitutes the parameters in the analysis, which in this case is the weather forecast. The results of the analysis show that the forecast rainfall would add large volumes of water to land that is already highly saturated. Wind speeds of 12–14 m/s are equivalent to strong breezes on the Beaufort scale which combined with a temperature of 0 °C, would lead to a wind-chill-effect temperature of minus 8 degrees Celsius. The results of the analysis would be useful, but without evaluation they would at best only be indications of the possible effects and/or limitations imposed by the weather on the operation. Evaluations convey validity, describing the assessed likelihood of an event occurring, such as a possible local flooding event that would probably lead to impaired manoeuvrability off-road in low-lying areas. In this case, an effective temperature of −8 °C would probably mean a moderate risk of frostbite. Such evaluations are subsequently linked to the impacts they will have on the military unit conducting the operation. In the example shown in Figure 6.2, it is concluded that it is highly likely that manoeuvring with wheeled vehicles will be limited to the roads in local areas. It is likely that frostbite could affect the endurance levels of military personnel operating on the ground. It is therefore possible to say something about the potential impact of the physical environment on the operation. From an analysis of the available information and an assessment of the results, the decision-maker would then be able to decide on an appropriate course of action.

As described in Chapter 4 (Analysis), it is important to evaluate the results of an analysis and the conclusions to ensure that relevant issues are adequately addressed. When the product is distributed with the conclusions, it is important to be able to account for any uncertainties and inaccuracies in the data on which the analysis is based. Also, documentation of the method is critical in order to be able to say something about the quality of the results.

6.3 Examples of terrain analysis. Flood data: NVE; basemap: © Norwegian Mapping Authority.

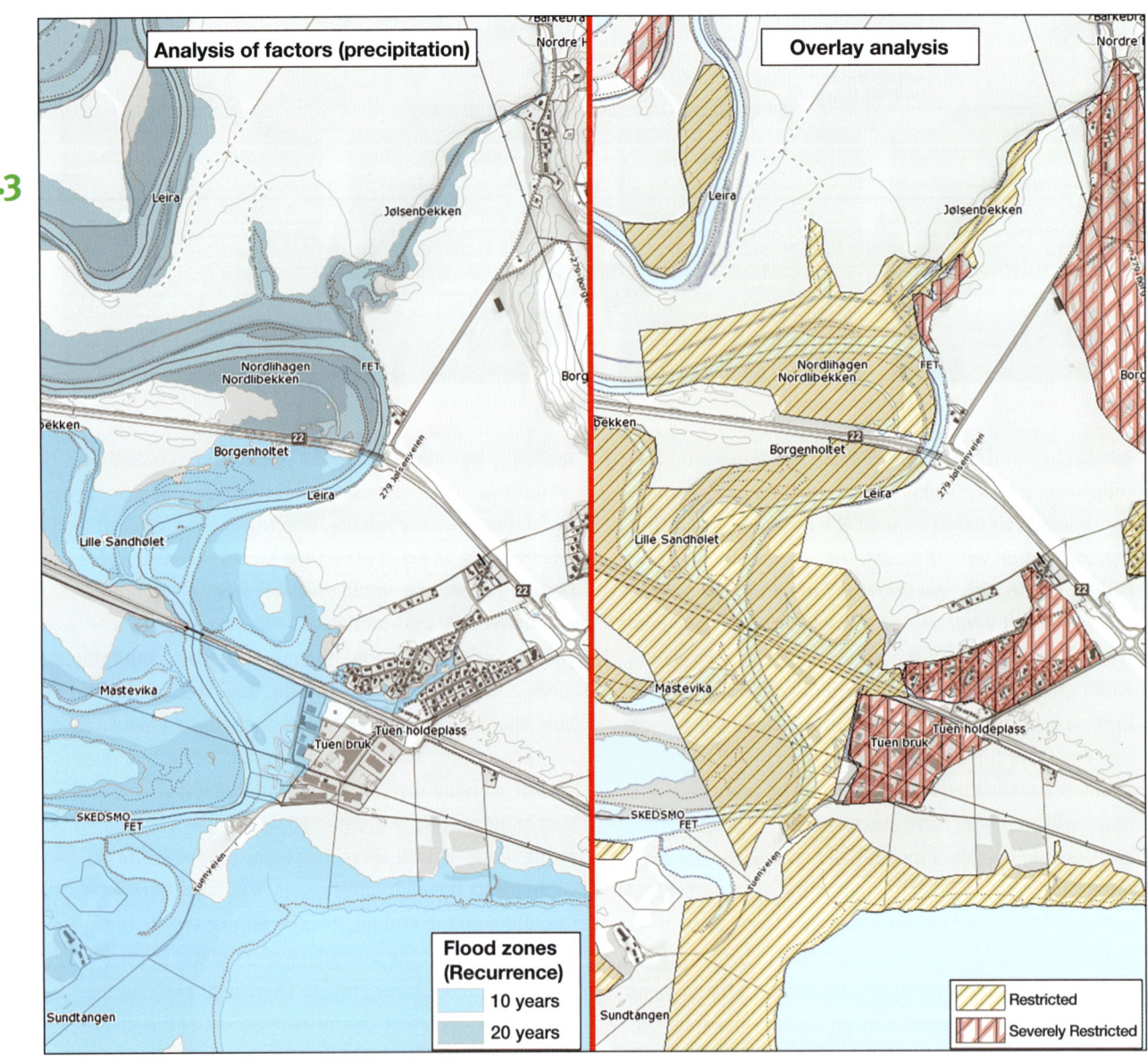

Point 4 in the GIS cycle covers distribution of the products, which may include information, analysis and/or evaluations, and conclusions. The examples from the terrain analysis in Figure 6.3 are partly based on the analysis in Figure 6.2. The overlay analysis summarizes the main conclusions from a number of analyses. Vegetation, buildings, and other factors are analysed in a similar manner as precipitation, in order to assess manoeuvrability in an area. Examples of products from a terrain analysis are shown in Figure 6.10. The preparation of a product is part of the distribution of that product. Chapter 5 (Presentation) describes cartographic theory and regulations, and how they are used to promote the results in the best possible way. In order to achieve an academically sound and straightforward product, it is essential to have knowledge of the theory of the visual variables combined with information levels. Figure 6.3 shows an overlay analysis that has been generalized using a GIS to improve its legibility and add clarification as results.

Contribution categories

MilGeo has its place as a recognized is an acknowledged authority for making contributions to many varied contexts at different levels. Its contributions, varying largely according to need, consist of products, simple analyses, and evaluations and assessments. The contributions fall into two fundamentally different categories: decision support and mission support (see Figure 6.4). The purpose of these categories is to show how different contributions can serve different purposes for different target groups and objectives. The terms 'decision support' and 'mission support' may be known to many readers from other military disciplines, but here they should be considered in relation to MilGeo.

Decision support is intended to provide decision-makers with the premises on which to base a decision and take appropriate action. To do this most effectively, a decision-maker will rely on the efficient utilization of their resources. In most military organizations and departments, staff are organized in a linear structure with standardized functions and specialist expertise in different disciplines. MilGeo analysts possess special expertise in the technical analysis of terrain and provide decision support in the form of products based on, for example, terrain evaluations. When it comes to distribution, decision support is treated separately from mission support.

Decision support is often distributed in the form of a written product presented jointly with decision-makers and other users. With the distribution of, for example, intelligence preparation of the operational environment (IPOE), the written product is considered the main product. Through presentations, decision-makers and other users have the opportunity to ask questions, make comments, and put forward suggestions for further work. In addition to knowledge of cartographic and geographic visual communication, the MilGeo analyst's presentation skills, both oral and written, are important for the *distribution* of decision support. MilGeo analysts help with decision support as an important part of a number of military work processes and methods. As a result, they depend on knowledge of existing plans and access to information, such as intelligence about the enemy and assessments of the conditions of friendly forces. This means that products can be tailored to provide the best possible decision support.

Products for mission support include more or less all products that support the accomplishment of any military operation, such as digital and analogue map products. Products in this category are often distributed without written or verbal monitoring. Thus, what the products convey largely depends on visual representation. Chapter 5 (Presentation) shows that, as a product, a good

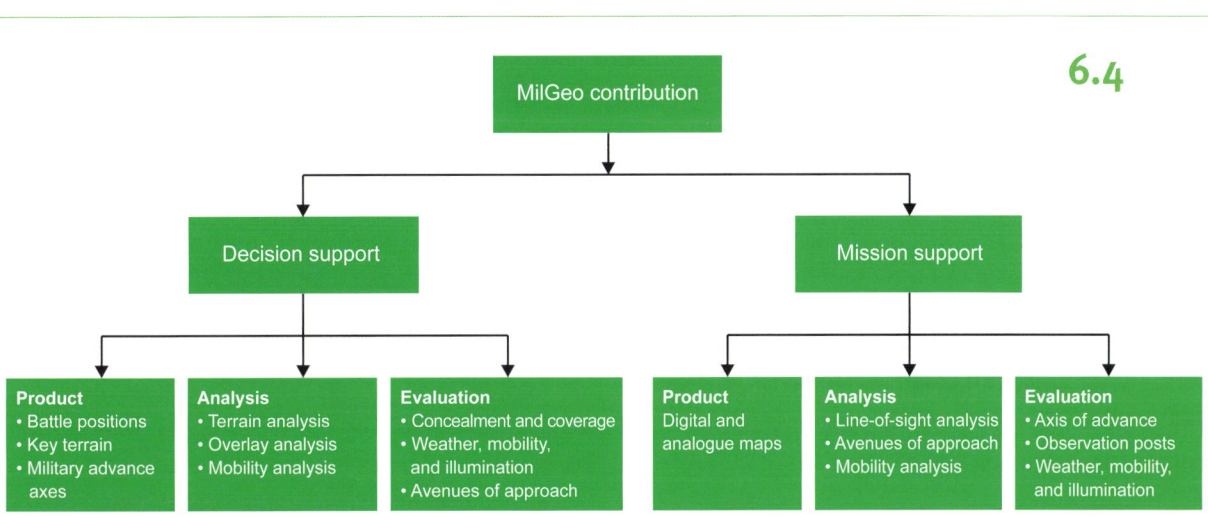

6.4 Flowchart of Milgeo's service provision, with examples.

6.5 Operations planning – part of the Norwegian Military Academy's engineering course.

Military methodology example – planning and decision-making process

The Norwegian Army's staff manual *Plan- og beslutningsprosessen* (PBP) (English translation: Planning and Decision-making Process, PDMP) describes an established and well-known military method that has existed in more or less the same form for a number of years. It states: 'The purpose of this staff manual is to create the basis for a unified approach to planning the Army's operations' (Hærens våpenskole, 2015. English translation: Norwegian Army Weapons School). Figure 6.6 shows the five steps of the PDMP process, together with one of the two subprocesses that helps to create the basis of a good PDMP product. A MilGeo analyst has an important role in the IPOE process, which, in addition to methodological efforts to achieve goals, is one of the two above-mentioned subprocesses in PDMP. Like PDMP, the two subprocesses are used at different levels by the Norwegian Armed Forces. In some departments in the Norwegian Armed Forces, MilGeo analysts have special responsibility for Step 2 in the IPOE process.

The Norwegian Army Weapons School states that '*Intelligence Preparation of the Operational Environment* (IPOE) is a systematic, continuous and cyclical, analytical process involving assessments of actors and their surroundings in a specific geographic area' (Hærens våpenskole 2015). The process is divided into four steps, each of which covers several different topics. Figure 6.7 shows the relationship between the four steps in IPOE process. Regardless of whether the operational area is already known, an IPOE product will help to increase situational awareness and a common standpoint for all involved parties. Figure 6.7 shows Step 1 as defining the area of operations. The area of operations defines, for example, the geographic framework for further work in the IPOE process. Step 1 also involves the identification of relevant factors and any immediate requirement for information. There is no emphasis placed on analysis in Step 1. To ensure an efficient process, it is important that each factor will have an impact on own and/or the opposing forces.

map can convey information in an efficient and transparent way. A user's ability to read, see, and interpret the information is largely determined by the use of the product, which further emphasizes the importance of good cartographic communication. Conventional cartographic rules are supplemented by the Norwegian Armed Forces' own cartographic descriptions, map symbols, and military exercise graphics. These rules can be regarded as aids to achieving continuity and interoperability in products intended for military use, and their usage is intended as far as possible. Different regulations govern the use of such map symbols across military disciplines, military departments, and nationalities.

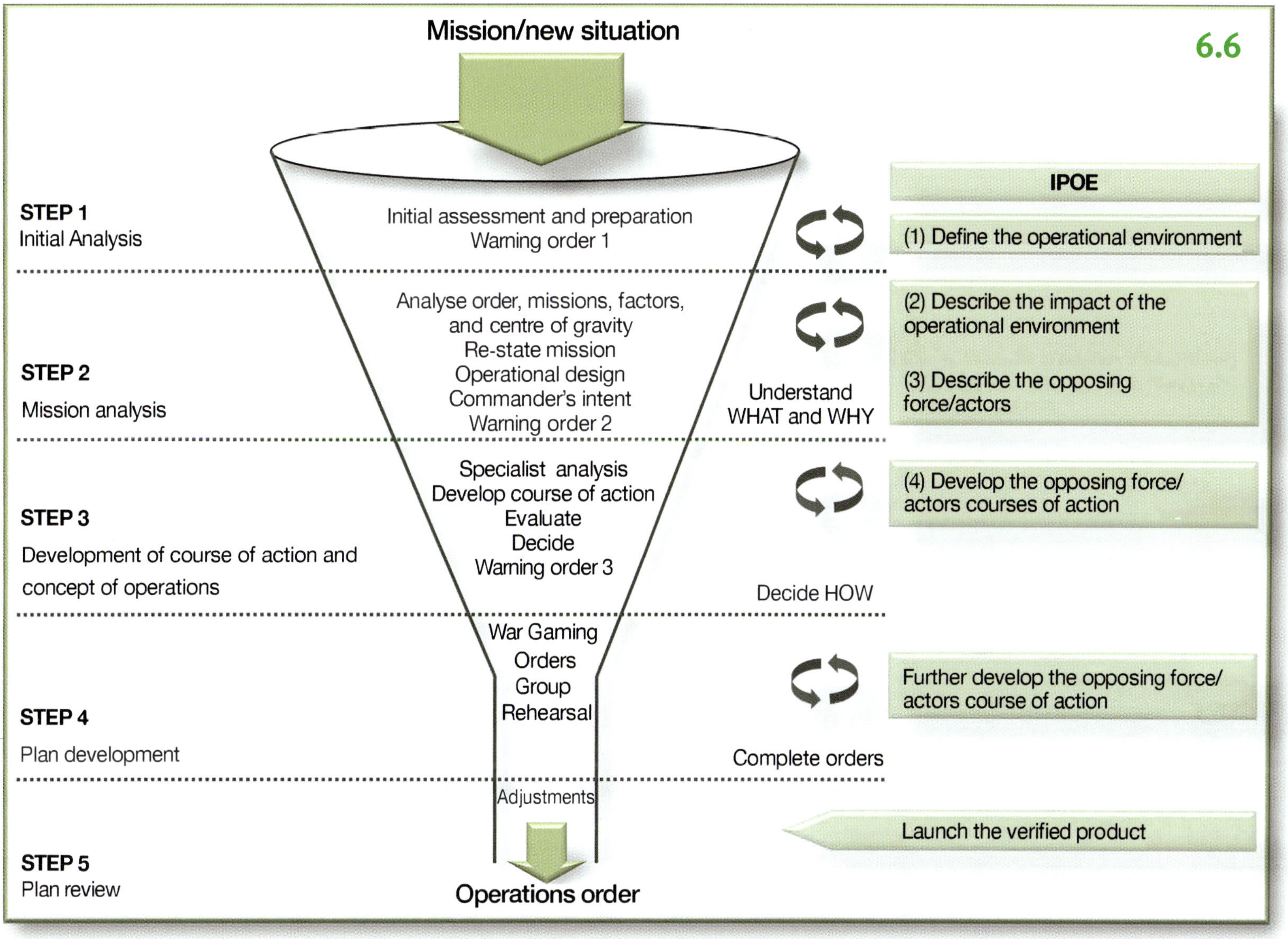

6 – MILITARY GEOPGRAPHY

6.6

	Mission/new situation ↓		**IPOE**
STEP 1 Initial Analysis	Initial assessment and preparation Warning order 1	⇄	(1) Define the operational environment
STEP 2 Mission analysis	Analyse order, missions, factors, and centre of gravity Re-state mission Operational design Commander's intent Warning order 2	⇄ Understand WHAT and WHY	(2) Describe the impact of the operational environment (3) Describe the opposing force/actors
STEP 3 Development of course of action and concept of operations	Specialist analysis Develop course of action Evaluate Decide Warning order 3	⇄ Decide HOW	(4) Develop the opposing force/ actors courses of action
STEP 4 Plan development	War Gaming Orders Group Rehearsal Adjustments	⇄ Complete orders	Further develop the opposing force/ actors course of action
STEP 5 Plan review	↓ **Operations order**		Launch the verified product

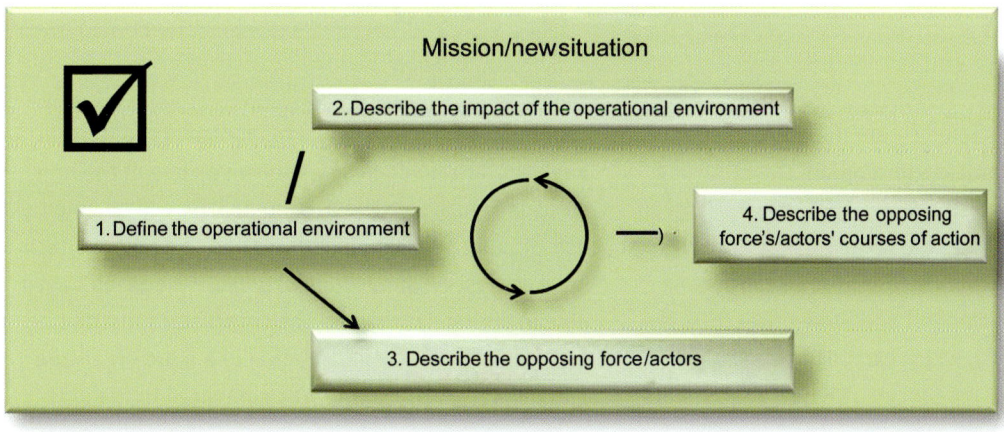

6.7

6.6 A simplified presentation of the steps in a planning and decision process. Source: Norwegian Army Weapons School (2015).

6.7 The relationship between the four steps of the IPOE process. Source: Norwegian Army Weapons School (2015).

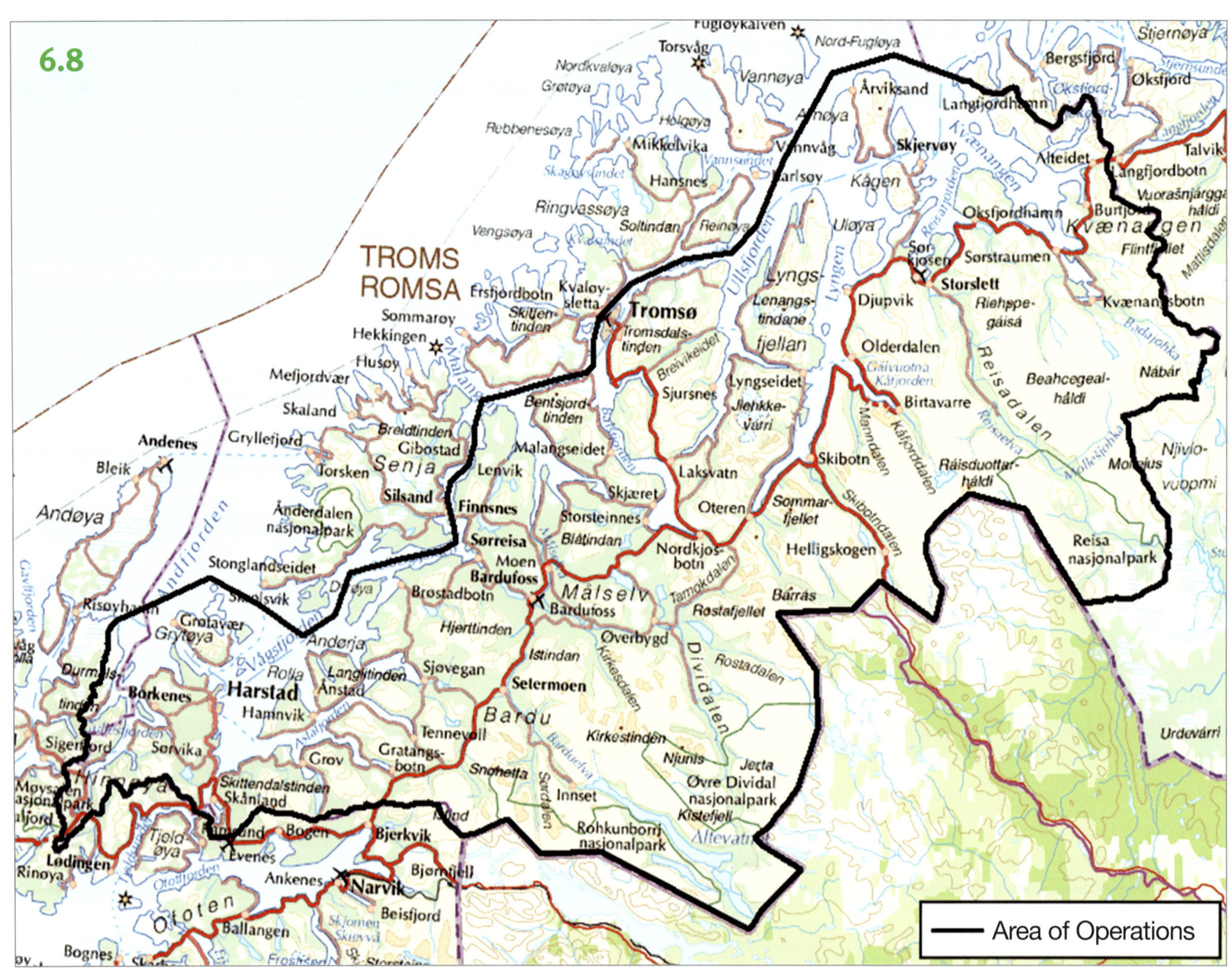

6.8 In this example, the area of operations (AO) largely coincides with Troms County, with the exception of the islands of Senja and Kvaløya. Background map: © Norwegian Mapping Authority.

Factor identification should create the prerequisites for a more effective process throughout the following steps in the IPOE process.

The factors from Step 1 are taken further in the IPOE process through both Step 2 (*Describe the impact of the operational environment*) and Step 3 (*Describe the opposing force/actors*), which can be done simultaneously in this phase of the process. MilGeo analysts are commonly assigned special responsibility for terrain analysis. Terrain analysis is shown as one of the four steps in Figure 6.9. The MilGeo analyst will be able to use GIS and geographic information to perform a technical analysis of the terrain and make the results available in digital databases. The conclusions distributed in a map will convey information in a more organized and pedagogically sound way compared to the pres-

Activity	Evaluation	Product
1 Weather analysis	• Wind • Temperature • Visibility • Cloud cover • Precipitation	Weather impact matrices
2 Terrain analysis	• Avenue of approch + man-made and natural obstacles • Observation and fields of fire • Concealment and coverage (linked to engagements areas) • Key terrain decided after evaluation of these factors	Mobility corridors Engagement areas Battle positions Key terrain
3 Civil relations (if required)	• Population/settlements calculation • Significant cultural relations • Calculation of civil infrastructure • Important terrain (human terrain) eliminated after evaluation of these factors)	Map overlay showing primary identities and loyalties, and local conflict lines Important terrain in relation to civil relations
4 Re-evaluate key terrain based on the conclusions	• Based on conclusions about key terrain, the relevant mobility corridors (physical and human access) can be identified and grouped into avenues of approach. These include recommendations, prioritizing, and clarification, as well as the affect of the weather in the relevant time perspective.	Key terrain Avenues of approch

6.9 The order of IPOE Step 2 (describe the impact of the operational environment). Source: Norwegian Army Weapons School (2015).

entation of texts, tables, and figures. The importance of the communicating characteristics of a cartographic product and human perception is emphasized in the section on cartography in Chapter 5 (Presentation). Together, these enable a far higher degree of detail in the analysis for reuse and better traceability compared to analysis performed directly using a hard copy of a map. The working method and products can be reused, rechecked, and quality assured.

The purpose of Step 2 in the IPOE process is to analyse and assess how topics within the four steps shown in Figure 6.9 both affect the operation and present conclusions for decision-makers and other users of the products. Throughout the IPOE process it is imperative that the conclusions drawn from the preceding steps are used as the basis for further work. Figure 6.10 shows examples of terrain analysis together with the conclusions from the last step in Figure 6.9 – key terrain and mobility corridors. Step 2 and Step 3 (Describe the opposing force/actors), form the basis for the final step of the IPOE process: Step 4 (*Develop the opposing force's/actors' courses of action*). As a specialist in military geography, a MilGeo analyst has a special role in conveying the interpretation and conclusions relating to the impact of the terrain on the operation through the work done in Step 4.

Step 4 ends with two courses of action: the opposing force's/actors' most likely course of action and their

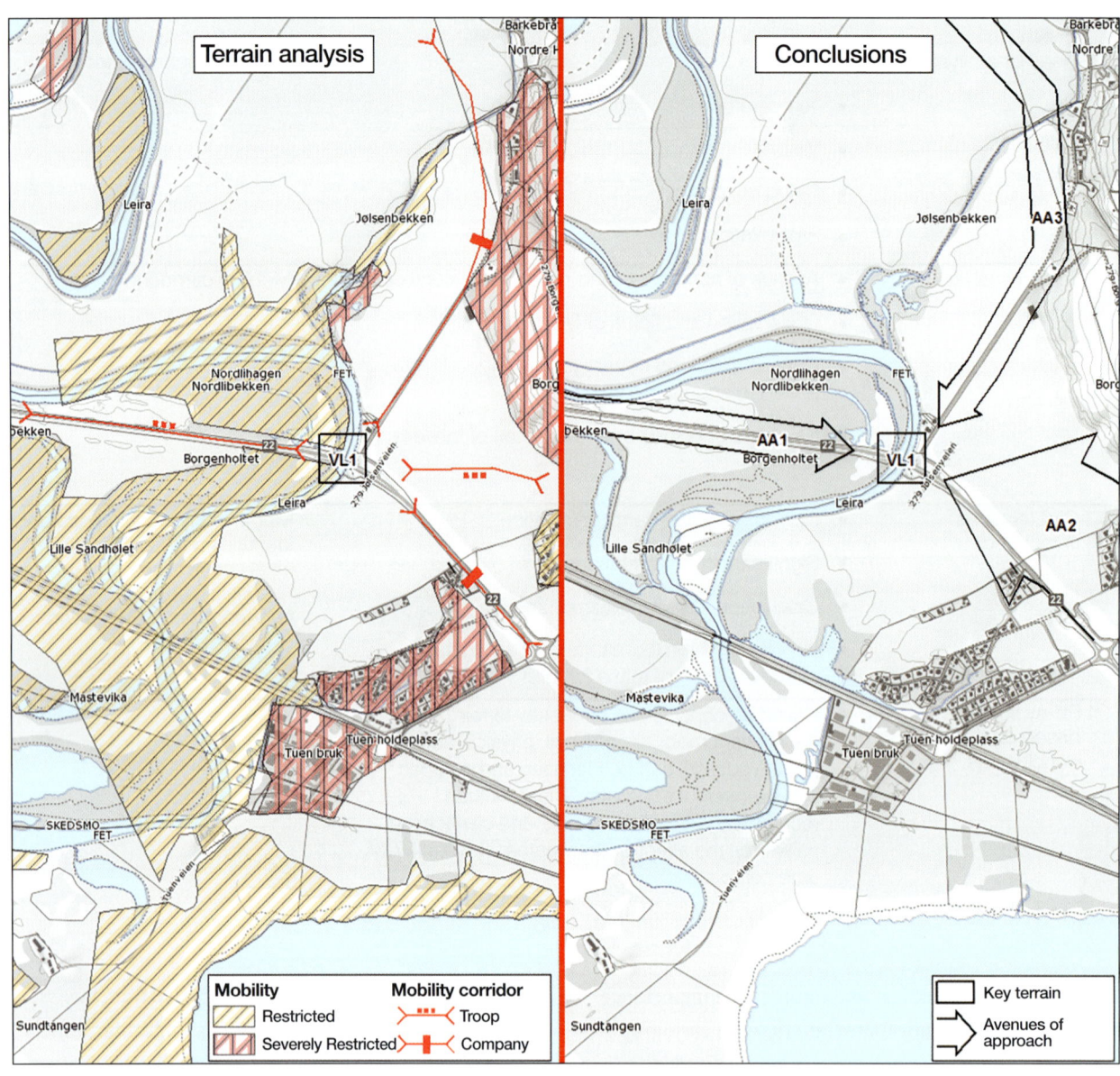

6.10 Examples of products from a terrain analysis. Background map: © Norwegian Mapping Authority. Left: a modified combined obstacle overlay; this is based on the overlay analysis in Figure 6.3 and is best suited for use in the analysis due to the large amount of information presented. Right: the conclusions presented as a separate product; in this map the conclusions include key terrain and avenues of approach.

most dangerous course of action. For practical purposes, an action is a description of how the opposing force/actors are expected to act during the operation. The two ways of acting are included in further work on the PDMP, such as the preparation of the decision-maker's own plan for conducting the mission. The IPOE product comprises the conclusions of the previous four steps and is distributed as decision support.

As already mentioned, the IPOE is a continuous and cyclical process that implies that products have to be regularly updated to maintain their relevance. The products are updated through evaluations based on the analysis of new information that is made available through the execution of the operation. Such information can range from information on ice thickness, destroyed roads, enemy movements, to changes in the decision-maker's ability to accomplish the mission. Thus, it can be said that the IPOE is a 'living' product that is only completed when the mission is completed.

Military use of GIS

This section looks at how various types of analyses can be combined with the IPOE method in order to be used in a military context. Modern GIS software has tools and features that offer the user great potential in terms of preparing and/or processing analysis, and various forms of presentation of geographic data. Today's software is becoming more adaptable due to its opportunities for modelling workflow, the development of its features and tools by using custom scripting or programming languages, and the ability to compile standard tools and functions. With a relatively simple level of understanding, users can customize software to suit different needs and working methods. Simultaneously, GIS technology is becoming increasingly user-friendly, making it more accessible to users at different levels, both in terms of expertise and needs. In sum, the technological developments enable us to resolve both old and new challenges in a more efficient manner. Today's challenge is primarily to identify how the use of technology can be optimized.

The following examples of analyses are suitable for use at a basic level in specialist fields. The examples mainly describe potential applications within the Norwegian Army, but are also relevant for the other branches of defence and for civilian actors. It should be noted that the examples used in the different analyses are fictitious and are intended to give readers a fundamental understanding of how tools and features may be used in a military context.

BUFFER ANALYSIS

In itself, buffer analysis is a relatively simple form of analysis but can be used for many different purposes. For example, buffers can be used to visualize the range of various sensors, communications, or weapons systems, or to calculate where a unit can cross a river. Conceptually, these examples are similar because buffer analysis is used to visualize the distance from one or more points, lines, or polygons.

Figure 6.12 shows an example of a buffer analysis in a military context. A unit wants to cross a river. In what places along the river will it be possible to cross by means of one or two bridges? The photo above shows a 26 m *Leguan* bridge. This type of bridge can be used both to reinforce existing bridges or to bridge dry and wet gaps when the enemy's operations have resulted in advances being interrupted and transport axes broken. The example analysis only deals with a quick way to identify possible crossing points based on the width of the river, but a number of existing conditions must be checked too. For instance, whether bridging is possible depends on the width of the river and whether the bridge can be transported, using specially produced wheel vehicles or an armoured vehicle launcher bridge (AVLB), to selected crossing point.

For the above-mentioned exercise, the buffer analysis is performed on a line segment representing the centre line of the river. By using two bridges (26 m), it will be possible to span a gap of 40 m, while one bridge will span up to 24 metres. A buffer distance of 20 m is used in the analysis, resulting in a buffer of 20 m on each side of the river's centre line.

The results of the analysis are presented in Figures 6.12 and show that theoretically the river is

6.11 **Leguan bridge layer.**
Photo: Torgeir Haugaard, Norwegian Armed Forces.

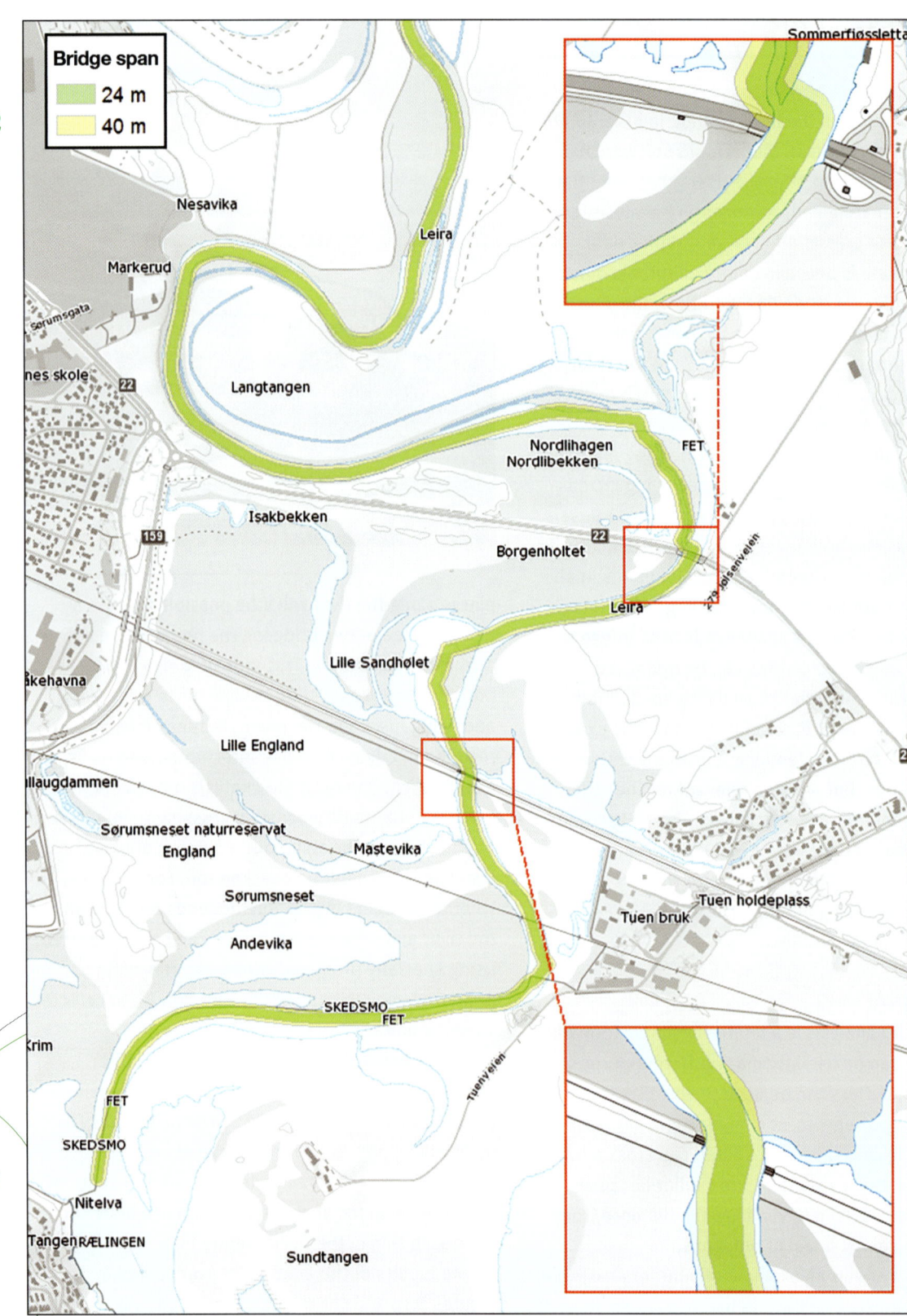

6.12 Buffer analysis – bridge span. River data: NVE; background map © Norwegian Mapping Authority.

sufficiently narrow for it to be crossed using one or two bridges. In the example, nowhere is it theoretically possible to use only one bridge (green buffer). The two current locations marked in red have naturally existing bridges, and there may be an opportunity to enhance or replace these by means of two bridges (forming a maximum bridge span of 40 m). It is crucial that whoever performs the analysis is aware of any limitations in the data, as the accuracy and degree of generalization are crucial for the result. The season, rate of flow in the river, and any other external factors that might affect the terrain should also be considered. In many cases, visual inspection of aerial photos can increase validity of the results. Figure 6.12 shows how the registered centre line of the main watercourse differs in some places from the watercourse shown in the more detailed background map.

Buffer analysis can also be a useful aid in connection with communications planning or the placement of other weapon or sensor systems. Figure 6.13 shows the position of a hypothetical weapons system. Based on technical data for the actual system, two different buffer distances are used: one representing practical range and the other maximum range. The position of the weapons system in the terrain is used as a basis for the analysis, the result of which visualizes respectively the weapon system's practical range (55 km) and maximum range (75 km). Various functions for creating a vector theme are described in the section on overlay analysis in Chapter 4 (Analysis). Based on the buffer analysis, these functions can be used to identify areas that are outside the weapon system's practical range in the actual area of operation. With already established battle positions, it can be helpful to identify any zones outside the range that should be reinforced by other means, such as finding the optimum position for the widest possible coverage when planning new battle positions.

Buffer analysis can also be used as a quick and efficient method for identifying the number of households or specific addresses located within a specific area. In the event of an evacuation it would then be possible to define a central point for the operation quickly and generate a buffer on this base. An appropriate buffer distance can be selected according to the type of incident, which then can be used to identify residential buildings to be evacuated.

ANALYSIS OF DIGITAL ELEVATION MODELS

Many of the analyses conducted in military contexts involve the use of digital elevation models. The nature of the terrain is crucial for planning military operations, such as the shape of the terrain or what lies above or below the surface. As described in Chapter 4 (Analysis), a distinction is normally made between *surface models* and *terrain models*. Essentially, many analyses fall into two principally different types of GIS technical analyses: analysis of the terrain slope (or of the direction of the inclination) and various forms of visibility analyses. In either case, numerous different models and types of data can be used as a basis for analysis. Some examples that demonstrate the range within a database are height contours in the form of vector data (isolines), different raster formats with different resolutions and levels of detail, and three-dimensional point clouds. In the military context, it is important to master the methods and techniques in order to use different types of data, as access to detailed surface models or high-resolution terrain models may be limited.

Slope analyses are mainly performed to make assessments and recommendations concerning mobility. When planning military manoeuvres, it is crucial that the decision-makers know which vehicles can be used and what restrictions the terrain imposes. Based on the limitations of different vehicle types, an analysis of the terrain's slope can indicate how it will be possible to use them during manoeuvres.

An evaluation of vehicle mobility is not only based on the gradient of the terrain but by a consideration of other factors, such as obstacles (both artificial and natural) present in the area of operation. Vegetation, water, buildings, and other types of infrastructure are examples of potential obstacles that represent more or less static challenges in terms of mobility. In addition,

218 Military use of GIS

6.13 Buffer analysis – weapon range. Background map © Norwegian Mapping Authority.

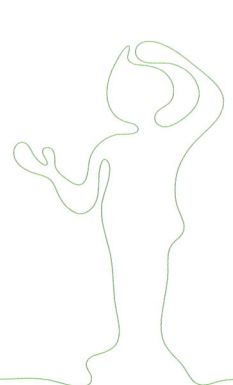

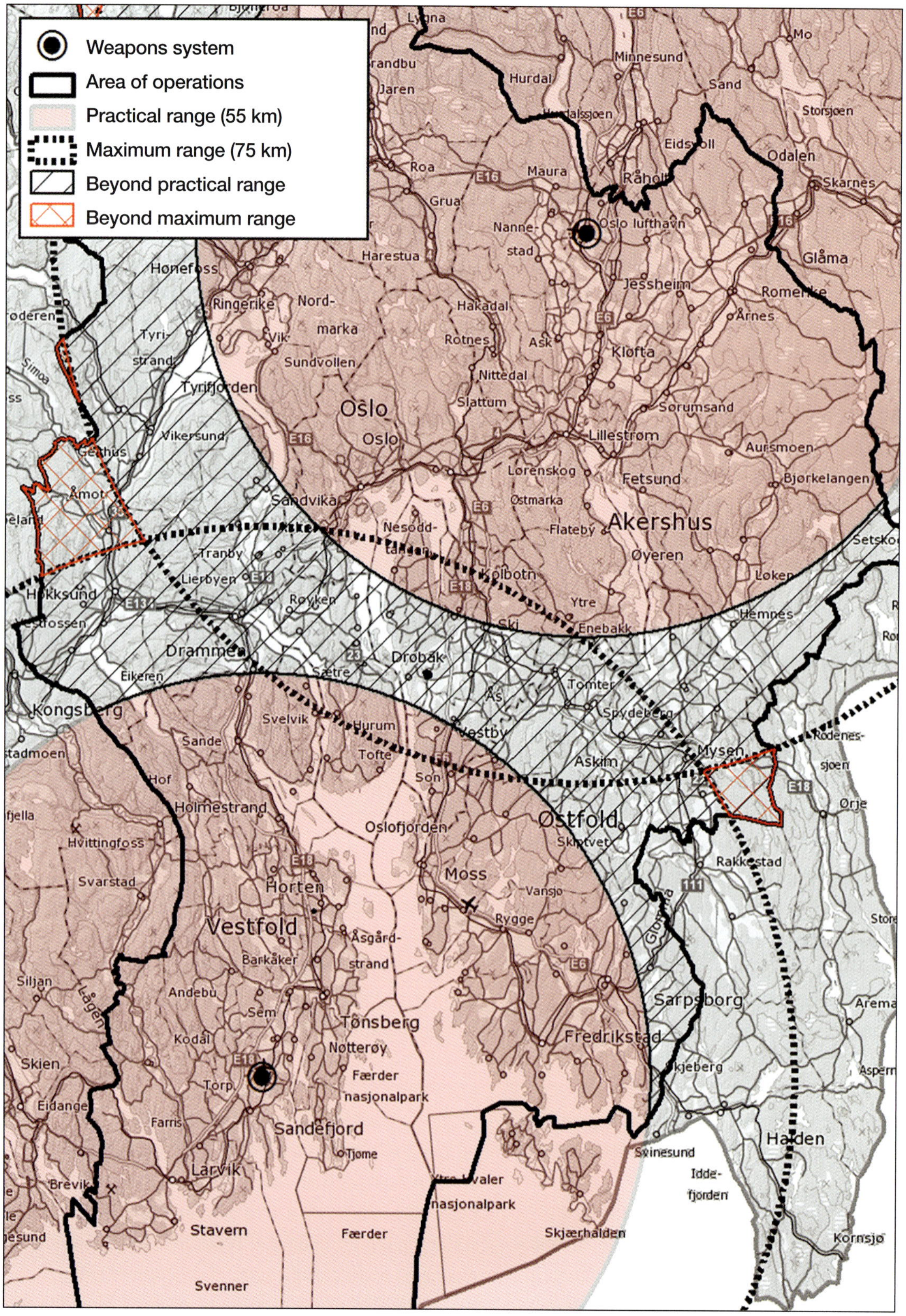

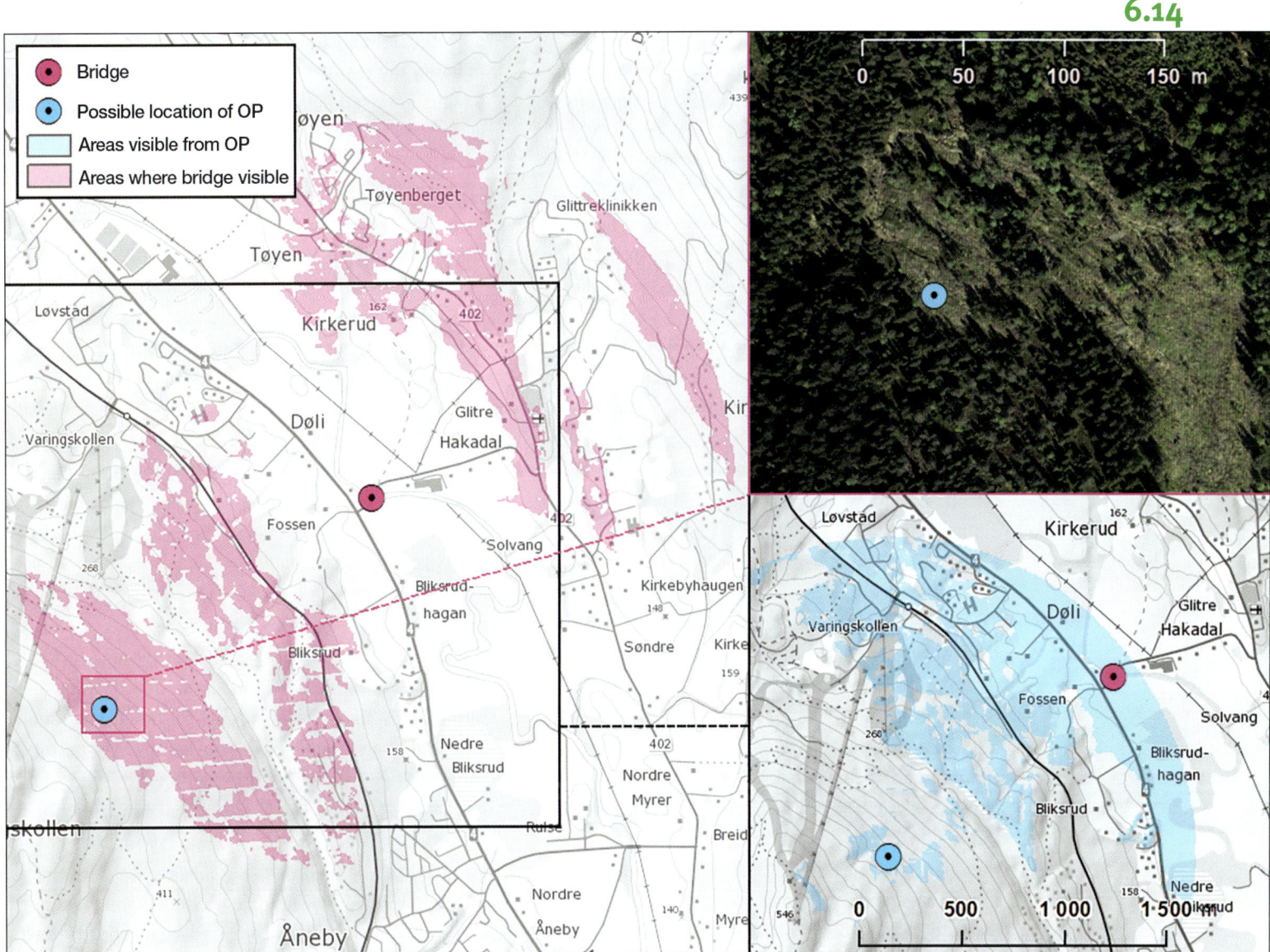

6.14 Visibility analysis. Background map © Norwegian Mapping Authority.

knowledge of the presence of any adversaries in an area and the type of threat this entails is crucial in military planning. Both static and more dynamic factors have to be taken into consideration. Therefore, the complexity of the analysis should be generalized as part of the evaluations before it can be presented together with concluding remarks, as shown in Figure 6.10.

There are many situations when visibility analysis can be used as a basis for assessments during the planning of military operations. Can the target be observed from this point? Where should an observation post be established to see as much as possible of a given stretch of road? Could enemy battle positions pose a threat to the planned movement? These are examples of questions for which relatively simple visibility analyses could contribute to more efficient planning and prioritization of resources.

In common with terrain slope analysis, the results of visibility analyses depend on what database is used. For example, to some extent it may be unhelpful to recommend the placement of an observation post based solely on the analysis of a rough terrain model, if in reality the area happens to be dense forest. Factors that are decisive for visibility in an area can range from local weather conditions and time of year to the type of vegetation found there. An updated surface model may often produce results that more closely correspond to the reality, but it will also have limitations. A surface model in raster format can include forests and buildings where, for example, the forest completely blocks the visibility. Sometimes this will correspond to the reality, while at other times the forest will represent a seasonal obstacle in terms of visibility. Figure 6.14 shows the result of an line-of-sight analysis conducted from the area that is to be observed (in this case the bridge). The analysis included parameters that determined the minimum distance from the bridge (500 m) and maximum distance of visibility (1500 m). Based on the results of the analysis and visual inspection of aerial photos, a possible location for an observation post (OP) was identified. The selected location is shown on the edge of an open area where the dominant height of the vegetation in relation to the bridge is good, and the gradient is very steep. This means it is possible to assume that the forest does not constitute a complete obstacle to visibility. In addition to having to observe the bridge itself, it is desirable to observe the road axis towards the bridge on the west side of the river. Therefore, an analysis of the visibility from the OP was performed, and the result showed that it should be possible to observe large section of the road in addition to the bridge.

In order to make the best and most realistic evaluations in terms of the mobility and visibility conditions of the terrain, a number of factors should be taken into consideration. The purely technical GIS analyses are still effective for quick refinements and generalizations. When planning military operations, it is critical that those who perform the analyses are aware of any limitations of the database; ultimately, lives could be lost if analyses and evaluations do not correspond sufficiently well with the reality. As already mentioned, in analyses of elevation models it may be advantageous to be able to verify the results. If aerial photographs or other images are used in this context, it is crucial to know when they were taken or produced.

NETWORK ANALYSIS

As described in Chapter 4 (Analysis), network analysis is used for many different purposes in GIS. Regardless of whether this concerns cable networks or roads, a certain amount of preparatory work is required to establish a usable network data set. In the military context, the capture of geographic data can vary depending on where in the world the preparatory work is done, and therefore network data sets tend to be established by those who are going to perform the analysis. This can be time-consuming, which means that evaluations often have to be made in the absence of a complete network.

Network analysis represents an effective aid in route planning, such as estimating response times or the optimal placement of vehicle checkpoints (VCPs). Other examples include immediate identification of the theoretically fastest route for the evacuation of wounded personnel or the optimization of patrol routes. When calculating driving times, it is also important to be aware of factors that can affect the speed along the road section in question. These are greatly affected by factors such as local traffic and weather, road, and lighting conditions. The planning and execution of military manoeuvre will be affected by the threats in the area. Such as improvised explosive devices (IEDs), or an enemy actors presence will affect and likely trigger countermeasures which will reduce speed of movement. The aforementioned examples represent more or less dynamic factors, which are virtually impossible to implement continuously in a network data set. Hence, when analysing relatively static networks, any evaluations and recommendations should take account of such factors.

Figure 1.12 (driving time analysis) in Chapter 1 (Application) shows an example of how a network analysis can be presented to give a comprehensive overview of the response time in a given area. The product

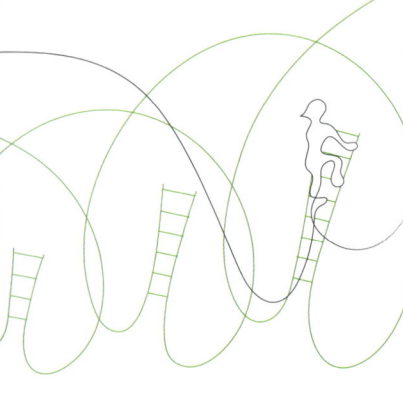

does not provide any information about which roads should be used but an indication of the response time from the starting point within the designated zones is readily apparent.

OVERLAY ANALYSIS

An overlay analysis focuses on how different themes and characteristics of the terrain can, for example, be combined to create a new theme, look at the connections between various terrain-related phenomena, or determine the suitability of something across multiple themes or features. Overlay analysis can be performed on both vector and raster data, as described in Chapter 4 (Analysis). In the section on the GIS cycle (near the start of this chapter) and military methods and processes, the process of starting with information and ending in conclusions is explained. The example in Figure 6.3 shows the results of terrain analysis, in which different factors have been weighted and combined to assess the potential for manoeuvres in an area. The example shows how more or less static theme data, dynamic factors (such as weather), and

6.15 Overlay analysis. Background map © Norwegian Mapping Authority, NRL, Geovekst.

6.15

6.16 An operation involving a Bell helicopter. Photo: Torbjørn Kjosvold, Norwegian Armed Forces.

human evaluations can be combined in a more complex overlay analysis.

Another concrete example of an overlay analysis is the identification of suitable areas for helicopter landing sites. Regardless of whether this concerns planned operations involving insertion of personnel by helicopter or possible landing places for medical evacuation, overlay analysis using GIS represent an effective aid in mission support. This type of analysis generally depends on a number of different data sets and different parameters, based on criteria of where it is physically possible, appropriate, safe, and permissible to land a helicopter. The following are examples of variable factors that may be included in an overlay analysis:

- Terrain slope
- Requirements for open area on solid ground and of a certain size
  (e.g. no forest, water, or buildings)
- Obstacles

(e.g. power lines, bridges, telecommunications masts, and electricity pylons)
- Operational considerations and current threats in the area

The parameters of the different factors will vary depending upon the situation, mission, and helicopter type. For example, requirements relating to the size of the area will depend on the total length or width of the helicopter, including its rotor blades. In peacetime or in conjunction with exercises, there will be other considerations, for example in relation to the noise-prone areas or other no-fly zones. In some cases, when personnel are only dropped or picked up, it may not be necessary for the helicopter to land. When it comes to planning military operations, plotting likely flight paths can be decisive in selecting which landing sites to recommend. The topography of the area, local weather conditions, and the general threat level can be crucial when planning flights, as these will also influence the choice of landing sites. As mentioned in previous examples, many factors affect the requirements for data and choice of analytical method. However, in many cases, relatively simple analyses of the more static and terrain-related factors could provide a better basis and streamline the planning process. Figure 6.15 shows part of an overlay analysis to identify helicopter landing sites.

A MilGeo analyst can help with analyses, assessments, and recommendations, but ultimately the helicopter crew will make the final decisions about whether and how they will land. It can be challenging to include factors such as threat assessments and weather and light conditions in a GIS analysis, but they provide significant added value for the decision-maker. In this way, decision support in the planning phase can end up as concrete mission support, such as in the form of map products, in this case with possible landing sites.

Summary

STRUVE'S GEODETIC ARC
Between 1826-1855 a chain of trigonometric points were established from Ishmael on the Black Sea to Hammerfest on the Arctic Ocean under the leadership of Fredrik Georg Wilhelm Struve. This was important work with respect to determining the earth's shape and size. The 34 measuring points in Struve's Geodetic Arc that are still intact were included as the first scientific cultural objects on UNESCO's World Heritage List in 2005. The photo shows the column erected at the meridian's most northerly point in Hammerfest. Photo: Bjørn Bergesen

In this book we describe and explain the thinking behind geographic information systems (GIS) and their applications. We start with the concept of the GIS cycle, which we introduce in Chapter 1 (Application). Chapter 1 also provides examples of how GIS can be applied to help resolve various problems. In addition, we present different planning methodologies that can be used to structure research objectives.

In Chapter 2 (Systems), we give a basic introduction to GIS and how data have to be processed before any analyses can be performed. We include GIS theory, which is fundamentally important to understand if one is to recreate the reality of a given situation as accurately as possible. Another very important aspect of GIS concerns data quality, since the results of GIS applications are dependent upon the quality of the original data.

In Chapter 3 we examine geographic data as data in the information society and what geographic data are available from different data sources. Readers gain an understanding of the content of data and what is regarded as critical information. At this stage in the GIS cycle the focus shifts from systems for gathering data to how to use of gathered data.

Chapter 4 (Analysis) examines how data are analysed methodically, as well as various ways of questioning and analysing. When asking such questions, the relevant keywords are: *where*, *what*, *when*, *how*, and *why*.

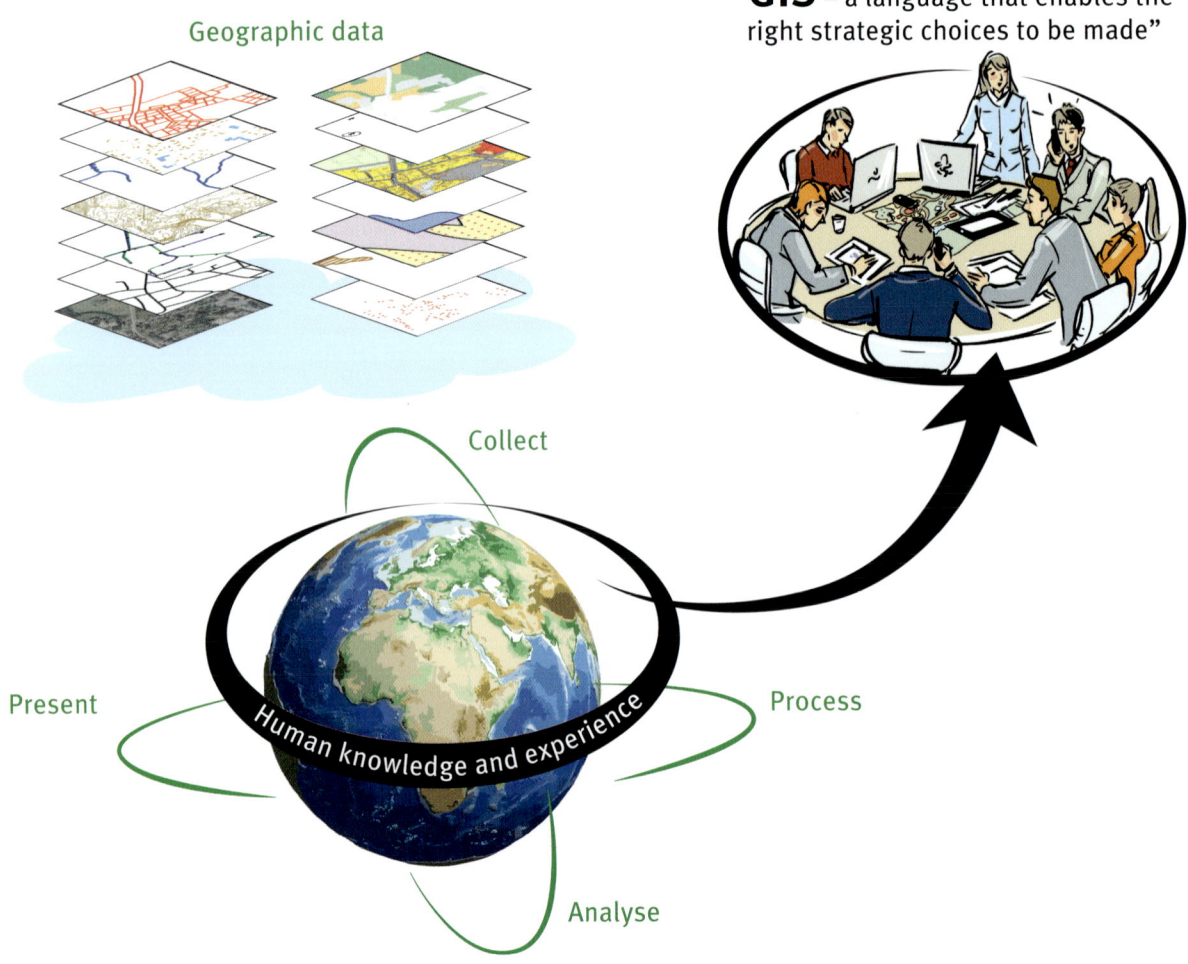

Chapter 5 (Presentation) addresses the visualization of geographic information and how the results can be presented in the best possible way. Cartography, or the knowledge of how information can be presented on a map so that it is easily understandable and readable, is a very important aspect of the use of GIS.

The last stage in GIS cycle concerns making strategic decisions based on the collected, processed, analysed, and presented data. We exemplify this through the inclusion of a separate chapter on military geography.

Chapter 6 provides insights into how the Norwegian Armed Forces use geographic information and GIS when making decisions. The Armed Forces' experienced-based methodology for learning, innovation, and development could be beneficial to other disciplines. Moreover, in military defence there are often situations when reliable evaluations of the terrain are crucial and therefore the collection, analysis, and presentation of geographic data are necessary for the right decisions to be made. However, since military objectives and the methods for achieving them are rarely known in advance, knowledge, experience, and the ability to understand, analyse, and present opportunities are crucial. Hence, in this book great emphasis is placed on human factors as important prerequisites when GIS are used as tools for making appropriate decisions.

We hope that this book will provide readers with a good understanding of GIS as tools for collecting, processing, analysing, and presenting geographic information. We also hope that it will contribute to the increased use of GIS as a geographic language in order to increase our knowledge and understanding of our differing environments.

A cadet at the Norwegian Military Academy.

Glossary

3D printing
A process for building up in layers a model from a digital file.

Abstraction
A conceptual simplification of the real world.

Associative symbols
Symbols used to visualise features of equal weight/importance. The variables used as associative symbols are value and size.

Back casting
A technique of stepping away from the current situation. You imagine yourself in the future and look back at the present.

Basemap
A map used as a common information base with general background information.

Bathymetry
Bathymetry can be likened to topography, but relates to the measurement of landscape features below water.

Biodiversity
The diversity of life in the natural environment.

Buffer
A zone created a given distance away from or around a feature. Buffers can be created around polygons, lines and points.

Buffer analysis
Calculating buffers in relation to a feature.

Buffer distance
The distance from a feature to the outer edge of the buffer zone around a feature. For example, the buffer around a point will be a circle with a radius equal to the buffer distance.

CAD
Computer Aided Design. Computer software for designing and drawing. A common tool in engineering and architecture circles.

Cartogram
On a cartogram the sizes of the areas are proportional to the data values – often used to visualise statistical data.

Cartography
The science of ensuring maps and map data have understandable content and an attractive appearance.

Classeless
A method whereby one uses the actual values of a data set in map production, instead of grouping the data into classes.

Clipping
Cutting out an area of a theme based on another theme's geographic area.

Cloud technology
Data and resources storage services provided via servers connected to the Internet. Such servers are often located in Asia or the United States. The service is also used to run data processing and software on remote servers. The technology is also known as net cloud, cloud computing, or cloud.

Contours
Provide information about elevations in the terrain. Contours pass through points in the terrain at the same elevation above see level (isoline).

Coordinate system
The reference system in which all geographic features are placed in relation to each other and in relation to a geographic origo.

Cyclical analysis process
A residual analysis process in which new information is continuously included in the analysis in order to ensure that it remains relevant.

Dasymetric maps
Surface symbolised maps in which the area units have undergone processing.

Datum
A datum is a set of parameters that describes an ellipsoid and how this is oriented in relation to a geoid.

Delaunay triangulation
A method for dividing a plane into triangles on the basis of points that mark the corners of the triangles.

DEM
An acronym for digital elevation model.

Det offentlige kartgrunnlaget (DOK) Public domain map data.
These data are public, quality assured spatial map data adapted for municipal planning and building works.

Digital elevation model
Digital elevation models are subdivided into digital terrain models (DTMs) and digital surface models (DSMs). DTMs describe the ground surface (terrain), and DSMs describe specific surfaces. DEMs are produced from a set of points with known x, y, and z coordinates, and there are rules for how the z value must be interpolated for the new x and y values in the model.

Doctrine
A doctrine is a procedural guide for dealing with a complex military operation. The doctrine explains what should be done, how it should be done, and with what resources it should be done. It helps to standardize operations.

Dot density analysis
An analytical method to determine the concentrations of particular phenomena in a given area.

Drone
A remote-controlled unmanned aircraft.

DSM
Abbreviation of digital surface model.

Elevation
A height in the terrain above a given level.

Ellipsoid
A model of the earth with an even surface without topography.

Equidistant
A method in which one chooses equally large intervals between the class thresholds in a data set. Is most suitable for visualising data with a regular, linear distribution on a map.

Foreign key
An attribute or set of attributes in one table that match the primary key attributes in another table. A foreign key thus refers to data in another database.

Fuzzy data set
A data set in which the boundaries between geographic features are unclear, e.g. a zone in which the land transitions from bogs to forest.

Geodata
An abbreviation of geographic data: information about geographic features with their coordinates, shape and attributes.

Geodesy
Geodesy is a scientific discipline that deals with the measurement and representation of the earth (i.e. its shape and size) and its gravitational field. Geodesy forms the basis for mapping and surveying.

Geographic data
Information about geographic features and their coordinates, shape and attributes.

Geographic information
Information about features or conditions that can be located geographically.

Geographic features
Features with known coordinates. Features in a vector format can be visualised either as points, lines or surfaces. In a raster format the features are visualised as a pattern using equally large pixels (cells).

Geoid
A geoid is a model of the earth's surface in which gravity is the same wherever you are. In a map context, elevation data often relates to the geoid since for practical reasons one will want to show which way water runs.

Geomatics
Geomatics is a generic term for disciplines that combine traditional mapping with information technology, and as a discipline it ranges from data collection and storage to the visualization and application of geographic information.

Geometric symbols
Geometric symbols are symbols such as circles, triangles and squares. When these are used on a map a legend is needed to explain what the geometric symbols are intended to represent.

Georeference
To give a georeference means to associate a physical map or raster image of a map with locations in physical space. In GIS aerial photographs with the same geometric properties as a map can be georeferenced because they are linked to a coordinate system.

GIS
A geographical information system (GIS) comprises geographical data mapping systems, methods, and human knowledge and experience that enable the collection, processing, analysis, and presentation of geographical data.

GIS cycle
A GIS cycle is working method that can be used for all types of analyses. The methodology consists of four steps: collection, processing, analysis, and presentation of geographical data.

GNSS
Global Navigation Satellite System – a generic term for satellite-based systems for determining locations.

Gradient analysis
A raster analysis used to calculate the gradient in an area.

Hybrid forms
Maps that are intended to meet specific needs and that are produced by combining a base map with special themes.

Identify
Overlay analysis in which the resulting area is the same as the area in the first input theme. The common area will inherit attributes from both themes.

INS
Internal Navigation System. An instrument system installed in an aircraft that monitors the aircraft's acceleration and heading as well as the aircraft's rotation in relation to its heading and the perpendicular.

INSPIRE
Infrastructure for **sp**atial **i**nformation in **E**urope is an European law that describes how the individual countries in Europe should organise infrastructure work, which types of data should be available, which types of services the countries should offer and how data shall be harmonised across national borders.

Interoperability
Interoperability refers to the capability of components or systems to exchange data with other components or systems. This capability is needed by GIS users when studying data using software that differs from the software used to compile the data.

Interpolation
A mathematical model for calculating values for new points based on points with known values.

Intersect
Overlay analysis in which the resulting surface is equal to the overlapping section of the surfaces of the input themes. The surfaces inherit the attributes of both themes.

Isoline map
A map in which the theme is visualised as lines that pass through points of equal value. Terrain contours are examples of such visualisation.

Isopleth map
A map in which the lines that visualise the data do not have the same value as in a isoline map, but contain derived data that is calculated in relation to a specific area, e.g. local authority boundaries.

Key (database)
In a database, a key is an attribute and/or set of attributes that identifies each record in the database.

Land Registry
The Land Registry (Matrikkelen) is the Norwegian Mapping Authority's public register of real estate, property boundaries, addresses and buildings.

Land surveying
Coordinate determination in the terrain using various instruments. Until recently this has mostly been carried out using angle and distance determining optical instruments (theodolites, total stations, etc). It is now mostly carried out using GPS-based instruments.

Laser surveying, Laser scanning
A method for surveying terrain formations and coordinates with the aid of laser technology installed beneath an aircraft or helicopter. Particularly used in the establishment of digital terrain models and when surveying the seabed.

Leguan
A type of military bridge construction.

LiDAR
LiDAR (light detection and ranging) data are collected using lasers and used to establish good elevation models. The lasers can be used to generate sets of data points with high point-density, where each point contains a pair of coordinates and an elevation value. In addition, the data are delivered with an intensity value.

Map projection
A method of reflecting the surface of the earth in two dimensions.

Metadata
Data about data. Data about the data's origin, accuracy, measurements, purpose, etc.

MilGeo
Abbreviation of military geography, including the military use of geographic information and GIS.

MilGeo analyst
A person who systematically collects, processes, interprets, analyses, and presents geographical data for military use.

National Reference Geodatabase (FKB)
The most detailed basic map data has been collated in the Feature Catalogue. This map data provides a basis for producing detailed technical and economic maps at scales of 1:500, 1:1000 and 1:5000.

Network analysis
A method of calculating the most appropriate route in a network.

Noise Zone Map
A thematic map showing the extent of noise pollution.

Norway Digital
A formalised partnership between municipalities, counties, state agencies and major users of geodata involving a financial cooperation concerning the obtaining, updating and use of geographic information.

Obeng's method
A method of modelling the real world in four stages depending on whether the goal and method is known or unknown.

Orthophoto
A vertical, georeferenced aerial photograph.

Overlay analysis
Forms new themes based on two or more themes with the same geographic area. The new theme can inherit the attributes of one or more of the input themes.

Pattern analysis
An analytical method for detecting and visualizing different spatial phenomena.

Percentile
A method in which classification of the data set is based on a specific percentage.

Perception
How we perceive and interpret what we see.

Photogrammetry
A measurement technique for determining a feature's coordinates by measuring photos.

Pictogram
A stylised drawing of the feature it is intended to represent. For example, a cross that is used as a symbol for a church.

Pictometry
Georeferenced aerial photos - oblique photos that are taken from five different angles meaning the user can see a building from all sides and from above, which means he or she can move around a house, measure the height of the house, and zoom out and look at the neighbourhood. Pictometry photos can also be used to drape a 3D model and thus make it more true to life.

Pixels
Pixels, picture elements, and image elements. The units a digital image is constructed from. Every pixel is assigned a colour and possibly other attributes.

Polygon
A plane figure that is bounded by a closed path or circuit, composed of a finite sequence of straight line segments.

Glossary

Polynomial
In mathematics a polynomial is the sum of a finite number of terms where each term is a constant multiplied by one or more variables raised to positive whole-number exponents. A polynomial is determined by using only the operators addition, subtraction, and multiplication.

Primary keys
An attribute or set of attributes in a database that is unique to each record (i.e. a unique identifier). The primary key is the preferred method for identifying a row in a table in a relational database.

Product specifications
SOSI product specifications specify in detail how basic geodata and thematic geodata should be established within defined fields.

Quantile
A value representing a class division, where the classes in the data set have an approximately equal number of observations.

Radar measurements
Remote sensing technology from, for example, a satellite that uses radio waves to determine the distance to geographic features. Frequently used for monitoring areas susceptible to landslides.

Raster calculator
A calculator that uses data from raster layers as input for mathematical calculations using operators.

Raster data
A raster file (bitmap) contains data in a digital image file format.

Real Time Information
Real Time Information is dynamic information that is made available via cloud services. For example, in Norway the National rail Administration displays real time train information that shows whether trains are on schedule.

Reference system
The coordinate system in which all geographic features are placed in relation to each other and in relation to a geographic origo.

Registration units
A system based on uniquely defined units of land. In Norway we use cadastral numbers, holding numbers, leasehold numbers, and section numbers, combined with municipality numbers.

Remote sensing
A surveying or analytical method in which one makes measurements or observations from a distance. The commonest methods of remote measurement are photogrammetry, laser surveying, satellite image data and radar.

Resistance
A factor used in network analyses to describe restrictions associated with a link or node.

Resource area map
Area resource maps provide information on the cultivation and use of agricultural land, the soil quality for forestry, and the land type, soil quality, and potential for cultivation in outlying fields.

Scenario thinking
Envisaging the different possible consequences of a pattern of action.

SOSI Feature Catalogue
A feature catalogue is a collection of standardised definitions of real worlds features in a geographical context, including definitions of feature types, their attributes and predefined code lists. In Norway the SOSI Feature Catalogue is a collection of definitions covering a variety of thematic fields, being a result of long-term collaboration between national agencies. SOSI is also used as a standard for exchange format, raw data formats and storage formats.

Statistics
A branch of mathematics that deals with the analysis of quantitative observations with the aim of drawing conclusions based on these data.

Synthesis
The combination of elements or components that form a connected whole.

Synthetic Aperture Radar (SAR)
A radar that can be mounted on an aircraft or a satellite, with a small antenna that moves over a controlled path. The method provides higher resolution than the use of conventional antenna.

Terrain analysis
Analysis of the features of the terrain to identify their affect on any potential scenarios.

TIN
See Triangular Irregular Network.

Topographic map
A map that represents a range of surface features without any emphasis on particular elements.

Topography
The terrain's characteristics that refer to the shape of the landscape or the physiogeographic characteristics of the landscape such as elevation above sea level, slope and spatial position.

Trend analysis
Another term for pattern analysis – an analytical method for detecting and visualizing different spatial phenomena.

Triangular irregular network (TIN) model
A terrain model in which the terrain is divided into triangles according to specific criteria from points with known coordinates (x, y, z).

Union
Overlay analysis in which the resulting surface is equal to the union of the surfaces in the input data. The surface inherits the attributes of both themes.

Vector data
Vector data consist of points, lines, and areas, and are used in map production.

Visual variables
Methods of symbolisation using variations in the shape.

WGS84
The World Geodetic System 1984 (WGS 84) defines a comprehensive global framework for geodesy and navigation. It is a geodetic datum.

WMS
Web Map Services (WMS) is a standard protocol for providing map-based services via the Internet.

Norwegian institutions and organizations – formal English names

ENGLISH NAME	NORWEGIAN NAME
BI Norwegian Business School	Handelshøyskolen BI
Cavalry squadron	Kavalerieskadronen
Directorate for Cultural Heritage	Riksantikvaren
Directorate of Fisheries	Fiskeridirektoratet
Directorate of Mining[1]	Direktoratet for mineralforvaltning
Directorate of Norwegian Customs	Tolldirektoratet
Directorate for Nature Management[2]	Direktoratet for naturforvaltning
Early Response Unit[3]	Beredskapstroppen
Geological Survey of Norway	Norges geologisk undersøkelse (NGU)
Home Guard	Heimevernet (HV)
Institute of Marine Research	Havforskningsinstituttet
Ministry of Climate and Environment	Klima-og miljødepartementet (KLD)
Ministry of Local Government and Modernisation	Kommunal- og moderniseringsdepartementet
NDEA	see Norwegian Defence Estates Agency
NDLO	see Norwegian Defence Logistics Organisation
NGU	see Geological Survey of Norway
NIBIO	see Norwegian Institute of Bioeconomy Research
NPRA	see Norwegian Public Roads Administration
NTNU	see Norwegian University of Science and Technology
NVDB	see National Road DataBase
NVE	see Norwegian Water Resources and Energy Directorate
National Road DataBase	Nasjonal vegdatabank (NVDB)
Norway's Institute of Technology[4]	Norges tekniske høgskole
Norwegian Armed Forces	Forsvaret
Norwegian Military Geographic Service	Forsvarets militærgeografiske tjeneste (FMGT)
Norwegian Armed Forces Media Centre	Forsvarets mediesenter
Norwegian Armed Forces Special Operations Command	Forsvarets spesialkommando
Norwegian Defence Command and Staff College	Forsvarets stabsskole (FSTS)
Norwegian Army Weapons School	Hærens våpenskole (HVS)
Norwegian Biodiversity Information Centre (NBIC)	Artsdatabanken
Norwegian Civil Defence	Sivilforsvaret
Norwegian Coast Guard	Kystvakten
Norwegian Coastal Administration (NCA)	Kystverket
Norwegian Defence Estates Agency (NDEA)	Forsvarsbygg
Norwegian Defence Logistics Organisation (NDLO)	Forsvarets logistikkorganisasjon
Norwegian Directorate for Civil Protection (DSB)	Direktoratet for samfunnssikkerhet og beredskap (DSB)

NORWEGIAN INSTITUTIONS AND ORGANIZATIONS – FORMAL ENGLISH NAMES

Norwegian Environment Agency	Miljødirektoratet
Norwegian Geotechnical Institute (NGI)	Norges Geotekniske Institutt (NGI)
Norwegian Institute for Air Research	Norsk institutt for luftforskning (NILU)
Norwegian Institute of Bioeconomy Research	Norsk institutt for bioøkonomi (NIBIO)
Norwegian Labour and Welfare Organization	Arbeids- og velferdsforvaltningen (NAV)
Norwegian Mapping Authority	Kartverket
Norwegian Meteorological Institute	Meteorologisk institutt
Norwegian Military Academy	Krigsskolen
Norwegian Pollution Control Authority[2]	Statens forurensningstilsyn
Norwegian Public Roads Administration (NPRA)	Statens vegvesen
Norwegian Space Centre	Norsk Romsenter
Norwegian University of Science and Technology (NTNU)	Norges teknisk-naturvitenskapelige universitet (NTNU)
Norwegian Veterinary Institute	Veterinærinstituttet
Norwegian Water Resources and Energy Directorate	Norges vassdrags- og energidirektorat (NVE)
Royal Norwegian Navy	Sjøforsvaret
Statistics Norway	Statistisk sentralbyrå (SSB)
The Norwegian Parliament	Stortinget
The Brønnøysund Register Centre	Brønnøysundregistrene

[1] The full name is: Directorate of Mining (with the Commissioner of Mines at Svalbard); the full Norwegian name is: Direktoratet for mineralforvaltning (med Bergmesteren for Svalbard)
[2] Now the Norwegian Environment Agency
[3] Part of Norwegian police force
[4] Now part of NTNU

References

Artsdatabanken (2015) *Natur i Norge*. Available from http://www.artsdatabanken.no/NaturiNorge (Accessed 29 June 2015).

Bernhardsen, T. (2000) *Geografiske informasjonssystemer*. Nesbru: Vett & Viten AS.

Bertin, J. (1983) *The Semiology of Graphics*. Madison, Wis.: University of Wisconsin Press.

Biography.com (2015) *Nadar Biography*. Available from http://www.biography.com/people/nadar-9419752 (Accessed 17 August 2015).

Bjørke, J.T. (1989) *Digitale terrengmodeller*. NLH/NTH.

Blankholm, H.P. (2004) *ArcGIS Manual til GIS i fortid, nåtid og fremtid*. Universitetet i Tromsø: Samfunnsvitenskapelig Fakultet.

Booth, B. (2000) *Using ArcGIS 3D Analyst Geografisk informasjonssystemer*. ESRI

Brodersen, L. (1999) *Kort som kommunikation*. Frederikshavn: Forlaget Kortgruppen.

Carcia-Molina, H., Ullman, J.D., & Widom, J. (2002) *Database Systems: The Complete Book*. Upper Saddle River, New Jersey, USA: Prenice Hall.

Cliff, A.D., & Hagget, P. (1988) *Atlas of Disease Distribution*. Oxford, UK: Blackwell.

Dale & McLaughlin (1999) *Land administration*. Oxford: Oxford University Press.

de Soto, H. (2000) *The Mystery of Capital*. New York: Basic Books.

Directorate for Nature Management (2003) *Grønn by – arealplanlegging og grønnstruktur*. DN-Håndbok 23–2003. Trodnheim: Direktoratet for naturforvaltning.

Directorate for Nature Management (2007) *Kartlegging av marint biologisk mangfold*. DN-Håndbok 19-2007. Trondheim: Miljødirektoratet.

Directorate for Nature Management (2007) *Kartlegging av naturtyper – verdisetting av biologisk mangfold*. DN-Håndbok 13-2007: Direktoratet for naturforvaltning.

Dueker, K.J. & Kjerne, D. (1989) *Multipurpose cadastre*. Falls Church: American Society for Photogrammetry and Remote Sensing.

Elmasri, R., & Navathe, S.B. (1994) *Fundamentals of Database Systems*, (2. ed.). Redwood City, California: The Benjamin/Cummings Publishing Company Inc.

European Environment Agency (EEA 1999) *Environment in the Euro- pean Union at the turn of the century*. Environmental assessment report No 2.

European Environment Agency (EEA 2003) *Environmental Indicators: Typology and Use in Reporting*. European Environment Agency. Available from http://didattica.ambra.unibo.it/didattica/att/456d.file.pdf (Accessed 15 April 2015).

European Environment Agency (EEA 2014) *Noise in Europe 2014*. Available from http://www.eea.europa.eu/publications/noise-in-europe-2014 (Accessed 1 April 2015).

Foundation for Design and Architecture (2010) *Barnetråkk veileder 2010: Registrering av barn og unges arealbruk*. Available from https://www.regjeringen.no/globalassets/upload/md/bilder/planlegging/veiledere/barn/barnetrakk_2010.pdf (Accessed 8 October 2015).

Frerichs, R.R. (undated) *John Snow – a historical giant in epidemiology*. Available from http://www.ph.ucla.edu/epi/snow.html (Accessed 2008).

Geological Survey of Norway (NGU) and Norwegian Directorate of Mineral Management (2015) *Mineralressurser i Norge 2014, Mineralstatistikk og bergindustriberetning*. Nr 1/2015. Available from http://www.ngu.no/publikasjon/mineralressurser-i-norge-i-2014-mineralstatistikk-og-bergindustriberetning (Accessed 28 September 2015).

Geological Survey of Norway (NGU) (2014) *Produktark jordsmonn*. Available from https://register.geonorge.no/data/documents/Jordsmonn_Produktark_S-L_Jordsmonn.pdf. (Accessed 28 September 2015).

Geological Survey of Norway (NGU) (2015) *Bruk av grunnvann*. Available from http://www.ngu.no/emne/bruk-av-grunnvann. (Accessed 25 June 2015).

Geological Survey of Norway (NGU) (2015) *Produktark Nasjonalt aktsomhetskart for Radon*. Available from https://register.geonorge.no/data/documents/produktark_Radon%20aktsomhet_produktark-nguradonaktsomhet_.pdf (Accessed 28 September 2015).

Hegstad, E. (1994/95) *Om eiendomsregistrering*. Doctor scientarium, NLH, Dept. of Landscape Architecture and Spatial Planning, Ås.

Hærens våpenskole (2015) *Stabshåndbok for Hæren – Plan -og beslutningsprosessen*.Produced by the Norwegian Army Weapons School (Hærens våpenskole).

Jacobsen, C.R. (1988) *Fundamentals of Data Storage*. NCGIA Core Curriculum in GIScience. Available from http://www.ncgia.ucsb.edu/giscc/units/u037/u037.html (Accessed 6 October 1998).

Kartverket (2014) *Nasjonal detaljert høydemodell*. Available from http://kartverket.no/Prosjekter/Nasjonal-detaljert-hoydemodell/ (Accessed 28 September 2015).

Kartverket (2015) *Arealstatistikk for Norge 2015*. Available from http://kartverket.no/globalassets/fakta-om-norge/arealer_fylker2015.pdf (Accessed 28 September 2015).

Kartverket (2015) *Det offentlige kartgrunnlaget*. Available from kartverket.no/kart/geodatasamarbeid/temadata/det-offentlige-kartgrunnlaget (Accessed 28 September 2015).

Kartverket (2015) *EUREF89 NTM (Norsk Transversal Mercator) sone 5–30*. Available from http://www.kartverket.no/globalassets/posisjonstjenester/euref89ntmbeskrivelse.pdf (Accessed 30 June 2015).

Kartverket (2015) *Inspire*. Available from kartverket.no/kart/geodatasamarbeid/inspire (Accessed 23 September 2015).

Kartverket (2015) *SOSI-standard. Versjon 4.5*. Available from http://kartverket.no/Standarder/SOSI/ (Accessed 28 September 2015).

Klima- og miljødepartementet (2012) *Veileder til retningslinje for behandling av støy i arealplanlegging (T1-442/2012)*. Available from https://www.regjeringen.no/no/dokumenter/retningslinje-stoy-arealplanlegging/id696317/ (Accessed 28 September 2015).

Kommunal- og moderniseringsdepartementet (2009/2015) *Nasjonal produktspesifikasjon for arealplan og digitalt planregister. Del 2 – Spesifikasjon for tegneregler*.

Kulturminneloven (1978) *Lov om kulturminner av 9. juni 1978 nr. 50*. Available from https://lovdata.no/dokument/NL/lov/1978-06-09-50 (Accessed 15 June 2015).

Longley P.A., Goodchild, M.F., Maguire D.J., & Rhind, D.W. (2010) *Geographic Information Systems and Science*. (3. ed.). Hoboken, NJ: Wiley.

Lovelock, J. (1979) *Gaia: A New Look at Life on Earth*. Oxford: Oxford University Press.

MacEachren, A.M. (1995) *How Maps Work*. New York: Guilford Press.

Magnus, O. (1539) *Carta marina*. James Ford Bell Library, University of Minnesota. Available from https://upload.wikimedia.org/wikipedia/commons/e/ea/Carta_Marina.jpeg (Accessed 11 September 2015).

Matrikkellova (2005) *Lov om eigedomsregistrering av 17.06.2005 nr. 101*. Available from https://lovdata.no/dokument/NL/lov/2005-06-17-101?q=matrikkellova (Accessed 20 September 2015).

Meyer, T.H. (1997) *Non-spatial Database Models*. NCGIA Core Curriculum in GIScience. Available from http://www.ncgia.ucsb.edu/giscc/units/u045/u045.html (Accessed 10 November 1997).

Ministry of Climate and Environment (2011) *Veileder T-1490 Reguleringsplan*. Available from https://www.regjeringen.no/no/dokumenter/reguleringsplanveileder/id613879/ (Accessed 15 May 2015).

Ministry of Climate and Environment (2012) *Veileder T-1491 Kommuneplanensarealdel*. Available from https://www.regjeringen.no/no/dokumenter/kommuneplanens-arealdel/id676377/ (Accessed 20 September 2015).

Ministry of Climate and Environment (2013) *T-1539 Strategi for åpnekart- og eiendomsdata*. Available from https://www.regjeringen.no/no/dokumenter/t-1539-strategiapne-kart-eiendomsdata/id736284 (Accessed 28 September 2015).

Mitchell, A. (1999) *The ESRI Guide to GIS Analysis*. Redlands, Calif.: ESRI Press.

Moen, A. (1998) *Nasjonalatlas for Norge: Vegetasjon*. Hønefoss: Statens kartverk.

Naturmangfoldloven (2009) *Lov om forvaltning av naturens mangfold av 19. juni 2009 nr. 100*. Available from https://lovdata.no/dokument/NL/lov/2009-06-19-100 (Accessed 28 September 2015).

Norwegian Directorate for Civil Protection (2011) *Samfunnssikkerhet i arealplanlegging*. Available from http://www.dsbinfo.no/DSBno/2011/Tema/temasamfunnssikkerhetareal/ (Accessed 28 September 2015).

Norwegian Environment Agency (2011) *TA2859/2011 Atmosfærisk nedfall av tungmetaller i Norge*. Available from http://www.miljodirektoratet.no/old/klif/publikasjoner/2859/ta2859.pdf (Accessed 25 May 2015).

Norwegian Environment Agency (2013) *M98 Kartlegging og verdsetting av friluftsområder*. Available from http://www.miljodirektoratet.no/Documents/publikasjoner/M98/M98.pdf (Accessed 25 September 2015).

Norwegian Environment Agency (2014) *M100 Planlegging av grønnstruktur i byer og tettsteder*. Available from http://www.miljodirektoratet.no/Documents/publikasjoner/M100/M100.pdf (Accessed 25 September 2015).

Norwegian Environment Agency (2014) *Miljøstatus i Norge – Forurenset grunn*. Available from http://www.miljostatus.no/Tema/Kjemikalier/Forurenset-grunn/ (Accessed 25 September 2015).

Norwegian Environment Agency (2015) *Miljøstatus i Norge – Inngrepsfrinatur (INON) 01.2013*. Available from http://www.miljostatus.no/Tema/Naturmangfold/inon/ (Accessed 28 September 2015).

Norwegian Mapping and Cadastre Authority (2002) *Grafisk utforming av kart i målestokk 1:500–1:10.000*. Versjon 1. Available at http://www.kartverket.no/globalassets/standard/bransjestandarder-utover-sosi/grafiskutformingf.pdf.

Norwegian Water Resources and Energy Directorate (NVE) (2011) *Retningslinjenr. 2/2011. Flaum- og skredfare i arealplanar*. Available from http://www.nve.no/no/Flom-ogskred/Arealplaner-i-fareomrader/ (Accessed 28 September 2015).

Norwegian Water Resources and Energy Directorate (NVE) (2011) *Samlet plan for vassdrag*. Available from http://www.nve.no/no/Energi1/Fornybar-energi/Vannkraft/Samletplan-for-vassdrag/ (Accessed 28 September 2015).

Norwegian Water Resources and Energy Directorate (NVE) (2015) *Kart og karttjenester*. Available from http://www.nve.no/no/Vann-og-vassdrag/Databaser-og-karttjenester/ (Accessed 14 September 2015).

NIBIO (2014) *Beitestatistikk*. Available from http://www.skogoglandskap.no/kart/beitestatistikk (Accessed 24 September 2015).

NIBIO (2014) *Vernskog*. Available from http://www.skogoglandskap.no/kart/Vernskog (Accessed 26 September 2015).

Norwegian Space Centre (2015) *Kort om jordobservasjon*. Available from http://www.romsenter.no/Fagomraader/Jordobservasjon/Kort-om-jordobservasjon (Accessed 16 September 2015).

Obeng, E. (1994) *All Change – The Project Leader's Secret Handbook*. London: Financial Times/Pitman.

Oreigningslova (1959) *Lov om oreigning av fast eigedom av 23. oktober 1959 nr. 3*. Available from https://lovdata.no/dokument/NL/lov/1959-10-23-3 (Accessed 20 May 2015).

Pickles, J. (1995) Representations in an Electronic Age: Geography, GIS, and Democracy. I J. Pickles (ed.), *Ground Truth: The Social Implications of Geographic Information Systems*. New York: Guilford Press.

Plan- og bygningsloven (2008) *Lov om planlegging og byggesaksbehandling (plan- og bygningsloven) av 27.06.2008 nr. 71*. Available from https://lovdata.no/dokument/NL/lov/2008-06-27-71 (Accessed 1 September 2015).

Robinson, A.H., Morrison, J.L., Muehrcke, P.C., Kimerling, A.J., & Guptill, S.C. (1995) *Elements of Cartography*, (6. ed.). New York: Wiley.

Sivilbeskyttelsesloven (2010) *Lov om kommunal beredskapsplikt, sivile beskyttelsestiltak og Sivilforsvaret av 25. juni 2010 nr. 45*. Available from https://lovdata.no/dokument/NL/lov/2010-06-25-45 (Accessed 17 September 2015).

Snow, J. (1855) *On the Mode of Communication of Cholera*. (2. ed.). London: John Churchill, New Burlington Street.

St.meld. nr. 30 (2002–2003) (2003) *«Norge digitalt» – et felles fundament for verdiskapning*. Available from https://www.regjeringen.no/no/dokumenter/stmeld-nr-30-2002-2003-/id196962 (Accessed 17 September 2015).

St.meld. nr. 33 (2012–2012) *Klimatilpasning i Norge*. Available from https://www.regjeringen.no/no/dokumenter/meld-st-33-20122013/id725930/ (Accessed 28 September 2015).

Stadnamnlova (1990) *Lov om stadnamn*. Available from https://lovdata.no/dokument/NL/lov/1990-05-18-11 (Accessed 29 September 2015).

Statens forurensningstilsyn (2009) *Veileder TA2553/2009 Tilstandsklasser for forurenset grunn*. Available from http://www.miljodirektoratet.no/old/klif/publikasjoner/2553/ta2553.pdf (Accessed 1 April 2015).

Statistics Norway (SSB) (1999) *Rapport 99/6 Standard for økonomiske regioner, Etablering av publiseringsnivå mellom fylke og kommune*. Available from http://www.ssb.no/offentlig-sektor/artikler-og-publikasjoner/standard-for-okonomiske-regioner (Accessed 6 October 2015).

Statistics Norway (SSB) (1999) *Regionale inndelinger: En oversikt over standarder i norsk offisiell statistikk*. Norges offisielle statistikk (NOS) C 513. Oslo and Kongsvinger: Statistisk sentralbyrå.

Statistics Norway (SSB) (2015) *Landsskogtakseringen, 2010–2014*. Available from www.ssb.no/lst (Accessed 25 September 2015).

Strømsholm, B. (2001) *Fjernanalyse og Geografiske informasjonssystemer*. NAROM AS.

Støeng, L.T., Jetlund, K., & Abelvik, T. (2014) *Vegdata for navigasjon*. Available from http://www.vegdata.no/2014/08/15/vegdata-for-navigasjon/ (Accessed 28 September 2015).

Tomlinson, R. (2003) *Thinking about GIS*. Redlands, Calif: ESRI Press.

Tenge, G. (2015) *Analysens plass i GIS*. Available from http://www.tenge.no/bilder_filer/analysens_plass_i_GIS.pdf (Accessed 16 September 2015).

Tenge, G. (2015) *Romlig datamanipulering*. Available from http://www.tenge.no/bilder_filer/Romlige_manipuleringsteknikker.pdf (Accessed 16 September 2015).

Tufte, E. (1990) *Envisioning Information*. Graphics Press.

Vaaje-Kolstad, T. (ed.) (2011) *Arealressurskart AR5, AR50, AR250, CLC*. Brochure produced by Skog og landskap, January 2011. Available from http://www.skogoglandskap.no/filearchive/arealressurskart_brosjyre.pdf (Accessed 10 Mars 2015).

Vannressursloven (2000) *Lov om vassdrag og grunnvann av 20. juni 2014 nr. 52*. Available from https://lovdata.no/dokument/NL/lov/2000-11-24-82 (Accessed 28 September 2015).

Vegdirektoratet (2014) *Håndbok V712 Konsekvensutredninger*. Available from http://www.vegvesen.no/_attachment/704540/binary/1056993?fast_title=H%C3%A5ndbok+V712+Konsekvensanalyser.pdf (Accessed 28 September 2015).

Visma.no (2015) *Gartners 10 teknologitrender for 2015*. Available from http://www.visma.no/blogg/gartners-10-teknologi-trender-2015/ (Accessed 25 September 2015).

Index

3D digital model 155

A
accuracy 40, 42, 48, 49, 50, 51, 52
administrative boundaries 134
administrative levels 79, 119
administrative units database (abas) 79
aerial photograph 220
aerotriangulation 55
agriculture 105, 106, 109, 133
air pollution 114
Antikva font 201
AR5 106
area coverage map 143
Askeladden 118
attribute data 180
attributes 46
average centre 149

B
backcasting 30
Barents Sea 35, 36
bar graph 198
barnetråkk 123
base station 52
basic geodata 77
basic statistical unit division 134
bathymetry 53
bedrock 74
bedrock data 91
benchmark 51
Bertin, Jacques 180
biodiversity 22, 93, 94, 97
brightness 182
buffer analysis 142, 143, 165, 170, 215
buffering 142, 143, 165
Buford, John 29
business risk 132

C
cables and pipelines 82
Cadastre, the 86
Canadian Geographic Information System (CGIS) 16
Cartesian coordinate system 44
cartographic communication 177
cartography 176
case studies 30
central point 149
central projection 55
children's paths 123
cholera 13
choropleth map 189, 190
civil protection 23
Civil Protection Law 127
classification 190
classification using natural break points 192
classless method 192
CMYK model 181
CMY model 181
cold colour series 190
colour 181, 182
communicable map 178
communication channel 52
COMPASS 51
complementary colours 199
computer-aided design (CAD) 16
conservation area 96, 102
constant visibility 180, 181, 184
coordinate system 42, 43, 201, 226
CPOS 52
cultural heritage 93, 116, 118
Cultural Heritage Act 102
cyan 181

D
dasymetric map 192
data capture 47, 51, 54, 59
data model 46, 65
data presentation 201
data quality 60, 224
data uniformity 146
datum 40, 64, 227
decision support 209
degree of change 203
descriptive statistics 148
development plans 124
differential GNSS 52
digital elevation model 155, 217, 226
digital terrain models 155, 226
Directorate for Cultural Heritage (Riksantikvaren) 116
doctrine 206, 226
dot density map 186, 190, 195
downstream analysis 154
DPOS 52
duration 203
dynamic zoom 203

E
earth observation 56
eKlima 113
election ward 134
electromagnetic radiation 56
Elektron incident 35
element level 178
ellipsoid 40
ellipsoid height 46
elvedatabasen (river database) 110
emergency preparedness 23, 81, 91, 113, 127, 129
energy 112
entities 46
environmental impact assessments 125
equidistant 42, 168, 193
equidistant classification 191
equipotential surface 40
equivalent 207
EUREF89 41, 43, 229
expert room 26, 32

F
false polygons 203
Felles Kartdatabase (FKB) 78, 201
filming 28
fixed property 85
FKB 78
footpaths 80, 122
forestry 105, 107, 108

G
geodesy 226
geographic coordinate system 43
geographic information system 14
Geological Survey of Norway (NGU) 90
geomatics 20
geoprocessing 68
georeferenced model 48
Geovekst 77, 78, 81, 82
germ theory 13
graphic communication 180
Graphic design of maps 201
gravity analysis 149
grid map 193

H
habitat 97, 98, 109
hiking routes, paths, and ski trails 122
hue 182, 195
hunting and fishing map 122
hydropower 96, 103, 112

I
inertial navigation system (INS) 54
information level 180, 185
information variable 185
Inngrepsfrie naturområder i Norge (INON) 104
Innsjødatabasen 112
Intelligence Preparation of the Operational Environment (IPOE) 210
interpolation 63, 155, 160
interpolation methods 160
intersect 144
intersection 45
interval data 179
isoline 193

J
jointly owned common land 85

K
kriging 161

L
landed property 85
landform 90
Land Resource Map 106
landscape 93
landslide 23, 25, 81, 146
land use planning 123, 127
laser scanning 50, 54, 55, 58, 81
latitude 42, 44, 45
Lee, Robert E. 28
legend 199
legionnaires' disease epidemic 15
Leguan 215
LiDAR 58, 81, 226
LiDAR data 81
light detection and ranging (LiDAR) 58
linear interpolation 161
line symbolization 187
links 47
local perception level 178, 197
longitude 35, 42, 44

M
magenta 181
manual digitizing 59
map 96
map design 199
map projection 41
map standard 201
MAREANO 119
masked polygons 203
matrikkelbrev 86
Meade, George G. 29
measuring at sea 53
median centre 149
meridian 35, 42, 43
metadata 61
miasma theory 13
MilGeo 206
MilGeo analysts 206
Military Grid Reference System (MGRS) 44
Miljødirektoratet 104, 114
Miljøstatus in Norge 113
mimetic symbol 183
mineral resources 91
mission analysis 206
mission support 209, 222, 223
municipality and county division 134
municipal master plans 124

N
national elevation model 81
national programme for repeated aerial photograph coverage of Norway 82
National Road DataBase (Nasjonal vegdatabank, NVDB) 132
natural neighbour 161
nature conservation 93, 103, 119
nature conservation area 102, 119, 125
Natur i Norge 96, 229
nautical chart 122
network analysis 154, 165, 170, 220
network RTK 52
NGO1948 41
NIBIO 96, 106, 108
nodes 47
noise 114, 223
Noise Zone Map 226
Norge digitalt 33, 50, 69
norgeibilder.no 82
Normalnull 1954 (NN1954) 46
north arrow 201
northern regions 35
Norwegian Biodiversity Information Centre 98
Norwegian Coast Guard 35, 36
Norwegian Directorate for Civil Protection 127
Norwegian Institute for Air Research (NILU) 15
Norwegian Meteorological Institute (MET Norway), the 113
Norwegian Public Roads Administration (NPRA) 132
Norwegian Transversal Mercator (NTM) 43
Norwegian Water Resources and Energy Directorate (NVE), the 110

O
oblique photograph 54
OpenStreetMap 60
operational environment (IPOE) 209
ordered data value 179
orientation 183
orthogonal projection 55
orthometric 46
orthophotos 81, 82
overlay analysis 142, 144, 145, 167, 208
overview map 201
ownership rights 86

P
painting by numbers 27
parallel circle 43
parishes 134
paths 121, 123
pattern analysis 149, 150
percentile classification 192
perception level 178
perceptual property 177, 180
phase 28, 203, 223
photogrammetry 54
Pickett's charge 30
pictogram 183
pie chart map 196
pixel 48
pixel size 48
pixel value 48
place names 79
planning 210
Planning and Building Act 125
Planning and Decision-making Process, PDMP 210
Plan- og beslutningsprosessen (PBP) 210
Plan- og bygningsloven 88
population data 135, 136
primary colours 181, 199
primary key 226
Probate Registry 86
properties 46
public map 88, 127
public map database 88

Q
qualitative data 179, 183
quantile classification 192
quantitative data 179, 181, 184, 185
quest 28
quick clay 25, 131

R
radar measurement 50, 57
radar (radio detection and ranging) 57
radon 127
raster 145, 146, 157, 160
raster calculator 170
rasterization 59
raster model 46, 48
rate of change 203
ratio data 179
readable map 178, 196
real estate 66, 83, 85, 86, 107, 226
reality model 46, 48
Real Time Kinematic (RTK) 52
reference surface 41, 43
REGINE 110
reindeer husbandry 109, 127
relational databases 65
Rett i kartet 60
RGB colour model 181
risk and vulnerability (RAV) 23
river 185, 187
road data 79, 137
Road Database 170
rough pasturelands 107
rover 52
rules for cartographic symbolization 180
rute 123

S
saddle points 162
satellite 156
satellite image 56, 82
saturation 182
scale 201, 203
scenario 25, 28, 165
scenario construction 30
school catchment areas 134
sector-specific administrative areas 135

SEFRAK register 119
selective point symbols 186
selectivity 186
seNorge.no 113
sensor 50, 215
sensor systems 217
Sentral stedsnavnregister (SSR) 79
sePlan 124
size 40, 48, 107, 146, 148, 151, 157, 180, 181
ski trail data 123
Skrednett (skrednett.no) 131
Snow, John 13
soil 58, 83, 90
soil quality 106, 227
sonar 53
SOSI 64, 226
source information 201
sources of errors 63
sources of statistical data 133
spaghetti data 47
spatial data manipulation 165
spatial planning 22
species data 98
spline 161
static linear zoom 203
static non-linear zoom 203
statistical data 133
Statistics Norway 133
Statistikkbanken 133
Statistisk Sentralbyrå 133
Statskog 20
St.meld. nr. 30 33
Stuart, James E.B. 29
surveying 40, 42, 50, 51
symmetrical difference 145
synthetic aperture radar 57
Synthetic Aperture Radar 227

T
terrain analysis 221, 227
texture 181
The Land Registry (Matrikkelen) 226
thematic data 187, 201
thematic map 176
theodolite 51
theoretical statistics 148
this 128
TIN model 160, 162
title 199
title and key 199
T-O map 12
topographic map 68, 74, 80, 81, 121, 176
topological accuracy 61, 62
topology 47, 48, 64, 80
topology model 47
total station 50, 51
touring map 122
traditional RTK 52
traffic light scale 182
transport network 151, 154
trend 161
triangulated irregular network (TIN) 160

U
uniform 40
union 144, 167
Universal Polar Stereographic (UPS) system 45
untouched nature 104
updates 63
updating 61, 63
upstream analysis 154
user competence 206
UTM 42, 43, 44
UTM system 44

V
valuable cultural landscapes 93
value 181
vector 46, 59
vectorization 59
vector model 47
vegetation map 96
vehicle launcher bridge (AVLB) 215
vertical photograph 54
viewable map 178
vigilance map 128
visibility analysis 219
visible light 56
visual perception 178, 180
visual variables 177, 180, 182, 184, 185, 187

W
walking in the fog 28
warm series 182
watercourse 103, 110
weather and climate 113
web coverage service 68
web feature service 68
web map 203
Web Map Services (WMS) 67, 227
web map tile service 68
WGS84 41, 227
wind power 112
wind power facilities 119

Z
zooming 203